Undergraduate Analysis

Undergraduate Analysis

A Working Textbook

Aisling McCluskey

Senior Lecturer in Mathematics
National University of Ireland, Galway

Brian McMaster

Honorary Senior Lecturer
Queen's University Belfast

OXFORD
UNIVERSITY PRESS

OXFORD
UNIVERSITY PRESS

Great Clarendon Street, Oxford, OX2 6DP,
United Kingdom

Oxford University Press is a department of the University of Oxford.
It furthers the University's objective of excellence in research, scholarship,
and education by publishing worldwide. Oxford is a registered trade mark of
Oxford University Press in the UK and in certain other countries

First Edition published in 2018

Impression: 1

Published in the United States of America by Oxford University Press
198 Madison Avenue, New York, NY 10016, United States of America

British Library Cataloguing in Publication Data

Data available

Library of Congress Control Number: 2017963197

ISBN 978-0-19-881756-7 (hbk.)
ISBN 978-0-19-881757-4 (pbk.)

Printed and bound by
CPI Group (UK) Ltd, Croydon, CR0 4YY

We dedicate this book to all those practitioners of the craft of analysis whose apprentices we have been in times long past, and to the colleagues who in more recent years have shared with us their insights and their enthusiasm.

In particular, we salute with gratitude and affection:

Samuel Verblunsky

Derek Burgess

Ralph Cooper

James McGrotty

David Armitage

Tony Wickstead

Ariel Blanco

Ray Ryan

John McDermott

AMcC, BMcM, October 2017

Preface

Mathematical analysis underpins calculus: it is the reason why calculus works, and it provides a toolkit for handling situations in which algorithmic calculus doesn't work. Since calculus in its turn underpins virtually the whole of the mathematical sciences, analytic ideas lie right at the heart of scientific endeavour, so that a confident understanding of the results and techniques that they inform is valuable for a wide range of disciplines, both within mathematics itself and beyond its traditional boundaries.

This has a challenging consequence for those who participate in third-level mathematics education: large numbers of students, many of whom do not regard themselves primarily as mathematicians, need to study analysis to some extent; and in many cases their programmes do not allow them enough time and exposure to grow confident in its ideas and techniques. This programme-time poverty is one of the circumstances that have given analysis the unfortunate reputation of being strikingly more difficult than other cognate disciplines.

Aspects of this perception of difficulty include the *lack of introductory gradualness* generally observed in the literature, and the *without loss of generality* factor: experienced analysts are continually simplifying their arguments by summoning up a battery of shortcuts, estimations and reductions-to-special-cases that are part of the discipline's folklore, but which there is seldom class time to teach in any formal sense: instead, students are expected to pick up these ideas through experience of working on examples. Yet the study time allocated to analysis in early undergraduate programmes is often insufficient for this kind of learning by osmosis. The ironic consequence is that basic analytic exercises are not only substantially harder for the beginner than for the professional, but substantially harder than they need to be.

This text, through its careful design, emphasis and pacing, sets out to develop understanding and confidence in analysis for first-year and second-year undergraduates embarked upon mathematics and mathematically related programmes. Keenly aware of contemporary students' diversity of motivation, background knowledge and time pressures, it consistently strives to blend beneficial aspects of the workbook, the formal teaching text and the informal and intuitive tutorial discussion. In particular:

1. It devotes ample space and time for development of insight and confidence in handling the fundamental ideas that – if imperfectly grasped – can make analysis seem more difficult than it actually is.

2. It focuses on learning through doing, presenting a comprehensive integrated range of examples and exercises, some worked through in full detail, some supported by sketch solutions and hints, some left open to the reader's initiative (and some with online solutions accessible through the publishers).

3. Without undervaluing the absolute necessity of secure logical argument, it legitimises the use of informal, heuristic, even imprecise initial explorations of problems aimed at deciding how to tackle them. In this respect it creates an atmosphere like that of an apprenticeship, in which the trainee analyst can look over the shoulder of the experienced practitioner, look under the bonnet of the problem and watch the roughwork develop, noting the occasional failures of opening gambits and the tricks of the trade that can be mobilised in order to circumvent them.

The price that has to be paid for such an approach is that the book is more verbose, sometimes positively long-winded, and certainly longer than one that would concentrate solely on finalised versions of standard proofs and slick model answers. Yet it appears to us that such a price is well worth paying: for one thing, it is our experience that a text principally consisting of streamlined, finalised demonstrations and solutions creates in the mind of many beginners a misleading and demoralising impression that this is how they are expected to create solutions at the first attempt; for another, the extra material – far from being just digressional – summarises what we find it necessary to say, time and time again, to students who ask us eminently reasonable questions such as: 'How do I start this?' 'How can we be expected to think of that?' 'Why is that step true, and why did you think of taking it?' An additional benefit is that the text will be easier and quicker to read, since the thoughtful reader will often find answers promptly supplied to the questions that would otherwise have impeded progress to the next step.

Especially because less-specialised learners will often need to deal with only some of the material covered here, we have streamed the presentation into basic and more advanced chapters and, within these, we have flagged up relatively specialised topics and sophisticated arguments that can reasonably be omitted without compromising overall comprehension. Analysis is more welcoming to the learner who has thoroughly grasped a modest amount of material than to one who has an imprecise understanding of a larger body of knowledge.

It is central to our teaching philosophy and to our classroom experience that students learn at a deeper level through doing than they ever could through reading alone: despite our intention to present here as full an account of basic analytic concepts, results and techniques as is reasonable to set before learners who have many other competing demands on their time and energy, it is only by active study, engaging in a broad range of exercises, that they will gain confidence and empowerment in acquiring useable, performable knowledge and the insight that directs it. Our account is therefore intended as a working textbook: each idea encountered is embedded in worked examples and in exercises – some with solutions, some with helpful hints encouraging the reader to explore and to internalise that idea.

Contents

A Note to the Instructor

The first twelve chapters present the ideas of analysis to which virtually everyone enrolled upon a degree pathway within mathematical sciences will require exposure. Those whose degree is explicitly in mathematics are likely to need most of the rest. Of course, how this material is divided across the years or across the semesters will vary from one institution to another.

Most of the exercises set out within the text are provided with specimen solutions either complete, outlined or hinted at, but in the final chapter we have also included a suite of over two hundred problems which are intended to assist you in creating assessments for your student groups. Specimen solutions to these are available to you, but not directly to your students, by application to the publishers: please see the webpage www.oup.co.uk/companion/McCluskey&McMaster for how to access them.

Prior knowledge that the reader should have before undertaking study of this material includes a familiarity with elementary calculus and basic manipulative algebra including the binomial theorem, a good intuitive understanding of the real number system including rational and irrational numbers, basic proof techniques including proof by contradiction and by contraposition, very basic set (and function) theory, and the use of simple inequalities including modulus. Substantial revision notes on several of these topics are provided within the text where appropriate.

A Note to the instructor

A Note to the Student Reader

If, as a student of the material that this book sets forth, you are enrolled on a course of study at a third-level institution, your instructors will guide and pace you through it. Careful consideration of the feedback they give you on the work you submit will be very profitable to you as you develop competence and confidence.

If you are an independent reader, not engaged with such an institution's programmes, we intend that you also will find that the text supports your endeavours through its design: in particular, through the expansive (almost leisurely) treatment of the initial ideas that really need to be thoroughly grasped before you proceed, through the informal and intuitive background discussions that seek to develop a feel for concepts that will work in parallel with their precise mathematical formulations, and through the explicit inclusion of roughwork paragraphs that allow you to look over the shoulder of the more experienced practitioner of the craft and under the bonnet of the problem being tackled.

In both cases, our strongest advice to you is to work through every exercise as you encounter it, and either check your answer against a specimen answer where available, see if it convinces a colleague or fellow student, or submit it for assessment or feedback as appropriate. Nobody learns analysis merely by reading it, any more than you can learn swimming or cycling just by reading a how-to book, however well-intentioned or knowledgably written it may be. No one can teach you analysis without your commitment; but you can choose to learn it and, if you do, this working textbook is designed to help you towards success.

Preliminaries

1.1 Real numbers

You can choose to think of the real numbers as being all the possible decimals – finite and infinite, recurring and non-recurring, positive and negative and zero, whole numbers and fractions and surds[1] and non-surds such as π and e, and every possible combination of such objects. Equally well, you can choose to think of them as being (or being represented by) all the points that lie on a continuous unbroken straight line (the *real line*, the *real axis*) that stretches away endlessly in both directions. Somewhere on that line is a point marked 0 (zero) which separates the positives (on its right) from the negatives (on its left), and pacing out from zero at regular intervals in both directions lie the whole numbers (the *integers*) like distance markers along that endless road.

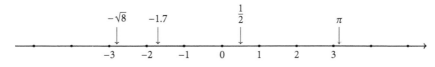

A naïve picture of the real line

This is not, of course, a proper definition of what real numbers are. We are taking what is sometimes called a *naïve* view of the system of real numbers: not having sufficient time to construct it – to dig deeply enough into the logical foundations of mathematics to come up with a guarantee of its existence – we are instead seeking to highlight the common consensus on how real numbers behave, combine and *compare*. This consensus will already be enough to let us start explaining some basic ideas in analysis (and we shall say more about the finer structure of the real numbers in Chapter 3).

Nothing in Section 1.2 is likely to strike the student reader as being much more than common sense, and nor should it at this stage of study. Nevertheless, it is all too easy to make mistakes in *comparisons between numbers – inequalities –* and it is consequently important to keep these apparently obvious rules in mind and to build up a good measure of confidence in their use, especially because so many arguments in analysis depend upon using inequalities. Sections 1.3 and 1.4 present a couple of useful operations on real numbers that are strongly connected with inequalities.

---

[1] that is, non-rational numbers involving roots, such as $\sqrt{2}$, $\dfrac{\sqrt[3]{5}}{1+\sqrt{2}}$, $\sqrt{10 - \sqrt[3]{2}}$.

Undergraduate Analysis: A Working Textbook, Aisling McCluskey and Brian McMaster 2018.
© Aisling McCluskey and Brian McMaster 2018. Published 2018 by Oxford University Press

1.2 The basic rules of inequalities – a checklist of things you probably know already

- Each real number is either positive or zero or negative. 'Non-negative' means positive or zero.
- $x > y$ and $y < x$ both mean $x - y$ is positive[2].
- $x \geq y$ and $y \leq x$ both mean $x - y$ is non-negative[3].
- $x < y < z$ means both $x < y$ and $y < z$. Likewise for $>$, $\leq$, $\geq$.
- If $x < y$ and $y < z$, then $x < z$. Likewise for $>$, $\leq$, $\geq$.
- If $x \leq y$ and $y \leq x$, then $x = y$.
- If x and y are different real numbers, then one of them is greater than the other, and is usually denoted[4] by $\max\{x, y\}$.
- You can add a number to an inequality without damaging it:
  $x < y \implies x + a < y + a$.
- You can add two inequalities:
  $(x < y$ and $a < b) \implies x + a < y + b$.
- Notice how to use the symbol '$\implies$' (pronounced *implies*): the last line is shorthand for 'if $x < y$ and $a < b$ then $x + a < y + b$'.
- You can multiply an inequality **by a positive number** without damaging it: provided $a > 0$, we have $x < y \implies ax < ay$.
- If you multiply an inequality **by a negative number**, the inequality becomes reversed:
  provided that $a < 0$, we have $x < y \implies ax > ay$.
- You can multiply two inequalities provided that all the numbers involved are positive:
  $(0 < a < b$ and $0 < x < y) \implies ax < by$;
  $(0 < a \leq b$ and $0 < x \leq y) \implies ax \leq by$.
- Provided that the numbers involved are positive, you can take reciprocals across an inequality, and the inequality becomes reversed:
  $x < y \implies 1/x > 1/y$ provided that x, y are positive.
- Provided that the numbers involved are positive, you can take square roots[5] across an inequality, and the inequality is preserved:
  $x < y \implies \sqrt{x} < \sqrt{y}$ provided that x, y are positive. Likewise for cube roots, fourth roots and so on.

---

[2] – and are pronounced as x is greater/larger/bigger than y, y is less/smaller than x.
[3] – and are pronounced as x is greater than or equal to y, y is less than or equal to x.
[4] If $x = y$ then $\max\{x, y\}$ means x (or y, which is the same thing).
[5] Recall that the symbol $\sqrt{x}$ always means the *non-negative* square root of x.

- 'There are large integers:' that is, for any given real number x we can find an integer n so that $n > x$.

1.3 Modulus

1.3.1 Definition If x is a real number, we define[6] its *modulus* (also called its *absolute value*) as $|x| =$ the greater of x and $-x$. That is:

- If $x \geq 0$ then $|x| = x$;
- If $x < 0$ then $|x| = -x$.

Since the effect of modulus is to 'throw away the minus from negative numbers', the following should be obvious:

1.3.2 Proposition For any real numbers x, y:

- $x \leq |x|, -x \leq |x|$,
- $|-x| = |x|$,
- $|xy| = |x||y|$,
- $\left|\frac{x}{y}\right| = \frac{|x|}{|y|}$ provided that $y \neq 0$,
- $\sqrt{x^2} = |x|$.

1.3.3 The triangle inequality For any real numbers x and y, we have $|x + y| \leq |x| + |y|$.

Proof

Since $x \leq |x|$ and $y \leq |y|$, adding gives us $x + y \leq |x| + |y|$.
Exactly the same reasoning gives us $-x + (-y) = -(x + y) \leq |x| + |y|$.
Now $|x + y|$ is either $x + y$ or $-(x + y)$. So whichever one it is, it is $\leq |x| + |y|$.

Note

It is easy to extend this by induction[7] to deal with any finite list of numbers, thus:

$$|x_1 + x_2 + x_3 + \ldots + x_n| \leq |x_1| + |x_2| + |x_3| + \ldots + |x_n|.$$

1.3.4 The reverse triangle inequality For any real numbers x and y, we have $||x| - |y|| \leq |x - y|$.

[6] More briefly: $|x| = \max\{x, -x\}$.
[7] We discuss this type of argument in detail later in the text.

Proof

Use the triangle inequality on $x = (x - y) + y$ and we get $|x| \le |x - y| + |y|$, from which $|x| - |y| \le |x - y|$.

Interchange x and y, and we also get $|y| - |x| \le |y - x| = |x - y|$.

Now $\big||x| - |y|\big|$ is either $|x| - |y|$ or $|y| - |x|$. So whichever one it is, it is $\le |x - y|$.

1.4 Floor

1.4.1 Definition When x is a real number, we define the *floor* of x (also called the *integer part* of x or, informally, *x rounded down to the nearest integer*) to be the largest integer that is $\le x$. The usual notation for the floor of x is $\lfloor x \rfloor$, although some books write it as $[x]$. For instance, $\lfloor 5.6 \rfloor = 5$, $\lfloor \pi \rfloor = 3$, $\lfloor 7 \rfloor = 7$, $\lfloor -8\frac{1}{2} \rfloor = -9$.

If you choose to imagine the real numbers as being set out along the real line, with the integers – marked here by heavier dots – embedded into it at regular intervals, then the following diagram should help you to picture the relationship between x and $\lfloor x \rfloor$.

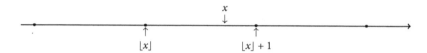

Case 1: when x is not an integer

Case 2: when x itself is an integer

In both cases, the essential inequality connecting x and $\lfloor x \rfloor$ is

$$\lfloor x \rfloor \le x < \lfloor x \rfloor + 1$$

or, equivalently

$$x - 1 < \lfloor x \rfloor \le x.$$

2 Limit of a sequence – an idea, a definition, a tool

2.1 Introduction

Mathematical analysis has acquired a reputation – not entirely justified – for seeming more difficult than other first-year undergraduate study areas. We shall begin our exploration of it by seeking to identify the factors that have contributed to this image, and what we can do to explain or address them.

Firstly, the study of mathematics is *cumulative* to a greater degree than that of most disciplines. Each new block of mathematics that a student encounters is built directly on other, underpinning, blocks, and it is practically impossible to achieve confidence in the new without having previously identified and grasped the older supporting material. No matter how well you can implement differentiation algorithms, your chance of successfully finding the second derivative of x^4 is very limited until you've learned your three-times table.

Secondly, mathematics is *hard*. By that we do not mean that it is intrinsically difficult: in this sense, 'hard' is the opposite of 'soft', not the opposite of 'easy'. Learning a piece of mathematics requires a precise understanding of the terms that it involves, of the arguments that it employs and of the questions that it seeks to answer. A broad appreciation, a solid general overview of the topic, will on its own be utterly insufficient for actual application. *Precision* of concept and of logical discourse, as well as the previously mentioned cumulativeness, are the hallmarks of a discipline that is 'hard' in this sense.

Yet these two factors are common to the whole of mathematics. Why does analysis in particular have such a daunting public image?

It seems to us that, thirdly, a *lack of introductory gradualness* comes into play here. Most topics, in mathematics and elsewhere, can be adequately explained to the beginner by working initially on simple special cases. So the usual arena for first steps in linear algebra is something like the coordinate plane, rather than an infinite-dimensional Banach space; French language lessons do not kick off by handing out a table of the complete tenses of common irregular verbs. In analysis, however, the very first concept that a beginner has to make sense of is one of the most demanding: until you have a crisp understanding of the notion of the limit of a sequence (or, a matter of similar difficulty, of the supremum of a set of real

Undergraduate Analysis: A Working Textbook, Aisling McCluskey and Brian McMaster 2018.
© Aisling McCluskey and Brian McMaster 2018. Published 2018 by Oxford University Press

numbers) you can neither read nor carry out any significant analytic activity. On the credit side, this means that we can honestly promise the beginner that the material gets easier once we are through most of Chapter 2 – an interesting contrast with many topics, both mathematical and otherwise – provided always that this first concept is fully and thoroughly understood before we go any further.

Fourthly – and this is another point that applies to the whole of the discipline, but is particularly relevant just here – mathematics as a subject and mathematicians as a breed are inclined to prefer *conciseness* over verbosity when they present final versions of their work, and to feel more at home with terse, lean, point-by-point arguments rather than expansive, wordy, descriptive accounts. There are, however, some key moments in analysis where expansive rather than compressed accounts actually help in delivering understanding, and the definition of sequence limits, right at the start of our study, is one of them. It is perfectly possible to write down that definition in one line: but if we do, most readers will not see the point of it, will not grasp the kind of problem that it is set up to address and will not be able to make effective use of it even in quite simple examples. So – with apologies to all those who don't like reading essays – we see no alternative to spending a fair bit of time and several hundred words filling in the background and 'thinking out loud' about how to use this idea in applications. We again reiterate that the concept itself is not intrinsically difficult; it is merely different from mathematical notions that you have already mastered, and needs a particular form of argument presentation in order to get the best out of it. We also commit to getting back to concise, un-wordy arguments as soon as and wherever possible.

With all this in mind, we shall devote most of Chapter 2 to a thorough and leisurely exploration of this one single idea that opens the path to analytic arguments in mathematics: limits of sequences – its intuitive meaning, some of the contexts in which it arises, how to define it in terms sufficiently precise to do serious mathematics with it, and how to handle that rigorous definition in a range of illustrative examples. Please keep in mind that, once the opening chapter is safely assimilated, most of the rest of the first-year analysis syllabus is easier. (By the way, there is a fifth factor contributing to the widespread perception of the difficulty of introductory analysis, but it concerns its logical structure rather than its narrowly mathematical content, so we shall set it aside until some familiarity with the basic idea has been gained – see Section 4.4.)

2.2 Sequences, and how to write them

A *real sequence* in mathematics, sometimes more properly called an *infinite real sequence*, is an unending list of real numbers in a particular order: a first one, a second, a third, and so on without end. In other topics within mathematics, it pays to look at unending lists of objects of other kinds – complex numbers, functions, sets – but for the present we shall restrict our focus to real numbers, and use the single word 'sequence' always to mean 'real sequence' (since no other varieties are under our attention). The sorts of symbols that we write down to identify a particular sequence that we want to work with look like one of the following:

$(a_1, a_2, a_3, a_4, \cdots, a_n, \cdots)$

$(a_1, a_2, a_3, a_4, \cdots)$

$(a_n)_{n \in \mathbb{N}}$

$(a_n)_{n \geq 1}$

(a_n)

– and in many cases we complete the description by setting down a formula for how to calculate each individual number a_n in the list (the so-called n^{th} term). For instance, if we wish to talk about the sequence of all perfect squares, that is, all the squares of positive integers in their natural order, then all of the following are acceptable symbols:

$(1, 4, 9, 16, \cdots, n^2, \cdots)$

$(1, 4, 9, 16, \cdots)$

$(n^2)_{n \in \mathbb{N}}$

$(n^2)_{n \geq 1}$

(n^2)

$(a_1, a_2, a_3, a_4, \cdots, a_n, \cdots)$ where $a_n = n^2$ for each positive integer n

$(a_1, a_2, a_3, a_4, \cdots)$ where $a_n = n^2$, each positive integer n

$(a_n)_{n \in \mathbb{N}}$ in which $a_n = n^2$ for each n

$(a_n)_{n \geq 1}$ with $a_n = n^2$ for each n

(a_n), $a_n = n^2$ for each positive integer n

It may seem a little irritating that so many different styles of symbol are allowed, but this is mostly to enable us to tailor the notation we use to the particular problem that we are working on without writing more than is necessary. For instance, if the formula for a_n is as simple as $a_n = n^2$, then we really have no need for a separate symbol for the nth term, and we might just as well write it as n^2 all the time; on the other hand, if the nth term is something as complicated as

$$\frac{n!(n+1)!(2n+3)!\,t^n}{((n+2)!)^3(4n-1)!}$$

then we shall certainly not want to write that out more often than is needful, and in such cases, having a brief symbol such as a_n to stand in for it will be a considerable benefit and relief.

Although the idea of denoting a sequence by a list of its first few terms or a formula for its general term, wrapped up in brackets, is little more than common sense, it will be important to use this notation consistently and correctly. So we now flag up a few *dos* and *don'ts* concerning how best to employ it:

• Whenever you use a notation like $(a_1, a_2, a_3, a_4, \cdots, a_n, \cdots)$ or $(a_1, a_2, a_3, a_4, \cdots)$, be careful not to leave out the final row of dots: because a symbol such as $(a_1, a_2, a_3, a_4, \cdots, a_n)$ or (a_1, a_2, a_3, a_4) is a standard way to write a *finite* list of numbers consisting of only n or, indeed, only four items, and you will confuse the person reading your work if you use it when you actually intend an infinite sequence.

- Also be cautious about using such a symbol as $(1, 4, 9, 16, \cdots)$: however obvious it may be to you that this intends the sequence of perfect squares, there are *other* perfectly good sequences whose first four terms are 1, 4, 9 and 16. Therefore, only use this style of notation if it is genuinely clear what the 'pattern' of the terms is. Note that the symbol $(1^2, 2^2, 3^2, 4^2, \cdots)$ makes this pattern quite unambiguous.

- Always take care not to leave off the enclosing brackets when writing down a sequence: if you write just n^2, your reader will think that you mean only the single number n^2 (for some particular n that you have in mind) rather than the whole endless list of the squares.

- There are some occasions when $n = 1$ is not the best starting point for a sequence. If, for instance, we need to discuss the sequence

$$\left(\frac{n}{(n-1)(n-3)} \right),$$

then we dare not use $n = 1$ or $n = 3$ because it would lead to division by zero (which is, of course, meaningless). The notation can be tweaked slightly to avoid this, for example, by writing

$$\left(\frac{n}{(n-1)(n-3)} \right)_{n \geq 4}$$

which starts the list off safely at $n = 4$.

- Again, if we want to work with the endless list of factorials, it may be useful to recall that zero-factorial is a perfectly good and useful number, and explicitly to include it in our discussion by using a notation such as

$$(n!)_{n \geq 0}.$$

Here are a few illustrative examples of sequences, some presented in more than one style of symbol. You may find it useful to 'translate them into English' in your head; for instance, the first is 'the sequence of odd positive integers', the fourth is 'the sequence of primes', the sixth is 'the sequence of reciprocals of the positive integers but with the sign alternating' and so on.

2.2.1 Example

1. $(1, 3, 5, 7, 9, \cdots) = (2n - 1)_{n \geq 1}$

2.
$$\left(\frac{3}{2}, \frac{3}{4}, \frac{9}{8}, \frac{15}{16}, \frac{33}{32}, \frac{63}{64}, \cdots \right) = \left(1 + \frac{1}{2}, 1 - \frac{1}{2^2}, 1 + \frac{1}{2^3}, 1 - \frac{1}{2^4}, \cdots \right)$$
$$= \left(\frac{2^n + (-1)^{n-1}}{2^n} \right)_{n \in \mathbb{N}}$$

3.

$$\left(5, \frac{1}{2}, 5, \frac{1}{4}, 5, \frac{1}{8}, 5, \frac{1}{16}, 5, \frac{1}{32}, \cdots\right)$$

$= (x_n)$ where $x_n = 5$ if n is odd but $x_n = 2^{-n/2}$ if n is even.

4. $(2, 3, 5, 7, 11, \cdots) = (y_n)_{n \geq 1}$ where y_n is the nth prime number. Notice how potentially misleading the first symbol was here: it could have meant several different sequences including, for example, 'two, and then the odd integers excluding the perfect squares'. The second symbol was free from any such ambiguity.

5.

$$(1, -8, 27, -64, 125, -216, \cdots) = (1, -8, 27, -64, \cdots, (-1)^{n-1} n^3, \cdots)$$

$$= ((-1)^{n-1} n^3)_{n \in \mathbb{N}}$$

Once again the first symbol might have been misunderstood, but the second and third left no room for confusion.

6.

$$\left(\frac{1}{1}, -\frac{1}{2}, \frac{1}{3}, -\frac{1}{4}, \frac{1}{5}, -\frac{1}{6}, \frac{1}{7}, \cdots\right) = \left(\frac{(-1)^{n-1}}{n}\right)_{n \geq 1}$$

7.

$$(1, \sqrt{2}, 3^{\frac{1}{3}}, 4^{\frac{1}{4}}, 5^{\frac{1}{5}}, \cdots)$$

8.

$$\left((1+1)^1, \left(1+\frac{1}{2}\right)^2, \left(1+\frac{1}{3}\right)^3, \left(1+\frac{1}{4}\right)^4, \cdots\right) = \left(\left(1+\frac{1}{n}\right)^n\right)_{n \in \mathbb{N}}$$

9.

$$\left(1, 1+\frac{1}{2}, 1+\frac{1}{2}+\frac{1}{3}, 1+\frac{1}{2}+\frac{1}{3}+\frac{1}{4}, \cdots\right)$$

10.

$$\left(1, 1+\frac{1}{2}, 1+\frac{1}{2}+\frac{1}{4}, 1+\frac{1}{2}+\frac{1}{4}+\frac{1}{8}, \cdots\right)$$

You should notice that some sequences, but by no means all of them, seem to be settling towards an 'equilibrium value', a 'steady state' as we scan further and further along the list. For instance, (2) above appears to be settling towards 1, and (6) towards 0; in contrast, (1) and (4) are so far showing no sign of settling, but are 'exploding towards infinity' (and of course we shall need to make that phrase a lot more precise before we do anything serious with it) while (5) is doing some kind of cosmic splits by exploding towards infinity and minus infinity at the same time (same comment). In the case of the last four sequences (7) to (10), it is much less clear – to unaided common sense – what is going to happen in the long run.

This feature of settling towards some limiting steady state is the most important property that a sequence can possess. Our major upcoming task is to seek ways of deciding whether a given sequence ultimately settles or not, and if so, to what steady state it does 'gravitate'. As a first step in tackling that task, we need to find a way to describe such a settling process that is crisp and precise enough that we can do proper mathematics with it. In this description, we shall need to avoid all vague and undefined phrases like 'gravitate' and 'gets extremely close to' and 'is as nearly equal to as makes no difference' without, of course, throwing away the valuable intuition that these phrases try to capture.

2.3 Approximation

Across the full expanse of science, engineering and mathematics, we find instances where some interesting constant is known not precisely but 'only' by estimation, by approximation. In most such scenarios, we expect to see not just one approximation, but several obtained at various times and by different procedures (hopefully with increasing accuracy over time) and, using if necessary a little imagination, we can conceive of an endless process of refinement (new experiments, wider data collection, more powerful computation, more sophisticated digital image enhancement ...) capable of generating better and better estimates for ever. Of course, in the best of all possible worlds it would be ideal if, at some point in the process, we should meet and recognise the *exact* value of the elusive constant...but this is unrealistic for several reasons (including the fact that no measuring device ever invented can operate to infinite precision) and, even within mathematics itself[1], one must normally be content with an endlessly refining approximation procedure.

'Your estimate is only as good as your assessment of its error' as the maxim puts it, so each approximation process has to focus on how bad the error term is...or, more precisely, on how bad the error term *could be*: because we are actually never going to know the exact size of the error since that presupposes knowing the exact value of the constant that we are struggling to estimate. The final piece in this jigsaw is: how good do we *need* the approximation to be? – for most estimations are carried out with a view to application, and different applications depend for their success or validity on different levels of accuracy. (If you are in the business of manufacturing ball bearings for use in cheap, disposable water pumps, then a radius accuracy of 0.1 mm may well be good enough since this also helps to hold the price down; but if your next customer is installing similar devices in a submarine lab environment where failure means transporting the device up through a thousand fathoms for replacement, you had better increase that accuracy by an order of magnitude or two; and if you want to seek a contract with a commercial aircraft manufacturer for whom pump failure places lives in jeopardy, another order of magnitude again...)

---

[1] The best of all possible worlds!

Proper understanding of an approximation procedure therefore entails aware-ness of six separate quantities: the ideal value; the stage we have reached in the approximation process (let's call that *stage n*); the current, n^{th} approximation; the (actual) error[2] at this stage; a calculable 'worst case scenario' (WCS) estimate for that error; and the tolerance (the amount of error that would be acceptable for the application that we presently have in mind). For a given tolerance, a good approximation process is one for which we can find a value of n that forces the worst case scenario error estimate at stage n to be smaller than the tolerance, because then the (necessarily smaller) actual error will also be less than the tolerance – and therefore satisfy the customer or the demands of the application. Preferably, we should also wish all the later-stage errors to be as small or smaller than that: so that if, in caution or for some other reason, we take a larger value of n than the one first found, we shall still be safely within the tolerance of error.

A perfect approximation process would be one that was capable of this for each and every value of the tolerance (except for the physically absurd suggestion of tolerance = zero).

2.4 Infinite decimals

One of the most basic situations in which endless lists of approximations turn up is that in which we write down – or rather, imagine writing down – an infinite decimal. For instance, let us consider the meaning of the statement that the fraction $\frac{13}{27}$ can be written in this way:

$$\frac{13}{27} = 0.48148148148\ldots\ .$$

For each positive integer n, let p_n stand for the decimal expansion, up to the nth decimal digit, of this number. So...

$$p_1 = 0.4$$
$$p_2 = 0.48$$
$$p_3 = 0.481$$
$$p_4 = 0.4814$$
$$p_5 = 0.48148$$

$$\vdots \quad \vdots$$

...and so on. None of these numbers equals $\frac{13}{27}$ exactly: if you take any one of them and multiply it by 27, you don't get 13; indeed, just from the way in

---

[2] that is, the difference between the nth approximation and the ideal value

which multiplication is carried out, you don't get a whole number. They are, however, approximations to $\frac{13}{27}$ and – broadly speaking – they provide better and better approximations as you work along the list: indeed, you could get 'as close to $\frac{13}{27}$ as you needed to be' just by going far enough.

It is that last, slightly vague, comment that we need to make precise and, in order to pin down its exact meaning, we shall look at the errors in the approximations, the differences $\frac{13}{27} - p_n$:

$$
\begin{aligned}
\tfrac{13}{27} - p_1 &= 0.08148148\ldots &<& \quad 0.1 &=& \quad 10^{-1} \\
\tfrac{13}{27} - p_2 &= 0.00148148\ldots &<& \quad 0.01 &=& \quad 10^{-2} \\
\tfrac{13}{27} - p_3 &= 0.00048148\ldots &<& \quad 0.001 &=& \quad 10^{-3} \\
&\;\;\vdots && \quad\;\;\vdots && \quad\;\;\vdots \\
\tfrac{13}{27} - p_n &= &<& \quad 0.00\ldots 01 &=& \quad 10^{-n} \\
&\;\;\vdots && \quad\;\;\vdots && \quad\;\;\vdots
\end{aligned}
$$

Notice that this display doesn't tell us explicitly the exact value of these errors[3], but that this is not going to matter: because, instead, we have an overestimate of the size of the typical n^{th}-stage error that is simple enough to work with. Look:

- If we are allowed a certain 'tolerance of error', that is, we've been asked to get some approximations whose actual errors are less than that tolerance, we can now easily see how to do it. Just find some positive integer N such that $\left(\frac{1}{10}\right)^N$ is smaller than the permitted tolerance, and then p_N will be a good enough approximation because $0 < \frac{13}{27} - p_N < \left(\frac{1}{10}\right)^N$, so $\frac{13}{27} - p_N$ is also smaller than the tolerance.

- Continuing…not only is *that* p_N good enough, but all the later ones p_{N+1}, $p_{N+2}, p_{N+3}, p_{N+4}$…will also be good enough in the same sense of 'good': they all have actual errors that are less than our allowed tolerance.

Incidentally, if we had approximated from above instead of from below, by opting for the list of numbers 0.5, 0.49, 0.482, 0.4815, 0.48149…, then the differences $\frac{13}{27} - p_n$ would have been negative. That would not have bothered us too much because the *size* of the error is usually more important than whether it is positive or negative. So we would have taken $p_n - \frac{13}{27}$ as the error measurement in this case instead of $\frac{13}{27} - p_n$, and the rest of the calculations would have worked out almost exactly the same.

The way to avoid worrying about whether our approximations are overestimates or underestimates is simply to define the error to mean $\left|\frac{13}{27} - p_n\right|$, so that error measurements are always counted as positive. We shall do this in future.

The last small step we take in order to compress our account of this string of improving approximations into a compact phrase is to agree on a standard symbol for what we called the tolerance of error. For historical reasons, the Greek letter ε

---

[3] for one thing, we are only working to so-many decimal places at this point

(pronounced '*EP-silon*') is used. Thus, our precise and concise reason for declaring the list of numbers (p_n) to be a 'perfect' approximation process for the fraction $\frac{13}{27}$ is:

> for each $\varepsilon > 0$, we can find a positive integer N such that
> for every $n \geq N$, we get $\left| p_n - \frac{13}{27} \right| < \varepsilon$.

You should probably read that last couple of lines several times in order to feel how it captures all the aspects of our lengthy discussion. Notice particularly that the N that we find depends on the particular ε that we are challenged with: if they change ε, we are free to change[4] the replying N that we find (and we shall usually need to do so). Sometimes it is denoted by $N(\varepsilon)$ or N_ε instead of plain N in order to make exactly this point, and other commonly employed symbols for it are n_ε, n_0, n_1 and n_2.

2.5 Approximating an area

For the moment, let's forget everything we ever knew about differential calculus and integration (we shall study these in detail later) and think how we might try to find the area A of the region in the coordinate plane that lies above the horizontal axis, below the curve $y = x^2$ and between the vertical lines $x = 0$ and $x = 2$.

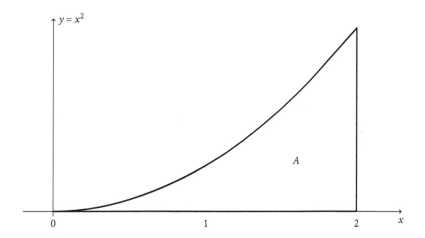

As a first attempt, we could divide this area into vertical strips by adding in extra vertical lines at $x = 0.2, 0.4, 0.6, 0.8, 1.0, 1.2, 1.4, 1.6$ and 1.8. Since (for positive numbers) $p < q$ implies $p^2 < q^2$, within each of these strips, the lowest point of the curve is at its left-hand edge and the highest point is at its right-hand edge. If we therefore imagine, inside each of the strips, the tallest rectangle that fits underneath

---

[4] For instance, in our '13/27' example, when ε is set at 0.001, $N = 3$ is a good enough choice; if the application requires ε to be reset to 0.000001, then N will have to alter to 6 at least. The relationship between ε and N is not always as simple as this, however.

the curve, it is easy to write down the area (*length times breadth*) of each of these rectangles. By adding these together, we get an estimate of the area under the curve.

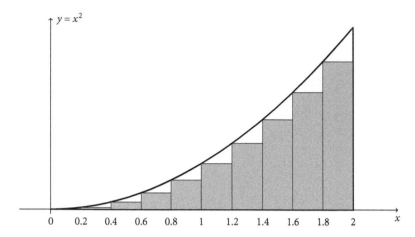

If we denote this estimate by U_{10} – since it is visibly an Underestimate of the intended area and we used ten strips in calculating it – we get

$$U_{10} = 0.2\{0.2^2 + 0.4^2 + 0.6^2 + 0.8^2 + 1^2 + 1.2^2 + 1.4^2 + 1.6^2 + 1.8^2\}$$

which calculates out as 2.28. In just the same way, we can find an overestimate (let us denote it by O_{10}) of the desired curved area by considering the shortest rectangle within each vertical strip that fits above the curve, as indicated in the next diagram:

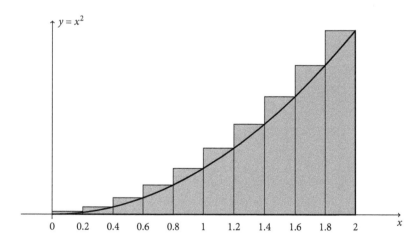

This time, the calculation is

$$O_{10} = 0.2\{0.2^2 + 0.4^2 + 0.6^2 + 0.8^2 + 1^2 + 1.2^2 + 1.4^2 + 1.6^2 + 1.8^2 + 2^2\} = 3.08.$$

It would, of course, be wrong to claim that U_{10} and O_{10} are *accurate* estimates of the area A that we set out to find: for one thing, our diagrams suggest that they are not; for another, the relatively large difference (0.8) between them makes it clear that they certainly cannot both have a high degree of accuracy. However, there are two very encouraging aspects of the discussion:

1. If we re-run the argument with more and narrower vertical strips, there are good prospects that the accuracy will improve.

2. We have control of the error: since the desired area A lies between U_{10} and O_{10}, the error we make in proposing either of these as an approximation to A cannot be more than the difference $O_{10} - U_{10}$. Therefore if, as we hope, the difference between the overestimate and the underestimate becomes smaller as we increase the number of strips, we have an improving sequence of approximations to A, just as in the previous illustration the decimals of increasing length provided an improving sequence of approximations to $13/27$.

Therefore, instead of using just ten vertical strips to slice up the area A, imagine that we choose a positive integer n and divide A into n strips (meeting at $\frac{2}{n}, \frac{4}{n}, \frac{6}{n}$ and so on up to $\frac{2n-2}{n}$). There are only very minor changes in the argument: we get the underestimate

$$U_n = \frac{2}{n} \left\{ \left(\frac{2}{n}\right)^2 + \left(\frac{4}{n}\right)^2 + \left(\frac{6}{n}\right)^2 + \cdots + \left(\frac{2n-2}{n}\right)^2 \right\}$$

$$= \frac{8}{n^3}(1^2 + 2^2 + 3^2 + \cdots + (n-1)^2)$$

and the overestimate

$$O_n = \frac{2}{n} \left\{ \left(\frac{2}{n}\right)^2 + \left(\frac{4}{n}\right)^2 + \left(\frac{6}{n}\right)^2 + \cdots + \left(\frac{2n}{n}\right)^2 \right\}$$

$$= \frac{8}{n^3}(1^2 + 2^2 + 3^2 + \cdots + n^2).$$

Now we can call in an algebraic identity for the sum of consecutive squares that you may have come across before:

$$1^2 + 2^2 + 3^2 + 4^2 + \cdots + k^2 = \frac{k(k+1)(2k+1)}{6}$$

(if this is not familiar to you, you will find a proof of it in the next chapter but one, as paragraph 4.2.2). Using this, the underestimate formula simplifies to

$$U_n = \frac{8}{n^3} \frac{(n-1)(n)(2n-1)}{6} = \frac{4(n-1)(2n-1)}{3n^2},$$

the overestimate formula to

$$O_n = \frac{8}{n^3} \frac{(n)(n+1)(2n+1)}{6} = \frac{4(n+1)(2n+1)}{3n^2}$$

and the difference between them to

$$O_n - U_n = \frac{4(n+1)(2n+1)}{3n^2} - \frac{4(n-1)(2n-1)}{3n^2} = \frac{4(6n)}{3n^2} = \frac{8}{n}.$$

At this point we are ready to obtain estimates for A that are as accurate as we choose to make them. For instance, if we need an approximation whose error is 0.01 or smaller, choosing $n = 800$ will be good enough since, at that point, $\frac{8}{n}$ is 0.01 and the error we make in claiming

$$`A = O_{800} = \frac{4(801)(1601)}{3(800)^2} = 2.67169 \text{ approximately}'$$

is less than that. If, instead, we need the error to be smaller than 0.0001, then choosing $n = 80,000$ (or, indeed, anything larger than 80,000 – for instance, it might make the arithmetic simpler if we opted for $n = 100,000$ instead[5]) will achieve it. Indeed, no matter how small the error is required to be, we now have a simple rule of thumb for choosing a positive integer N so that *any* value of n that exceeds N will give us an approximation O_n (or U_n or, indeed, anything in between) whose actual error is smaller: we have, in the language of Section 2.4, set up a 'perfect' approximation procedure for A.

2.6 A small slice of π

We conclude this group of illustrations with one that has more underlying theory than the previous two, so we shall consider it only briefly. If you add together an initial block of terms in the list

$$1, \quad -\frac{1}{3}, \quad +\frac{1}{5}, \quad -\frac{1}{7}, \quad +\frac{1}{9}, \quad -\frac{1}{11}, \quad +\frac{1}{13}, \quad -\frac{1}{15}, \quad \cdots$$

then (firstly) the total is an estimate for $\pi/4$ and (secondly) the error in that estimate is smaller than the modulus of the next number in the list – the first one that you decided not to take. Exactly why this is true is not at all obvious, but we shall investigate it later in the text (see 18.3.17).

If, for instance, we add the first five numbers, then the running total so far is $263/315$ which is really not all that good as an estimate, but does at least come with an assessed error: the error must be smaller than $1/11$, the modulus of the sixth term. Likewise, the total of the first twenty fractions in the list (which would

---

[5] Working to five decimal places, $O_{100,000}$ calculates out at 2.66671 and $U_{100,000}$ at 2.66663

be very tedious but routine to calculate by hand) will provide an approximation that differs from $\pi/4$ by less than $1/41$, the (modulus of the) twenty-first term. Continuing, if we needed an estimate whose error was less than 0.01, notice that term number n here is $\pm\frac{1}{2n-1}$, so by choosing $n = 51$ or more we would get the modulus of the nth term to be at most $1/101$: in other words, the sum of the first 50 terms will differ from $\pi/4$ by less than 0.01.

It should be becoming clear that we can force the error term here to be smaller than any given tolerance, merely by taking an initial block with enough terms in it. So, although the laborious arithmetic that it entails severely limits the usefulness of this approximation process, it is nevertheless 'perfect' in the technical sense that we are using here.

2.7 Testing limits by the definition

The case studies on which we have spent the last few pages should, at least, have removed every element of surprise from the following definition.

2.7.1 Definition   1. A sequence $(x_n)_{n\in\mathbb{N}}$ is said to *converge* to a *limit* ℓ (or to *tend* to ℓ), where ℓ is a real number, if:

for each $\varepsilon > 0$ there is some positive integer n_ε such that $|x_n - \ell| < \varepsilon$ for every $n \geq n_\varepsilon$.

2. When this is so, we write $x_n \to \ell$ (as $n \to \infty$) or, equivalently, $\lim_{n\to\infty} x_n = \ell$ or

$$\lim_{n\to\infty} x_n = \ell.$$

(The phrases '(as) $n \to \infty$' are often omitted, especially in situations where we might otherwise be writing them many times over. When not omitted, these phrases are commonly spoken as 'as n tends to infinity'.)

3. Not all sequences have limits. Those that do are called *convergent*, whereas those that do not are called *divergent*.

It will probably help you to keep on thinking of x_n as the nth item in a succession of approximations to ℓ, and of ε as the tolerance for some intended application. In that sense the open interval $(\ell - \varepsilon, \ell + \varepsilon)$, which consists of exactly the numbers whose distances from ℓ are smaller than ε, is where to find the 'good' approximations – those whose errors are smaller than the current tolerance. Keep in mind that the physical distance between numbers x and y on the real line is $|x - y|$, so that the phrase $|x_n - \ell| < \varepsilon$ simply says 'the distance between x_n and ℓ is smaller than ε'.

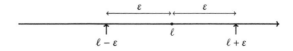

The 'good' approximations to ℓ lie between ℓ–*tolerance* and ℓ+*tolerance*

2.7.2 Example   To show that the harmonic sequence $\left(\frac{1}{n}\right)_{n\geq 1}$ converges[6] to 0.

Draft solution

Our task is – given any positive tolerance ε – to find a value of n beyond which all terms of the sequence lie within that tolerance of 0. Such a task usually needs a piece of roughwork first. We want $\left|\frac{1}{n} - 0\right| < \varepsilon$, that is, $\frac{1}{n} < \varepsilon$, that is, $n > \frac{1}{\varepsilon}$. This shows how big n needs to be. However, $\frac{1}{\varepsilon}$ is probably not an integer so, in order to line up with the definition, we had better round it up to the next whole number (or one greater still, if you prefer) and call *that* n_ε. Now we are ready:

Solution

Given $\varepsilon > 0$, let n_ε be any integer larger than $\frac{1}{\varepsilon}$. Then for every integer $n \geq n_\varepsilon$ we have $n > \frac{1}{\varepsilon}$, therefore $\frac{1}{n} < \varepsilon$, that is, $\left|\frac{1}{n} - 0\right| < \varepsilon$. By the definition, $\lim_{n\to\infty}\frac{1}{n} = 0$.

2.7.3 Example   Put $x_n = 4 - 3n^{-2}$ for each positive integer n. We show that (x_n) is a convergent sequence, and that its limit is 4.

Draft solution

We need to arrange $|x_n - 4| < \varepsilon$ for each given positive tolerance ε – or, more precisely, to decide how big n needs to be in order to force this to happen. Now $|x_n - 4|$ simplifies to $\frac{3}{n^2}$ and this will be less than ε just when $\frac{n^2}{3} > \frac{1}{\varepsilon}$, that is, when $n^2 > \frac{3}{\varepsilon}$, that is, when $n > \sqrt{\frac{3}{\varepsilon}}$. We can locate a suitable n_ε for the definition by rounding up that last expression to a whole number.

Solution

Given $\varepsilon > 0$, let n_ε be any integer larger than $\sqrt{\frac{3}{\varepsilon}}$. Then for every integer $n \geq n_\varepsilon$ we have $n > \sqrt{\frac{3}{\varepsilon}}$ and therefore $n^2 > \frac{3}{\varepsilon}$ and therefore $\frac{1}{n^2} < \frac{\varepsilon}{3}$ and therefore $\frac{3}{n^2} < \varepsilon$, that is,

$$|x_n - 4| = |4 - 3n^{-2} - 4| = \frac{3}{n^2} < \varepsilon.$$

By definition, the sequence (x_n) converges to 4.

2.7.4 Example   To show that

$$\lim_{n\to\infty} \frac{n(3n - 1)}{n^2 + 1} = 3.$$

[6] While we begin to build up experience and confidence in using this definition, we shall often practise on sequences (such as this one) for which it is possible to guess fairly easily the *exact* numerical value of the limit.

Draft solution

Let x_n stand for the typical term in this sequence and let ε denote a given positive tolerance. We want to arrange that $|x_n - 3| < \varepsilon$ for sufficiently big values of n, that is,

$$\left| \frac{n(3n-1)}{n^2+1} - 3 \right| = \left| \frac{3n^2 - n - 3n^2 - 3}{n^2+1} \right| = \left| \frac{-n-3}{n^2+1} \right| = \frac{n+3}{n^2+1} < \varepsilon.$$

This time it is not straightforward to determine exactly how big n must be to force this, but we do not need to do so exactly: we can look for a WCS[7] overestimate that is easier to work with, and use that instead to decide where n_ε can be safely placed. Look carefully at the following overestimation:[8]

$$\frac{n+3}{n^2+1} < \frac{n+3}{n^2} \leq \frac{n+3n}{n^2} = \frac{4n}{n^2} = \frac{4}{n}.$$

Now it is easy to make $\frac{4}{n}$ less than ε: just ensure that n exceeds $\frac{4}{\varepsilon}$ or, rather, an integer larger than that.

Solution

Given $\varepsilon > 0$, let n_ε be any integer larger than $\frac{4}{\varepsilon}$. Then for every integer $n \geq n_\varepsilon$ we have $n > \frac{4}{\varepsilon}$ and therefore $\frac{n}{4} > \frac{1}{\varepsilon}$ and therefore $\frac{4}{n} < \varepsilon$. But

$$|x_n - 3| = \left| \frac{n(3n-1)}{n^2+1} - 3 \right| = \frac{n+3}{n^2+1} < \frac{n+3}{n^2} \leq \frac{n+3n}{n^2} = \frac{4n}{n^2} = \frac{4}{n}$$

so also $|x_n - 3| < \varepsilon$. By definition, $x_n \to 3$.

2.7.5 **EXERCISE**   Show that the sequence $\left(\dfrac{n+5}{2n+13} \right)_{n \geq 1}$ converges to $\frac{1}{2}$.

Partial draft solution

Given $\varepsilon > 0$, we need to arrange that $\left| \dfrac{n+5}{2n+13} - \dfrac{1}{2} \right| < \varepsilon$. That simplifies to $\dfrac{3}{2(2n+13)} < \varepsilon$. Either calculate[9] how big n needs to be in order to ensure

---

[7] 'worst case scenario'
[8] and keep in mind that, to increase a fraction of positive numbers, you can increase the numerator, or *decrease* the denominator, or both.
[9] The inequality we desire is equivalent to $\dfrac{2(2n+13)}{3} > \dfrac{1}{\varepsilon}$, and to $2n + 13 > \dfrac{3}{2\varepsilon}$, and to $2n > \dfrac{3-26\varepsilon}{2\varepsilon}$, and to $n > \dfrac{3-26\varepsilon}{4\varepsilon}$. So if we choose n_ε to be a positive integer that is bigger than $\dfrac{3-26\varepsilon}{4\varepsilon}$, then $n \geq n_\varepsilon$ will guarantee that we get it.

that this happens, *or else* (preferably) use some (WCS) overestimation to make your task easier; for instance:

$$\frac{3}{2(2n+13)} < \frac{4}{2(2n+13)} = \frac{2}{2n+13} < \frac{2}{2n} = \frac{1}{n}.$$

2.7.6 EXERCISE Show that the sequence $(17n^{-3} - 2)_{n\geq 1}$ converges.

Partial draft solution

They haven't told us what the limit is, but it will not be difficult to make an informed guess. Roughly what is the value of the nth term if n is very big? What if $n = 1,000$? What if $n = 1,000,000$? What if $n = 1,000,000,000$? Once you have correctly guessed what the limit is going to turn out to be, it should be simple enough to calculate how big n must be taken in order to make the nth stage error less than any given $\varepsilon > 0$.

2.7.7 EXERCISE Prove the convergence (and evaluate the limit) of the sequence (a_n) described by

$$a_n = \frac{15n^2 + n + 1}{5n^2 - n - 2}.$$

Partial draft solution

You should again be able to guess the limit pretty certainly just by trying huge values of n (but better methods are coming). The tricky point this time comes in the (WCS) overestimating of the error term. You ought to find that the error simplifies to $\dfrac{4n+7}{5n^2 - n - 2}$, and it is then tempting to argue as follows:

$$\frac{4n+7}{5n^2 - n - 2} \leq \frac{4n+7n}{5n^2 - n - 2} = \frac{11n}{5n^2 - n - 2} < \frac{11n}{5n^2} \cdots$$

but the last step is *wrong*: by changing $5n^2 - n - 2$ into $5n^2$ we have actually increased the denominator and therefore decreased the fraction, which is the exact opposite of what we intended and needed. Instead, try this: $5n^2 - n - 2 \geq 5n^2 - n^2 - 2n^2$ so

$$\frac{11n}{5n^2 - n - 2} \leq \frac{11n}{5n^2 - n^2 - 2n^2} = \frac{11n}{2n^2} < \frac{12n}{2n^2} = \frac{6}{n}$$

and so on.

2.7.8 Remark How a sequence converges (or not) is not influenced in any way by the first few terms – nor, indeed, by the first trillion terms. For imagine that we take a convergent sequence (a_n) with limit ℓ, and alter the first trillion ($= 10^{12}$) terms in some fashion. Given positive ε, we can find n_ε so that $n \geq n_\varepsilon$ makes the error terms $|a_n - \ell|$ *in the original sequence* less than ε. If it happens that n_ε is more than a trillion, this remains true for the modified sequence (since the modifications only affected the early terms). Yet if n_ε is a trillion or less, we see that $n \geq 10^{12}$ forces the nth stage errors in both the modified and the unmodified sequences to be smaller than ε once again. In both cases, the limit has not been affected.

This allows us, when exploring the limit of a sequence, to ignore the first few (or the first many – but never *infinitely* many) terms if it simplifies our argument. Here is an illustration:

2.7.9 Example To show that $\dfrac{3n}{7n^2 - 6n - 12} \to 0$.

Roughwork and partial draft solution

Given $\varepsilon > 0$, we want to get $\left| \dfrac{3n}{7n^2 - 6n - 12} \right| < \varepsilon$. When $n = 1$, $7n^2 - 6n - 12$ is actually negative so the modulus signs are important; but their importance disappears once $n \geq 2$ since then $7n^2 - 6n - 12$ is positive. Thus, provided we deal only with $n \geq 2$, our problem becomes the slightly simpler condition $\dfrac{3n}{7n^2 - 6n - 12} < \varepsilon$.

As before, we would like to replace that fractional expression with a *larger but simpler* overestimate, so that (i) it will become easy to see how big n needs to be to make the overestimate less than ε, and (ii) then the original (smaller) fraction will automatically be less than ε also. We dare not simplify by throwing away the $-6n - 12$ from the denominator because that would increase the denominator and thus decrease the fraction – the opposite of what we need. Nor can we easily replace the $-6n - 12$ by $-6n^2 - 12n^2$ this time because it would make the denominator go negative, and oblige us to use modulus signs again. Instead, and keeping in mind that the $7n^2$ in that denominator is (for big values of n) more important than the $6n$ or the 12, consider replacing the $6n$ by a 'slice' of $7n^2$ that is small compared with $7n^2$ but big compared with $6n$. (The intention here is to simplify the algebra in the denominator, while ensuring that it remains positive but becomes smaller than it was.)

Let us be a bit more specific: provided that n is 7 or more, $6n$ will be smaller than n^2, so $7n^2 - 6n - 12$ will be larger than $7n^2 - n^2 - 12 = 6n^2 - 12$. Now do something similar to get rid of the -12: provided that n is 4 or more, 12 will be smaller than n^2, so $6n^2 - 12$ will be larger than $6n^2 - n^2 = 5n^2$. Therefore (gathering up all the restrictions on the value of n that we found useful), if n is at least 7, we shall get $7n^2 - 6n - 12 > 5n^2$, and $\dfrac{3n}{7n^2 - 6n - 12} < \dfrac{3n}{5n^2} = \dfrac{0.6}{n}$. Now the rest of the proof will run like that of earlier examples because we have a simple (WCS) overestimate of the nth stage error *and because our decision to ignore the first 6 terms will not alter our conclusion.*

Before we become too complacent about using the phrase 'the limit of a sequence', we ought to take the trouble to check that no sequence can ever possess two or more limits.

2.7.10 Theorem: uniqueness of limit of a convergent sequence   If a sequence (a_n) converges to a limit ℓ_1, and also converges to a limit ℓ_2, then $\ell_1 = \ell_2$.

Proof

If not, then one of the two is larger. Without loss of generality we'll assume $\ell_1 < \ell_2$ (otherwise, just change the labels of the two alleged limits). Put $\varepsilon = \frac{1}{2}(\ell_2 - \ell_1) > 0$.

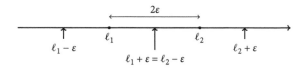

ε is half the difference between ℓ_1 and ℓ_2

From the definition of limit there must be a positive integer n_1 such that

$$\ell_1 - \varepsilon < a_n < \ell_1 + \varepsilon$$

for every $n \geq n_1$. Then again, there must be another positive integer n_2 such that

$$\ell_2 - \varepsilon < a_n < \ell_2 + \varepsilon$$

for every $n \geq n_2$. Choose any integer n that is bigger than both n_1 and n_2, and we have all of these inequalities working for us simultaneously. In particular, and holding in mind that the way we chose ε made $\ell_1 + \varepsilon$ and $\ell_2 - \varepsilon$ be the same number:

$$a_n < \ell_1 + \varepsilon = \ell_2 - \varepsilon < a_n$$

which produces the **contradiction** $a_n < a_n$.

2.7.11 Example   To show that each constant sequence converges (and that its limit is that constant).

Solution

Consider a sequence (x_n) in which *every* x_n is the same number: that is, a sequence of the form $(c, c, c, c, c, \cdots)$. We show that the limit of (x_n) is also c. In other notation, $\lim_{n \to \infty} c = c$.

Given $\varepsilon > 0$, let us choose $n_\varepsilon = 1$...yes, with a constant sequence, we can get away with a constant choice of n_ε also. Then for every $n \geq n_\varepsilon$, we get $|x_n - c| = |c - c| = 0 < \varepsilon$ and the demonstration is complete.

2.7.12 EXERCISE

- Show by example that the following statement is not true: if the sequence (a_n^2) converges to ℓ^2, then the sequence (a_n) must converge to ℓ or to $-\ell$.
- Prove that if the sequence (a_n^2) converges to 0, then (a_n) converges to 0.

Roughwork

- For questions like the first part, the thing to keep in mind is that if you only know about the value of x^2 (x being a real number) then, generally, you don't know whether x itself is positive or negative.

- The second part of this exercise highlights a small trick that frequently turns out to be extremely useful. Suppose we know that a particular sequence (x_n) converges to a limit ℓ, and we wish to use this information to show that a different but related sequence (y_n) converges to a limit m. Our task is to demonstrate that, for any given $\varepsilon > 0$, we can force $|y_n - m| < \varepsilon$ (for sufficiently large values of n). The available information is that $|x_n - \ell|$ *actually can* be made less than any given positive number, such as ε...*but not only ε*: absolutely any positive quantity can be used instead if it helps us solve the problem – for the basic given information is that $|x_n - \ell|$ can be forced to be less than any positive tolerance whatever.

  Here, we need to show that (given $\varepsilon > 0$) we can make $|a_n - 0| < \varepsilon$, and the given information is that $|a_n^2 - 0|$ can be made less than any tolerance. We ask ourselves: what should that tolerance be, in order to be able to show that $|a_n - 0| < \varepsilon$? Another reading of the sentence (if needed) should show you that we can choose the 'missing' tolerance as ε^2: because $|a_n^2 - 0| < \varepsilon^2$ certainly implies that $|a_n - 0| < \varepsilon$. So a formal solution to the second part can begin as follows:

Partial solution

Given that (a_n^2) converges to 0, and given $\varepsilon > 0$, notice that ε^2 is also greater than 0, so there is a positive integer n_0 such that $|a_n^2 - 0| < \varepsilon^2$ for all $n \geq n_0$ ...

2.7.13 **EXERCISE**   Let $(x_n)_{n \geq 1}$ be any sequence. Verify that $x_n \to 0$ if and only if $|x_n| \to 0$.

Roughwork

The cautious approach to an 'if and only if' claim is to break the argument into two parts: the 'if' and the 'only if'. That is, we set out to show the following:

1. if $x_n \to 0$ then $|x_n| \to 0$, and

2. if $|x_n| \to 0$ then $x_n \to 0$.

To set up part (1), assume that $x_n \to 0$. Then (for a given value of $\varepsilon > 0$) what we know is that $|x_n - 0| < \varepsilon$ for all $n \geq$ some n_0. What we need to know is that $||x_n| - 0| < \varepsilon$ for all sufficiently large values of n. Now compare what we know with what we want to know.

Part (2) should work in a very similar way.

2.7.14 EXERCISE Let $(a_n)_{n \geq 1}$ be any sequence. Show (directly from the definition of the limit) that $a_n \to \ell$ if and only if $a_n - \ell \to 0$ if and only if $|a_n - \ell| \to 0$.

Roughwork

The *second* if-and-only-if is something that we know already: just put $x_n = a_n - \ell$ in the previous exercise (2.7.13). So we need only focus on the first one: that $a_n \to \ell$ if and only if $a_n - \ell \to 0$. For any given $\varepsilon > 0$, write down the definitions of what $a_n \to \ell$ and $a_n - \ell \to 0$ require, and compare them.

2.8 Combining sequences; the algebra of limits

This section offers us the first of a number of quicker ways to establish limits for relatively complicated sequences. The basic idea of these results is that whenever you can see how to build up a sequence from simpler ones whose limits you already know, it is possible to write down the limit of the complicated one just by combining the limits of its simpler 'components'. Furthermore, the proof via the definition that this is legitimate is done once and for all in the verification of the results, and there is no need to repeat that proof for each suitable example that you meet from then on.

2.8.1 Theorem Suppose that (a_n) and (b_n) are convergent sequences, with limits ℓ and m respectively. Then:

1. $a_n + b_n \to \ell + m$,
2. $a_n - b_n \to \ell - m$,
3. $a_n b_n \to \ell m$,
4. For each constant k, $k a_n \to k\ell$,
5. $|a_n| \to |\ell|$,
6. Provided that $m \neq 0$ and that no b_n is zero, also $\dfrac{a_n}{b_n} \to \dfrac{\ell}{m}$.

REMARK: There are several other ways of expressing this collection of results, including the following version:

1. $\lim(a_n + b_n) = \lim a_n + \lim b_n$,
2. $\lim(a_n - b_n) = \lim a_n - \lim b_n$,
3. $\lim(a_n b_n) = \lim a_n \lim b_n$,
4. For each constant k, $\lim(k a_n) = k \lim a_n$,
5. $\lim |a_n| = |\lim a_n|$,
6. Provided that no division by zero occurs, $\lim \dfrac{a_n}{b_n} = \dfrac{\lim a_n}{\lim b_n}$.

(The entire result can even be turned into English, thus: taking limits of convergent sequences is compatible with addition, with subtraction, with multiplication, with 'scaling' (that is, multiplying by constants), with taking modulus, and with division provided always that no illegal division by zero is attempted.)

We shall eventually provide proofs of all six parts of this highly useful result, but not all at once since we are keener just now to show how to use it. Here is a start on that project:

Proof of part (1)

Let $\varepsilon > 0$ be given. Then (of course) $\varepsilon/2$ is also a positive number.[10] Since we know that (a_n) converges to ℓ, there is some positive integer (let us call it n_1) such that

$$n \geq n_1 \ \text{ forces } \ |a_n - \ell| < \frac{\varepsilon}{2}.$$

For similar reasons, there is another positive integer (call this one n_2) such that

$$n \geq n_2 \ \text{ forces } \ |b_n - m| < \frac{\varepsilon}{2}.$$

We cannot know which of n_1, n_2 is the greater, but one of them is.[11] Write $n_0 = \max\{n_1, n_2\}$ and notice that whenever $n \geq n_0$, we get both the displayed lines working for us at the same time. Therefore, for every $n \geq n_0$:

$$|a_n + b_n - (\ell + m)| = |(a_n - \ell) + (b_n - m)| \leq |a_n - \ell| + |b_n - m| < \frac{\varepsilon}{2} + \frac{\varepsilon}{2} = \varepsilon.$$

Thus the proof of (1) is complete. (Notice the use of the triangle inequality 1.3.3 in the last line here.)

Proof of part (4)

In the special case $k = 0$, we know this already: because then (ka_n) is a constant sequence of zeroes, $k\ell$ is exactly zero and an earlier example told us that $\lim 0 = 0$. So for the rest of the proof we can assume that $k \neq 0$.

Given any $\varepsilon > 0$, notice that $\dfrac{\varepsilon}{|k|}$ is also positive (and that we needed the modulus signs on k to make that true). Since $a_n \to \ell$, there must exist a positive integer (call it n_0, for instance) such that

$$n \geq n_0 \ \text{ forces } \ |a_n - \ell| < \frac{\varepsilon}{|k|}.$$

---

[10] Why did we make that choice? Well, to show that $a_n + b_n$ converges to $\ell + m$, we need to arrange that $|a_n + b_n - (\ell + m)|$ shall be less than ε. Rearrange that desired punch-line into $|a_n - \ell + b_n - m| < \varepsilon$. Each of $|a_n - \ell|$ and $|b_n - m|$ can be made as small as we please...and if we make each of them less than one half of ε, then their combined total will be smaller than two halves of ε, which is exactly what we needed. Now, rejoin the main text.

[11] Unless, of course, they happen to be equal, in which case it really doesn't matter which one you choose.

Then also

$$n \geq n_0 \text{ forces } |ka_n - k\ell| = |k(a_n - \ell)| = |k||a_n - \ell| < |k|\frac{\varepsilon}{|k|} = \varepsilon$$

which is exactly what the definition asked in order to show that $ka_n \to k\ell$.

2.8.2 **EXERCISE** Construct a proof of part (2) of the theorem. (You should find that an argument very like that given for part (1) will be convincing; after all, addition and subtraction are quite similar operations.)

Moving towards the *use* of this theorem now, notice for a start that (according to part (3), and remembering from Example 2.7.2 that $\frac{1}{n}$ tends to zero)

$$\lim\left(\frac{1}{n^2}\right) = \lim\left(\frac{1}{n}\frac{1}{n}\right) = \lim\left(\frac{1}{n}\right)\lim\left(\frac{1}{n}\right) = 0.0 = 0$$

and, consequently, that

$$\lim\left(\frac{1}{n^3}\right) = \lim\left(\frac{1}{n^2}\frac{1}{n}\right) = \lim\left(\frac{1}{n^2}\right)\lim\left(\frac{1}{n}\right) = 0.0 = 0$$

and so on. Consequently $\frac{1}{n^k} \to 0$ for each positive integer k. (It is easy to prove this formally by induction,[12] if you wish to try it.) This observation, combined with pieces of our main theorem, allows us to deal with a large number of limit problems quickly and painlessly. Consider, for instance:

2.8.3 **Example** To establish the convergence of the sequence (a_n) described by

$$a_n = \frac{15n^2 + n + 1}{5n^2 - n - 2}.$$

Solution

Begin by dividing the numerator and denominator of the fraction formula by the largest power of n appearing, n^2 (which dividing, of course, doesn't change the fraction in the least):

$$a_n = \frac{15n^2 + n + 1}{5n^2 - n - 2} = \frac{15 + n^{-1} + n^{-2}}{5 - n^{-1} - 2n^{-2}} \to \frac{15 + 0 + 0}{5 - 0 - 2(0)} = \frac{15}{5} = 3.$$

We are finished already! Notice, however, how many aspects of the key theorem were used in that last line: part (6) to let us deal with the numerator and

---

[12] If this technique is not familiar to you, wait until Section 4.2 where we shall discuss it in detail.

denominator separately, parts (1) and (2) to add/subtract the separate parts of each and part (4) to see that $\lim 2n^{-2} = 2 \lim n^{-2}$. It is not considered necessary to write out all of these moves, but be aware of them.

2.8.4 **Example**   To find the limit (as $n \to \infty$) of:

$$\left| \frac{n^2 - 2n + 5}{n^3 + n^2 + 7} - \left| \frac{n - \left| \frac{2-3n}{4+5n} \right|}{3n + \pi} \right| \right|.$$

Solution

(Despite its forbidding appearance, this expression has been built up layer by layer from simple pieces, and we only need to 'shadow' its construction to get the answer.) To begin with, dividing top and bottom by n:

$$\frac{2 - 3n}{4 + 5n} = \frac{2n^{-1} - 3}{4n^{-1} + 5} \to \frac{2(0) - 3}{4(0) + 5} = \frac{-3}{5}$$

and therefore (using part 5 of the theorem)

$$\left| \frac{2 - 3n}{4 + 5n} \right| \to \left| \frac{-3}{5} \right| = \frac{3}{5}.$$

Next, again dividing top and bottom by n:

$$\frac{n - \left| \frac{2-3n}{4+5n} \right|}{3n + \pi} = \frac{1 - n^{-1} \left| \frac{2-3n}{4+5n} \right|}{3 + \pi n^{-1}} \to \frac{1 - 0 \left(\frac{3}{5} \right)}{3 + \pi(0)} = \frac{1}{3}$$

and consequently

$$\left| \frac{n - \left| \frac{2-3n}{4+5n} \right|}{3n + \pi} \right| \to \left| \frac{1}{3} \right| = \frac{1}{3}.$$

Moving on (but still digesting fractional formulas by dividing top and bottom by the largest power involved, currently n^3):

$$\frac{n^2 - 2n + 5}{n^3 + n^2 + 7} = \frac{n^{-1} - 2n^{-2} + 5n^{-3}}{1 + n^{-1} + 7n^{-3}} \to \frac{0 - 2(0) + 5(0)}{1 + 0 + 7(0)} = \frac{0}{1} = 0.$$

Finally, we put the whole construction together:

$$\left| \frac{n^2 - 2n + 5}{n^3 + n^2 + 7} - \left| \frac{n - \left| \frac{2-3n}{4+5n} \right|}{3n + \pi} \right| \right| \to \left| 0 - \frac{1}{3} \right| = \left| -\frac{1}{3} \right| = \frac{1}{3}.$$

2.8.5 **EXERCISE**   Use the algebra of limits to find the limit of each of the sequences whose n^{th} terms are given by the following formulae:

1.
$$\frac{1 - 2n + 3n^2 - 4n^3}{-5 - 6n + 7n^2 + 8n^3}$$

2.
$$\left(\frac{11 - 6n}{5n + 2}\right)^5$$

3.
$$\left|\frac{9 - 4n}{1 + n + 7n^2} - \frac{9 - 4n + n^2}{1 - n + 7n^2}\right|$$

4.
$$\left|\left(\frac{\pi^2 n^2}{\pi n^2 + \pi^3}\right)\left|\frac{3 - n^4}{3 + n^4} + \left|\frac{1 - \pi n}{2 + \pi^2 n}\right|\right| - 3\right|$$

Remark

In the first of these, be sure to check that the denominator of the given fraction is never zero (because otherwise the relevant part of the key theorem could not be used). In the fourth of these exercises, keep in mind the traditional 'order of precedence' of the arithmetical operations: for example, that multiplications and divisions are always done *before* additions and subtractions, except where brackets or other 'enclosing' operations dictate otherwise (since material inside brackets and the like needs to be evaluated first). In this context, pairs of modulus signs behave like brackets.

2.8.6 **EXERCISE**   Prove part (5) of the algebra of limits theorem. That is, given that $a_n \to \ell$, show that $|a_n| \to |\ell|$.

Draft solution

For a slick solution, almost all you will need is the reverse triangle inequality 1.3.4

$$\left|\,|x| - |y|\,\right| \le |x - y|$$

(where x, y are any real numbers).

2.8.7 **A look forward**   What we have done so far about detecting and proving limits of sequences works fine provided that *either* we can sensibly guess the limit and then come up with an overestimate of the error term that is simple enough to work with, *or else* we can break the typical term down into simpler pieces whose separate limits we already know. Unfortunately, many important and useful sequences don't fall into either of those categories. For instance (look back to a previous example 2.2.1 (7)), although it is not very hard to guess the limit of $n^{1/n}$, it is then far from obvious how we could estimate the error-gap between that guess and the typical term. Again, in the cases of $\left(\,(1 + \frac{1}{n})^n\,\right)_{n \ge 1}$ and of

$$\left(1 + \frac{1}{2} + \frac{1}{3} + \frac{1}{4} + \cdots + \frac{1}{n}\right)_{n \geq 1},$$

although a few minutes' roughwork with a calculator will strongly suggest that the nth term of each is getting steadily bigger as n increases, it is hardly self-evident whether or not they are getting steadily closer to some limiting 'ceiling' or, if so, what number that ceiling might be. In short, we need more techniques, more analytic technology, to tackle such questions. The aim of the next chapter is to develop and deploy some of that technology.

2.9 POSTSCRIPT: to infinity

It is built into our definition of convergence of a sequence $(x_n)_{n \in \mathbb{N}}$ to a limit, as $n \to \infty$, that the limit is always a real number: it is not a mysterious symbol such as ∞ or $-\infty$. Indeed, many useful theorems would collapse if we did not define convergence in this fashion: in particular, several of the algebra of limits results would have fallen into chaos by inviting us to calculate such expressions as $\infty/\infty, \infty - \infty$ and 0 times ∞. Yet there are circumstances in which the idea of a sequence 'moving towards infinity or minus infinity' can be useful and can be made precise. Indeed, we have used one such scenario many times already, for in every sequential convergence question the label n that serves to identify the nth term x_n does indeed '$\to \infty$' as the familiar notation says.[13] Hopefully it has been clear that the phrase '$n \to \infty$' simply means that n becomes enormously big, bigger than any possible upper bound or ceiling n_0 that could have been suggested. It is this insight that allows us to give a clear definition of a sequence tending to infinity also:

2.9.1 **Definition**   A sequence $(x_n)_{n \in \mathbb{N}}$ is said to *tend to infinity* or *diverge to infinity* if:

> for each $K > 0$ there is some positive integer n_K such that
> $x_n > K$ for every $n \geq n_K$.

When this is so, we write $x_n \to \infty$ or even $\lim_{n \to \infty} x_n = \infty$, but we must take care not to treat $\lim_{n \to \infty} x_n$ in such cases as if it were a number (for instance, by using it in an arithmetical calculation). Avoid saying that '(x_n) *converges to infinity*' since, in fact, (x_n) is not convergent at all.

In a similar way, we can define the idea of a sequence tending to minus infinity:

---

[13] Likewise, it is unlikely to surprise anyone if we claim that $n^2 \to \infty$ and $n^3 \to \infty$ (as $n \to \infty$).

2.9.2 Definition   A sequence $(x_n)_{n \in \mathbb{N}}$ is said to *tend to minus infinity* or *diverge to minus infinity* if:

for each $K < 0$ there is some positive integer n_K such that $x_n < K$ for every $n \geq n_K$.

When this is so, we write $x_n \to -\infty$ or even $\lim_{n \to \infty} x_n = -\infty$.

2.9.3 Example   To show that $n^3 - 5n^2 - 2n - 17 \to \infty$ as $n \to \infty$.

Roughwork

Given $K > 0$, we need to guarantee that, for all values of n that are big enough, $n^3 - 5n^2 - 2n - 17 > K$. Now (to simplify the algebra, just as we did for convergent sequences)[14]

1. $5n^2 \leq \frac{1}{10}n^3$ provided that $n \geq 50$,

2. $2n \leq \frac{1}{10}n^3$ provided that $20 \leq n^2$ and therefore surely if $n \geq 5$,

3. $17 \leq \frac{1}{10}n^3$ provided that $170 \leq n^3$, and therefore surely if $n \geq 6$.

Hence, $n \geq 50$ will force

$$n^3 - 5n^2 - 2n - 17 \geq n^3 - \frac{1}{10}n^3 - \frac{1}{10}n^3 - \frac{1}{10}n^3 = \frac{7}{10}n^3.$$

Then, to force $\frac{7}{10}n^3 > K$, it is good enough to take $n > \sqrt[3]{\frac{10K}{7}}$ ...

Solution

Given $K > 0$, choose an integer $n_K > \max \left\{ 50, \sqrt[3]{\frac{10K}{7}} \right\}$. Then (as shown in the roughwork) $n \geq n_K$ will guarantee that

$$n^3 - 5n^2 - 2n - 17 \geq \frac{7}{10}n^3 > K.$$

2.9.4 EXERCISE   Show that

$$\frac{3n^4 - 7n^2}{2n^2 + n - 1} \to \infty \quad \text{as} \quad n \to \infty.$$

---

[14] Sometimes, even the roughwork needs roughwork. Here, as n becomes bigger, the n^3 will become much bigger and more important than the other pieces. To bring this out, we ought to force each of $5n^2$, $2n$ and 17 to be less than a small fraction of the n^3...for instance, less than one tenth of it. How do we make $5n^2$ less than a tenth of n^3? By ensuring that $50n^2 < n^3$...and any value of n over 50 will do that. Then the same kind of discussion will handle $2n$ and 17. At this point you can rejoin the text and see why we restrict n as we do.

2.9.5 **EXERCISE**   Write out a formal proof of the following (almost immediate) consequence of the definitions: $x_n \to -\infty$ if and only if $-x_n \to \infty$.

2.9.6 **Example**   To verify that

$$\frac{100 - n^5}{200 + n^2 + n^4} \to -\infty \text{ as } n \to \infty.$$

2.9.7 **Roughwork**   By the previous exercise, it is equivalent to show that

$$\frac{n^5 - 100}{200 + n^2 + n^4} \to \infty \text{ as } n \to \infty$$

(which will be slightly easier since there are fewer minuses involved). Given $K > 0$, we therefore need to arrange (just by taking large enough values of n) that

$$\frac{n^5 - 100}{200 + n^2 + n^4} > K.$$

This is too hard an inequality to solve directly, so we should try to replace $\dfrac{n^5 - 100}{200 + n^2 + n^4}$ by a *smaller but simpler* fraction that we can still make greater than K: and this will involve replacing the numerator $n^5 - 100$ by something *smaller*, but the denominator $200 + n^2 + n^4$ by something *bigger*. As n becomes large, it will be the n^5 that matters more in the numerator – for instance, if n is at least 4, then the 100 will be smaller than one tenth of n^5, and the numerator will then be bigger than nine tenths of n^5.

In the denominator, it will be the n^4 (again, the biggest power of n appearing) that matters most – for instance, if n is at least 2 then n^2 will be smaller than n^4, and if n is at least 4 then 200 will be smaller than n^4. So $n \geq 4$ will guarantee that $200 + n^2 + n^4$ is less than $n^4 + n^4 + n^4$.

Summary so far: provided than $n \geq 4$, we have that

$$\frac{n^5 - 100}{200 + n^2 + n^4} > \frac{0.9n^5}{n^4 + n^4 + n^4} = \frac{0.9n^5}{3n^4} = \frac{3n}{10}.$$

It will be easy to ensure that this final fraction is greater than K: we need only make n exceed $10K/3$. A formal solution can now be compiled as follows:

2.9.8 **Solution**   Given $K > 0$, we choose n_K to be any positive integer that is greater than both 4 and $\frac{10K}{3}$. Then (as seen in the roughwork) for any $n \geq n_K$, we shall have

$$\frac{n^5 - 100}{200 + n^2 + n^4} > \frac{0.9n^5}{n^4 + n^4 + n^4} = \frac{0.9n^5}{3n^4} = \frac{3n}{10} \geq \frac{3}{10}n_K > K.$$

This shows

$$\frac{n^5 - 100}{200 + n^2 + n^4} \to \infty$$

which, by the preceding exercise, is equivalent to the question posed.

2.9.9 **EXERCISE** Show from the definitions that:

1. if $x_n \geq y_n$ for all $n \geq$ some n_0, and $\lim_{n \to \infty} y_n = \infty$, then $\lim_{n \to \infty} x_n = \infty$ also.

2. if $x_n \leq y_n$ for all $n \geq$ some n_0, and $\lim_{n \to \infty} y_n = -\infty$, then $\lim_{n \to \infty} x_n = -\infty$ also.

Roughwork towards part 1

The task is, given $K > 0$, to obtain evidence that $x_n > K$ for all sufficiently large values of n. What we know is that $y_n > K$ for all sufficiently large values of n, and also that $x_n \geq y_n$ for every $n \geq n_0 \dots$

Some fragments of the algebra of limits are still valid for sequences that diverge to ∞ or $-\infty$; for instance:

2.9.10 **Theorem**

1. if $x_n \to \infty$ and $y_n \to \infty$ (as $n \to \infty$) then $x_n + y_n \to \infty$ also;

2. if $x_n \to \infty$ (as $n \to \infty$) and there is some positive constant A such that $-A \leq y_n \leq A$ for every $n \in \mathbb{N}$ [15] then $x_n + y_n \to \infty$ also;

3. if $x_n \to \infty$ and A is a positive constant then $A x_n \to \infty$;

4. if $x_n \to \infty$ and A is a negative constant then $A x_n \to -\infty$;

5. if $x_n \to \infty$ then $\dfrac{1}{x_n} \to 0$ (as $n \to \infty$);

6. if $x_n \to -\infty$ then $\dfrac{1}{x_n} \to 0$ (as $n \to \infty$);

7. if $x_n > 0$ for all $n \geq$ some n_0 and $x_n \to 0$ then $\dfrac{1}{x_n} \to \infty$ (as $n \to \infty$);

8. if $x_n < 0$ for all $n \geq$ some n_0 and $x_n \to 0$ then $\dfrac{1}{x_n} \to -\infty$ (as $n \to \infty$).

(Notice that in 5 and 6 we have been slightly cavalier about never dividing by zero. If some of the x_n were zero then the corresponding $1/x_n$ would not be defined. However, in each case x_n is eventually becoming very big (whether positive or negative) and therefore certainly non-zero; to recover the result in all cases, we only have to ignore the first few terms where zero might have occurred and, as usual, ignoring finitely many terms has no effect upon a limit.)

---

[15] that is, $(y_n)_{n \in \mathbb{N}}$ is bounded: see paragraph 4.1.4 for more on this.

Sample proofs

All of the above are proved in a routine[16] way from the definitions, so we shall only demonstrate a few.

Roughwork towards 1.

Given $K > 0$, we need to show that $x_n + y_n > K$ for all $n \geq$ some threshold value. Now we are told that $x_n > K$ for all $n \geq$ some n_0, and also that $y_n > K$ for all $n \geq$ some possibly different n_1. What if we combine these inequalities while $n \geq \max\{n_0, n_1\}$? Remember that K is positive, so $2K > K$.[17]

Proof of 2.

Begin by choosing A as in 2., and let K be a given positive constant.

(*Roughwork*: we need to arrange that $x_n + y_n > K$ (for large values of n) and, knowing that $y_n \geq -A$ in any case, it will be enough to get $x_n > K + A$; but this we can do, because $x_n \to \infty$.)

Since $x_n \to \infty$, we can find a positive integer n_0 such that, for each $n \geq n_0$, $x_n > K + A$. Since we also know that $y_n \geq -A$, adding the two inequalities gives $x_n + y_n > K$. Therefore $x_n + y_n \to \infty$.

Proof of 4.

Suppose that $x_n \to \infty$ and that A is a negative constant. Given $K < 0$, we need to think how to ensure that $Ax_n < K$ and, because *negative* multipliers can cause confusion in inequalities, it is safer to write that as $-Ax_n > -K$, that is, as $|A|x_n > |K|$, or as $x_n > \frac{|K|}{|A|}$.

Since $x_n \to \infty$ and $\frac{|K|}{|A|}$ is positive, there exists[18] $n_0 \in \mathbb{N}$ such that $x_n > \frac{|K|}{|A|}$ for all $n \geq n_0$: that is, $|A|x_n > |K|$ or $-Ax_n > -K$ or $Ax_n < K$. Therefore $Ax_n \to -\infty$.

---

[16] Incidentally, *routine* is not the same as *easy*! By calling an argument routine, all we mean is that it is built up by putting together the definitions and the 'obvious' results in a rather predictable fashion, without depending on surprise insights. That may or may not be brief or easy – in some cases, it is neither.

[17] Incidentally, a slightly more elegant solution could begin with $x_n > K/2$ for all $n \geq$ some n_0 and $y_n > K/2$ for all $n \geq$ some n_1.

[18] This is similar to the trick we discussed in the roughwork to 2.7.12; once we are told that $x_n \to \infty$, then x_n can be made larger not only than a particular positive constant such as $|K|$ that we are challenged with, but also than any positive expression built from that constant that suits our purpose, such as $|K|/|A|$.

Proof of 7.

Supposing that $x_n > 0$ for all $n \geq$ some n_0 and that $x_n \to 0$; let K be any given positive constant. Then $\frac{1}{K}$ is positive[19] so there exists $n_1 \in \mathbb{N}$ such that $|x_n - 0| < \frac{1}{K}$ for all $n \geq n_1$: indeed, if $n \geq n_0$ as well, then $0 < x_n < \frac{1}{K}$ for all such n. Therefore

$$n \geq \max\{n_0, n_1\} \Rightarrow 0 < x_n < \frac{1}{K} \Rightarrow \frac{1}{x_n} > K$$

which, since K can be arbitrarily big, delivers $1/x_n \to \infty$.

2.9.11 **EXERCISE** Choose any other two parts of this theorem and write out proofs for them.

2.9.12 **Example** To obtain an easier proof that

$$\frac{100 - n^5}{200 + n^2 + n^4} \to -\infty \ as \ n \to \infty.$$

2.9.13 **Solution** Once we divide the top and bottom lines by n^5, the biggest power of n appearing, it is immediate from the algebra of limits that

$$\frac{200 + n^2 + n^4}{100 - n^5} \to 0 \ as \ n \to \infty$$

and we also notice that the expression is negative for all $n \geq 3$. The conclusion follows from part 8 above.

2.9.14 **EXERCISE** Decide (with proof) whether each of the following is true or false:

- If $x_n \to 0$ and no x_n is exactly zero, then either $1/x_n \to \infty$ or $1/x_n \to -\infty$.
- If $x_n \to 0$ and no x_n is exactly zero, then $1/|x_n| \to \infty$.

2.9.15 **Remark** Once we had checked that $1/n \to 0$, we learned from the algebra of limits that its various positive-integer powers

$$\frac{1}{n^2}, \frac{1}{n^3}, \frac{1}{n^4}, \frac{1}{n^5},$$

and so on, also converged to zero. This conclusion is not, however, restricted to *integer* powers:

---

[19] Where did that come from? Once again, it comes from thinking about what we need to show. In order to prove that $\frac{1}{x_n} \to \infty$, we need to get $\frac{1}{x_n} > K$ for big values of n and, provided that x_n is positive, that is the same as asking for $x_n < \frac{1}{K}$. Can we make that happen? Yes, because $x_n \to 0$.

2.9.16 Example   To use part (5) of the preceding theorem to show that, for any positive real number a,

$$\frac{1}{n^a} \to 0.$$

Solution

By the referenced part (5), it is good enough to show that $n^a \to \infty$.

(*Roughwork*: for each given positive K, we need to arrange that $n^a > K$, that is, that $n > \sqrt[a]{K} \ldots$)

Given $K > 0$, we note that $\sqrt[a]{K}$ is also merely a real number, so we can find a positive integer n_0 that is larger: $n_0 > \sqrt[a]{K}$. Then for any $n \geq n_0$ we have $n > \sqrt[a]{K}$ and therefore[20] $n^a > K$. In other words, the sequence (n^a) diverges to ∞, as desired.

2.10 Important note on 'elementary functions'

The ideal way to develop any mathematical text is, of course, to begin with elementary material that everyone already agrees upon and, using that as basis, to work step-wise through more sophisticated matters, always establishing the truth of any newly encountered result as a consequence of earlier ones. While it would be possible to do precisely this in an account of real analysis, and while it is indeed essential as regards the evolution of the *theory*, the suite of *examples* that rightfully illustrate that theory would be somewhat sterile and flavourless if we adhered too rigidly to the policy.

The reason is that the so-called elementary functions[21] such as $\sin x, \cos x, \ln x$ and e^x, and their basic properties, will already be familiar to anyone who sets out to read this text, and we can greatly increase the illustrative power of our examples[22] by using them: but a logically sound definition of these functions requires a surprising deal of preparation, and so therefore does proper mathematical proof that they do behave, in all circumstances, in the ways in which those readers believe that they do. It will not, in fact, be until Chapter 18 that we set out definitions of the four functions just named.

---

[20] Here we are taking it for granted that larger positive numbers have larger a^{th} powers which, although true, will not be amenable to formal demonstration until we have properly defined x^a for all real a and positive real x.

[21] Note that there are a number of notations in use for the natural logarithm function in particular; we have chosen to use 'ln', but 'log' and '$\log_e$' are also widespread.

[22] We should stress that expressions such as $\sin x$ and e^x are *functions* rather than *sequences*, since it is to be understood that their control variable x is a real number rather than just a positive integer. However, we can use such functions to build a wide variety of sequences, such as $(\sin n), \left(n \sin \left(\frac{1}{n}\right)\right), \left(e^{1+\sin(\pi n/4)}\right), (\ln(1 + 1/n))$ and so on, n being a positive integer in each case.

Pending our future encounter with those definitions and proofs, it seems right that we allow ourselves to mobilise certain basic information concerning $\sin x$, $\cos x$, $\ln x$ and e^x provided that we take care not to use it in the development of the theory, but only in examples, and provided that we do eventually get around to showing that this information is reliable. The following summary lists explicitly the details, concerning the four functions, that we temporarily accept for use in examples (and that we promise, in the long run, to establish).

1. • $\sin : \mathbb{R} \to [-1, 1]$ is an odd, periodic function in the sense that

$$\sin(-x) = -\sin x \quad \text{and} \quad \sin(x + 2\pi) = \sin x \quad \text{in all cases.}$$

- $-1 \leq \sin x \leq 1$ for all real x.
- The derivative of $\sin x$ is $\cos x$.
- $\sin 0 = \sin \pi = 0, \sin(\pi/2) = 1, \sin(-\pi/2) = -1$.

2. • $\cos : \mathbb{R} \to [-1, 1]$ is an even, periodic function in the sense that

$$\cos(-x) = \cos x \quad \text{and} \quad \cos(x + 2\pi) = \cos x \quad \text{in all cases.}$$

- $-1 \leq \cos x \leq 1$ for all real x.
- The derivative of $\cos x$ is $-\sin x$.
- $\cos 0 = 1, \cos \pi = -1, \cos(\pi/2) = \cos(-\pi/2) = 0$.
- $\sin^2 x + \cos^2 x = 1$.
- $\cos x = \sin\left(\frac{\pi}{2} - x\right)$ for all x.

3. • $\ln : (0, \infty) \to \mathbb{R}$ is an increasing function.
- $\ln 1 = 0, \ln(xy) = \ln x + \ln y, \ln(x/y) = \ln x - \ln y, \ln(x^y) = y \ln x$.
- The derivative of $\ln x$ is $1/x$.
- For positive values of x very close to 0, $\ln x$ is enormous but negative.
- For sufficiently big positive values of x, $\ln x$ is enormous and positive.

4. • The exponential function e^x is an increasing function from $\mathbb{R}$ to $(0, \infty)$: all its values are *strictly* positive.
- It equals its own derivative.
- $e^0 = 1, e^{x+y} = e^x e^y, e^{x-y} = e^x/e^y, (e^x)^y = e^{xy}$.
- For very large negative x, e^x is extremely small (but still positive).
- For large positive x, e^x is extremely big (and, of course, positive).
- $\ln(e^x) = x$ for all real x, $e^{\ln x} = x$ for all positive x.

3 Interlude: different kinds of numbers

3.1 Sets

It is convenient for us to use a little of the language and symbolism of set theory, although the theory itself lies beyond the scope of this text. By the term **set** we mean a well-defined collection of distinct objects that are called its **elements**. By **well-defined** we mean that each object either definitely is an element of the set in question, or definitely is not: there must be no borderline cases. (So, for instance, we have to avoid ideas such as 'the set of all very large integers' or 'the set of numbers that are extremely close to 3': for it would be a matter of opinion and context which objects were to belong to such ill-defined collections.) By **distinct** we mean in practice that repetitions among the elements of a set are not allowed, so that, for instance, the *set* of prime factors of 360 ($= 2^3 \times 3^2 \times 5$) has only the *three* elements 2, 3 and 5, although the *list* (2, 2, 2, 3, 3, 5) of its prime factors comprises *six* items.

3.1.1 Notation   If S is a set then

- $x \in S$ says that x is one of the elements of S,
- $x \notin S$ says that x is not one of the elements of S.

Enclosing a list of its elements within curly brackets is a quick way to create a symbol for a small, simple set. For instance, $\{2, 3, 5\}$ denotes the set of prime factors of 360, and $\{1, 4, 9, 16, 25, 36, 49, 64, 81\}$ is a reasonably tidy notation for the set of perfect squares that are less than one hundred. The same style can be used for larger sets, even including some infinite sets, provided that the pattern within the listing is really obvious: most people will accept $\{1, 2, 3, 4, 5, 6, \cdots 98, 99, 100\}$ as 'clearly' meaning the set of the first one hundred positive integers, and $\{2, 4, 8, 16, 32, 64, \cdots\}$ as the (infinite) set of all the powers of 2.[1] For more complicated sets, however, this type of notation really doesn't work.

The sets we particularly work with are sets of real numbers – either real numbers of a particular type, or a *selection* of real numbers lifted out for some particular discussion. Some of these have turned out to be useful so often in the past that there are now standard symbols for them that should be known and recognised:

---

[1] More precisely, the powers of two in which the index is a positive whole number.

Undergraduate Analysis: A Working Textbook, Aisling McCluskey and Brian McMaster 2018.
© Aisling McCluskey and Brian McMaster 2018. Published 2018 by Oxford University Press

3.1.2 Number systems

1. $\mathbb{N}$ is the set of **positive integers**,[2] the set of whole numbers that are greater than zero. In this case, the curly bracket notation is clear enough:
$\mathbb{N} = \{1, 2, 3, 4, 5, \cdots\}$.

2. $\mathbb{Z}$ is the set of all **integers** – positive, negative and zero whole numbers. The symbol comes from the German word *Zahl* meaning *number*. Sometimes we stretch the list-in-curly-brackets notation and display it as
$\mathbb{Z} = \{\cdots -3, -2, -1, 0, 1, 2, 3, 4, \cdots\}$.

3. $\mathbb{Q}$ is the set of **rational** numbers, those that can be expressed exactly as a fraction in the usual sense of the word, that is, as an integer divided by a non-zero integer. For instance, $\frac{2}{3} \in \mathbb{Q}$, $1.4 \in \mathbb{Q}$ because $1.4 = \frac{7}{5}$ exactly, $-3 \in \mathbb{Q}$ because $-3 = \frac{-3}{1}$ (or, if you feel it is somehow cheating to have 1 as the unnecessary denominator of a fraction, then $-3 = \frac{-6}{2}$ is an alternative reason).

4. $\mathbb{R}$ is the set of all **real** numbers. (So, for instance, $\pi/e \in \mathbb{R}$ but $\sqrt{-1} \notin \mathbb{R}$.)

It is not obvious to common sense alone that there are any real numbers that are not rational. You may have seen proofs that surds such as $\sqrt{2}, \sqrt{5}, \sqrt[3]{12}$ and so on are not rational but, if not, here is a sample argument that is worth reading carefully as an illustration of proof by contradiction. We shall take it as given that every positive integer n (except 1) can be expressed as a product of powers of prime numbers, and that (apart from the order in which these are written) such a prime decomposition is unique for each n.

3.1.3 Proposition   The real number $\sqrt{35}$ is not a rational number.

Proof

Suppose it were. Then it must be possible to find two integers p and q such that $\sqrt{35} = \frac{p}{q}$. Then $35 = \frac{p^2}{q^2}$ or, more simply,

$$p^2 = 35q^2.$$

(Now pick on one of the primes that appear to be involved in that last equation– say, the 5. The 7 would do equally well.)

Let 5^a be the power of 5 that appears in the prime decomposition of p, and 5^b be the power of 5 that appears in the prime decomposition of q. (We are not ruling out

---

[2] We have avoided using the phrase 'natural numbers'. Some writers use the term *natural numbers* as a synonym for the positive integers, and others take it to mean the set comprising the positive integers and 0 itself. In addition, some writers use the symbol $\mathbb{N}$ to mean the set of natural numbers rather than the set of positive integers. So $0 \in \mathbb{N}$ in some books and $0 \notin \mathbb{N}$ in others! Be aware of this possibly confusing point if you are reading a range of textbooks.

that 5 may not be a prime factor at all of p or of q, for a or b might be zero.) Then 5^{2a} is the power appearing in (the decomposition of) p^2, and 5^{2b} the power appearing in q^2. Also the power appearing in $35q^2 = 5 \times 7 \times q^2$ will be 5^{2b+1}. However, p^2 and $35q^2$ are the same number so, from the uniqueness of prime decomposition, 5^{2a} and 5^{2b+1} must be the same thing – which tells us that $2a = 2b + 1$ and that

$$a - b = \frac{1}{2}.$$

Since a and b are integers, that is impossible. The contradiction completes the proof (for it shows that it is impossible for $\sqrt{35}$ to be rational).

Real numbers that are not rational are called **irrational**. We shall explore a few aspects of the relationship between reals, rationals and irrationals in the third section of this chapter.

3.1.4 Definition If A and B are sets, and if every element of A is also an element of B, then A is called a **subset** of B, and we write $A \subseteq B$. For example, $\mathbb{N}, \mathbb{Z}$ and $\mathbb{Q}$ are subsets of $\mathbb{R}$, while $\mathbb{N}$ and $\mathbb{Z}$ are subsets of $\mathbb{Q}$, and $\mathbb{N}$ is a subset of $\mathbb{Z}$. Notice that the wording of the definition makes every set a subset of itself: $A \subseteq A$ merely because every element of A is (of course) an element of A.

There are ways of combining two (or more) sets that will also help us to discuss some matters in analysis:

3.1.5 Definition Let A and B be any sets. We define

1. their **union** $A \cup B$ to be the set of all objects that are elements of A or of B (or of both A and B),

2. their **intersection**[3] $A \cap B$ to be the set of all objects that are elements both of A and of B,

3. the **set difference** $A \setminus B$ to be the set of all objects that are elements of A but not elements of B.

We can usefully re-write those definitions, concentrating on which objects are elements of the three defined sets:

1. $x \in A \cup B$ says precisely that $x \in A$ or $x \in B$ (or both),

2. $x \in A \cap B$ says precisely that $x \in A$ and $x \in B$,

3. $x \in A \setminus B$ says precisely that $x \in A$ but $x \notin B$.

It is perfectly possible for the intersection of two sets not to include any elements at all. For this reason among others, it is useful to admit the idea of an **empty set** – a set that does not include any objects as elements. The standard symbol for such a set is $\emptyset$. Naturally, most of the sets we deal with in practice are **non-empty**.

---

[3] (sometimes informally called their *overlap*)

3.1.6 Selection   Whenever A is a set and $P(x)$ is a statement that makes sense[4] for each element x of A, the notation

$$\{x \in A : P(x)\} \quad or \quad \{x \in A \mid P(x)\}$$

means the subset of A comprising just those elements of A for which the statement in question is true. This is a much more versatile style of notation than the list-in-curly-brackets that we previously presented. The whole symbol is usually pronounced as 'the set of all x in A such that $P(x)$ is true' (and the words 'is true' can be left out). Whether you use a colon (:) or a vertical (|) half-way through the symbol is a matter of taste and readability; for instance, if $P(x)$ begins with something like $|x|$ or already involves a colon, then using the *other* divider will help the eye to take in quickly what is written. Of course, this *selection* notation only applies to sets that are subsets of some pre-existing set, but that is not a problem for us since virtually all the sets we need to work with are subsets of the real number system $\mathbb{R}$. Here are a few illustrations:

- $$\left\{x \in \mathbb{R} \mid x = \frac{p}{q} \text{ for some integers } p, q \text{ where } q \neq 0\right\}$$

  – is the definition of the set $\mathbb{Q}$ of rationals.

- $$\{x \in \mathbb{R} : x \notin \mathbb{Q}\}$$

  – is the set $\mathbb{R} \setminus \mathbb{Q}$ of irrationals.

- $$\{x \in \mathbb{R} \mid x = 2^n, \text{ some } n \in \mathbb{N}\}$$

  – is the set we previously (and a little clumsily) wrote as $\{2, 4, 8, 16, 32, 64, \cdots\}$.

- $$\{x \in \mathbb{R} : |x| < 3\}$$

  – is the 'solid block' of real numbers lying between -3 and 3. We shall next turn our attention towards such unbroken ranges of real numbers.

3.2 Intervals, max and min, sup and inf

An *interval* is a subset of the real line $\mathbb{R}$ that – intuitively speaking – has no gaps in it, but stretches unbroken across a connected region of $\mathbb{R}$. It is easy to turn that intuitive impression into a sharp definition:

3.2.1 Definition   A non-empty subset I of $\mathbb{R}$ is an *interval* if $x \in I, z \in I, x < y < z$ together imply that $y \in I$.

---

[4] That is to say, for each individual $x \in A$, $P(x)$ is either true or false – once again, there must be no borderline cases.

There are several different kinds of interval, depending on whether the subset extends limitlessly far up or down the real line, and on whether it includes or excludes points 'at the edge' (properly called *endpoints* of the interval), and we list all of them here, together with an exact set-theoretic description of each type (in each case, a and b denote arbitrary real numbers, with $a < b$ if appropriate):

1. $(a, b) = \{x \in \mathbb{R} : a < x < b\}$
2. $[a, b] = \{x \in \mathbb{R} : a \leq x \leq b\}$
3. $(a, b] = \{x \in \mathbb{R} : a < x \leq b\}$
4. $[a, b) = \{x \in \mathbb{R} : a \leq x < b\}$
5. $[a, a] = \{x \in \mathbb{R} : a \leq x \leq a\} = \{a\}$
6. $(a, \infty) = \{x \in \mathbb{R} : a < x\}$
7. $[a, \infty) = \{x \in \mathbb{R} : a \leq x\}$
8. $(-\infty, b) = \{x \in \mathbb{R} : x < b\}$
9. $(-\infty, b] = \{x \in \mathbb{R} : x \leq b\}$
10. $(-\infty, \infty) = \mathbb{R}$

(Incidentally, some texts do not class case (5) as an interval at all, and some others call it a *degenerate interval*.)

The numbers a and b appearing in these descriptions are referred to as the *endpoints* of the relevant interval, and all the other points of an interval are called its *interior points*. It is important to bear in mind that an endpoint of an interval may or may not itself belong to that interval. Also notice that the symbols ∞ and $-\infty$ are not counted as endpoints, mainly because (whatever they are) they are not real numbers: their purpose in these notations is just to draw attention to the *absence* of a right-hand or a left-hand endpoint.

The first five cases in our list are called *bounded* intervals. Cases (6) and (7) are called *bounded below* (but not bounded above) while cases (8) and (9) are called *bounded above* (but not bounded below). These ideas can be extended to apply to *any* subsets of the real line:

3.2.2 Definition   An *upper bound* of a set A of real numbers means a number u such that $a \leq u$ for *every* $a \in A$. The set A is called *bounded above* if it has an upper bound. Of course, if u is an upper bound of A then any number bigger than u is also an upper bound of A.

3.2.3 Definition   A *lower bound* of a set A of real numbers means a number l such that $l \leq a$ for *every* $a \in A$. The set A is called *bounded below* if it has a lower bound. Of course, if l is a lower bound of A then any number smaller than l is also a lower bound of A.

3.2.4 Definition   The set A is called *bounded* if it is both bounded above and bounded below: that is, if it has an upper bound and a lower bound.

You should check back to our list of types of interval and see that the way we used the terms 'bounded', 'bounded above' and 'bounded below' there is consistent with the definitions that we have just given.

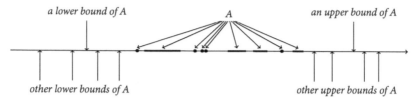

a lower bound of A   A   *an upper bound of A*

other lower bounds of A   *other upper bounds of A*

Bounds for a set

3.2.5 EXERCISE   Verify that a subset A of the real numbers is:

- bounded if and only if it is contained in some bounded interval,
- bounded above if and only if it is contained in some bounded-above interval,
- bounded below if and only if it is contained in some bounded-below interval,
- bounded if and only if there is some constant K such that, for every $a \in A$, $|a| \leq K$.

Specimen solution

Consider the last of these four assertions. If there does exist a constant K as described, then $-K \leq a \leq K$ for every $a \in A$, so $-K$ and K are respectively lower and upper bounds for A, and A is therefore bounded. Conversely, if A is indeed bounded, and we then choose a lower bound l and an upper bound u for it, then for each $a \in A$ we have

$$l \leq a \leq u \leq |u|, \quad -a \leq -l \leq |-l| = |l|$$

so both a and $-a$ are less than or equal to $\max\{|u|, |l|\}$. Put $K = \max\{|u|, |l|\}$ and we have (for each $a \in A$) $|a| \leq K$.

When I is an interval that is bounded above, that is, one of the form (a, b) or $[a, b]$ or $[a, b)$ or $(a, b]$ or $(-\infty, b)$ or $(-\infty, b]$, it is obvious where the 'right-hand edge' of I is (namely, the endpoint b) and it is obvious from the shape of the closing bracket whether that edge point belongs to the interval or not. These are things also worth asking about sets that are more complicated than intervals, but we need to be more careful about defining 'right-hand edge' for a set that is not just an interval.

3.2.6 Definition   If A is a set of real numbers and m is a particular real number then we say that m is the *maximum* element of A if

- $m \in A$ and
- $x \leq m$ for every $x \in A$

(in other words, if m is both an element of A and an upper bound of A). Notice immediately that many sets do not possess a maximum element: for instance, (a, b), $(0, 1) \cup (2, 4)$, $\mathbb{N}$ and $\{\frac{1}{2}, \frac{3}{4}, \frac{7}{8}, \frac{15}{16}, \frac{31}{32}, \cdots\}$ do not. On the other hand, $(a, b]$, $(0, 1) \cup (2, 4]$, $\{-n : n \in \mathbb{N}\}$ and $\{\frac{1}{2^n} : n \in \mathbb{N}\}$ do have (fairly obvious) maximum elements. Informally, we often use terms such as *biggest*, *largest*, *greatest* or *top* element instead of *maximum* element.

3.2.7 **Definition**   The *supremum* of a non-empty set A of real numbers that is bounded above means the least[5] upper bound of A. It is often written briefly as sup A.

This definition will be better understood if we unpack it a little. To say that a number t is the supremum of a set A says two things: firstly, that t is one of A's upper bounds and, secondly, that no smaller number can be. That is, $t \geq x$ for every $x \in A$ but, for any positive number ε, $t - \varepsilon$ fails to be greater than or equal to all of the elements of A, that is, it is strictly less than at least one element of A. In summary, $t = \sup A$ says:

- $t \geq x$ for every $x \in A$, and
- for each $\varepsilon > 0$ there exists $x_\varepsilon \in A$ such that $t - \varepsilon < x_\varepsilon$.

3.2.8 **EXERCISE**

- Show that if a set A possesses a maximum element, then this maximum element is the supremum of A.
- Show that if sup A belongs to A as an element, then it is the maximum element of A.
- For each interval that has a right-hand endpoint, show that the supremum is that right-hand endpoint.
- For the set
$$A = \left\{ \frac{-3}{n + 1} + \frac{-5}{m + 4} : n \in \mathbb{N}, m \in \mathbb{N} \right\}$$
prove that sup $A = 0$.

Specimen solution

Consider, for example, the fourth of these assertions. It is clear that every element of the set in question is negative, so 0 is an upper bound of the set. If ε is any given positive number, then we can choose (after some roughwork[6]) an integer p that is larger than $\frac{8}{\varepsilon}$. Now choose n, m both greater than p and we see that

---

[5] We are using the word 'least' in its common-sense meaning here; if in any doubt, please refer forward to paragraph 3.2.9.

[6] We need to find an element of A that is greater than $-\varepsilon$. Looking at what a typical element of A is, and getting rid of the complicating minuses, that says we want $\frac{3}{n+1} + \frac{5}{m+4} < \varepsilon$. We shall achieve that if both $\frac{1}{n+1}$ and $\frac{1}{m+4}$ are less than $\frac{\varepsilon}{8}$, so those bottom-line integers will need to be greater than $\frac{8}{\varepsilon}$ ...

$$n + 1 > n > p > \frac{8}{\varepsilon}, \frac{1}{n+1} < \frac{\varepsilon}{8}, \frac{-3}{n+1} > \frac{-3\varepsilon}{8},$$

$$m + 4 > m > p > \frac{8}{\varepsilon}, \frac{1}{m+4} < \frac{\varepsilon}{8}, \frac{-5}{m+4} > \frac{-5\varepsilon}{8}$$

and by adding these two lines we find a particular element of A:

$$\frac{-3}{n+1} + \frac{-5}{m+4} > \frac{(-3-5)\varepsilon}{8} = 0 - \varepsilon$$

thus confirming that 0 is the supremum (because no smaller number $0 - \varepsilon$ exceeds all of A's elements).

The preceding three paragraphs can be replicated 'in the opposite direction' to investigate left-hand edges of bounded-below sets. Here are the appropriate definitions:

3.2.9 Definition If A is a set of real numbers and l is a particular real number then we say that l is the *minimum* element of A if

- $l \in A$ and
- $l \leq x$ for every $x \in A$

(in other words, if l is both an element of A and a lower bound of A). Notice that many sets do not possess a minimum element: for instance, (a, b), $(0, 1) \cup (2, 4)$, $\{-n : n \in \mathbb{N}\}$ and $\{0.1, 0.01, 0.001, 0.0001, \cdots\}$ do not. On the other hand, $[a, b)$, $[0, 1) \cup (2, 4)$, $\mathbb{N}$ and $\{-1, \frac{1}{2}, -\frac{1}{3}, \frac{1}{4}, -\frac{1}{5}, \frac{1}{6}, -\frac{1}{7}, \cdots\}$ do have (fairly obvious) minimum elements. Informally, we often use terms such as *smallest*, *least* or *bottom* element instead of *minimum* element.

3.2.10 Definition The *infimum* of a non-empty set A of real numbers that is bounded below means the greatest lower bound of A. It is often written briefly as inf A.

We shall again unpack that definition a little. To say that a number t is the infimum of a set A says two things: firstly, that t is one of A's lower bounds and, secondly, that no greater number can be. That is, $t \leq x$ for every $x \in A$ but, for any positive number ε, $t + \varepsilon$ fails to be less than or equal to all of the elements of A, that is, it is strictly bigger than at least one element of A. In summary, $t = \inf A$ says:

- $t \leq x$ for every $x \in A$, and
- for each $\varepsilon > 0$ there exists $x_\varepsilon \in A$ such that $t + \varepsilon > x_\varepsilon$.

3.2.11 EXERCISE

- Show that if a set A possesses a minimum element, then this minimum element is the infimum of A.

- Show that if inf A belongs to A as an element, then it is the minimum element of A.

- For each interval that has a left-hand endpoint, show that the infimum is that left-hand endpoint.

- For the set

$$A = \left\{ \frac{7}{n+3} + \frac{2}{m+12} : n \in \mathbb{N}, m \in \mathbb{N} \right\}$$

prove that inf $A = 0$.

Partial solution

For instance, let us consider the first of these assertions.

Let z denote the minimum element of the set A. That is, z belongs to A as an element, and $z \leq x$ for every x in A. The second of those observations tells us that z is one of the lower bounds of A. On the other hand, for any $\varepsilon > 0$ we can indeed find an element x_ε of A that is less than $z + \varepsilon$, namely $x_\varepsilon = z$. So z is the infimum of A.

We have made the point several times that many sets do not have maximum or minimum elements. **The vital point about sups and infs[7] is that, in contrast, these virtually always exist within the real numbers** – provided only that the set in question does not 'stretch off towards infinity or minus infinity' and is not merely the empty set. This is, in many ways, the most critical property of $\mathbb{R}$:[8]

3.2.12 The completeness principle for the real number system

Every non-empty set of real numbers that is bounded above has a supremum. Every non-empty set of real numbers that is bounded below has an infimum.

It is possible to construct the real number system within a framework of set theory and to establish this key completeness property, but such a construction lies outside the scope of this text so we must ask you to take it on trust at present. When you have seen how powerful it is, you will have better reasons for going deeper into set theory with a view to understanding such a construction.

Note once again that the sup and the inf of a set A might or might not belong to A as elements. Note also, as a point of interest, that each of the two sentences in our statement of the completeness principle logically implies the other,[9] so we did not really need to state both of them.

---

[7] The official Latin plurals are *suprema* and *infima*, but it is common practice to speak of sups and infs, and also of sup and inf in less formal discussions.

[8] But not a property of $\mathbb{Q}$! We shall return to this issue in 3.3.9.

[9] Supposing that each non-empty bounded-above subset of $\mathbb{R}$ has a supremum, and that $A \subseteq \mathbb{R}$ is non-empty and bounded below, put $B =$ the set of lower bounds of A. Then B is non-empty and bounded above (by any element of A that you choose to consider) so it possesses a supremum s. Now it is routine to confirm that s is the infimum of A.

We'll finish off this section with a few results that combine or compare sup and inf of different sets.

3.2.13 Lemma Suppose that A and B are two non-empty subsets of $\mathbb{R}$, each bounded above. Let $A + B$ mean the set $\{a + b : a \in A, b \in B\}$. Then $A + B$ is also bounded above, and $\sup(A + B) = \sup A + \sup B$.

3.2.14 Lemma Suppose that A is a non-empty subset of $\mathbb{R}$ and is bounded above, and that k is a positive real number. Let kA mean the set $\{ka : a \in A\}$. Then kA is bounded above, and $\sup(kA) = k \sup A$.

Proof

Let s be a temporary symbol for $\sup A$. We know that (for each $a \in A$) $a \leq s$ and therefore $ka \leq ks$, so at least ks is one of the upper bounds of the set kA. Given $\varepsilon > 0$, we see that $\frac{\varepsilon}{k}$ is also positive, therefore $s - \frac{\varepsilon}{k} < a'$ for some $a' \in A$. Hence $ks - \varepsilon < ka'$ where ka' is an element of kA. This establishes ks as the supremum of kA.

3.2.15 Lemma Suppose that A is a non-empty subset of $\mathbb{R}$ and is bounded above, and that k is a negative real number. Let kA mean the set $\{ka : a \in A\}$. Then kA is bounded below, and $\inf(kA) = k \sup A$.

3.2.16 EXERCISE

1. Prove 3.2.13.
2. State and prove modifications of 3.2.13 and 3.2.14 for infima instead of suprema.
3. Prove 3.2.15.
4. State and prove a modification of 3.2.15 for a set kA where A is bounded below (and k is negative).
5. Use these lemmata[10] to determine the supremum of the set

$$C = \left\{ \frac{3}{n+1} - \frac{4}{m+2} : m \in \mathbb{N}, n \in \mathbb{N} \right\}.$$

6. Determine the supremum and the infimum of the set

$$D = \left\{ 5 + \frac{12}{n^2 + 1} - \frac{2}{m^2 + 3m + 5} : m \in \mathbb{N}, n \in \mathbb{N} \right\}.$$

---

[10] The official plural of the Greek word *lemma* is *lemmata*, but it is perfectly ok to use the anglicised plural 'lemmas' instead.

Partial solutions

In part (1), it is purely routine to check that $\sup A + \sup B$ is *an* upper bound of the set $A + B$.

Now if $\varepsilon > 0$ is given, note that $\varepsilon/2$ is also positive, so there are elements $a' \in A, b' \in B$ greater than $\sup A - \varepsilon/2$ and $\sup B - \varepsilon/2$ respectively. Combine these observations.

For part (5), the notation set up in the lemmata lets us express the set C as $3A + (-4)B$ where

$$A = \left\{ \frac{1}{n+1} : n \in \mathbb{N} \right\}, \quad B = \left\{ \frac{1}{m+2} : m \in \mathbb{N} \right\}.$$

It is easy to see that the biggest element (and therefore the supremum) of A is $\frac{1}{2}$ and that the infimum of B is 0. Using the machinery set up by the lemmata, we therefore find

$$\sup C = \sup(3A + (-4)B) = \sup(3A) + \sup(-4B)$$

$$= 3 \sup A - 4 \inf B = 3 \left(\frac{1}{2} \right) - 4(0) = \frac{3}{2}.$$

3.3 **Denseness**

Our main objective in this section is to establish (and to use) the fact that between each two distinct real numbers, there is a rational number. We should begin, however, by looking a little more closely at (what appears to be) the simplest number system of all,[11] that of the positive integers $\mathbb{N}$.

Suppose that a is a particular positive integer and that b_1 is another that is less than a. Since the differences between integers have to be integers, it follows that b_1 is at most $a - 1$. Likewise, if $a > b_1 > b_2$ (all three being positive integers) then b_2 is at most $b_1 - 1$ and, consequently, at most $a - 2$. Continuing this argument, $a > b_1 > b_2 > b_3$ will guarantee that $b_3 \le a - 3$ provided that all these numbers are positive integers.

Repeating this argument a times, we find that $a > b_1 > b_2 > b_3 > \cdots > b_a$ will guarantee that $b_a \le a - a = 0$ if all the numbers involved are positive integers: but this is impossible, since the *positive* integer b_a cannot be ≤ 0. The contradiction shows that no strictly decreasing succession of positive integers, starting with a, can contain more than a terms.

This insight can be presented as a statement about *sets* of positive integers, as follows:

---

[11] Actually, $\mathbb{N}$ is not as simple as it appears to be. In particular, a complete logical account of the positive integer system would need to justify the idea of carrying out some procedure an *arbitrary* positive-integer-number of times, as we describe in this discussion and in the next proof. But this is not a textbook on mathematical logic, and we shall accept some intuitive input into our view of $\mathbb{N}$, just as we did – and continue to do – concerning the real number system $\mathbb{R}$.

3.3.1 **Proposition**   Every non-empty subset of $\mathbb{N}$ possesses a least element.

Proof

Given non-empty $A \subseteq \mathbb{N}$, suppose that A does not possess a least element – that is, for each element of A that we look at, there will always be a smaller element of A. Since A is not empty, we can choose an element a somewhere in A.

Since a is not the least element of A, we can find some a_1 in A that is smaller.

Since a_1 is not the least element of A, we can find some a_2 in A smaller than a_1.

Since a_2 is not the least element of A, we can find a_3 in A smaller than a_2, and so on.

Run that argument a times, and we shall have created a strictly decreasing succession of $a + 1$ positive integers beginning with a. This contradicts what we observed above.

3.3.2 **Theorem: $\mathbb{Q}$ is dense in $\mathbb{R}$**   If $c < d$ are any two distinct real numbers, then there is a rational number q such that $c < q < d$.

Roughwork

The informal idea is this. We choose a positive integer n so big that $\frac{1}{n}$ is smaller than the gap $d - c$ between c and d, and we think about all the rational fractions whose denominator is n. These are evenly spaced out across the entire real line at intervals of $\frac{1}{n}$, and some of them lie to the left of c, and some of them lie to the right of d. If we imagine switching our attention from one that lies $> d$, step by step toward the left with strides of length $\frac{1}{n}$ until we eventually reach one that lies $< c$ then, because each step that we took was shorter than the gap between c and d, one of them must have fallen into that gap. The first one that does this is the rational q that we were looking for.

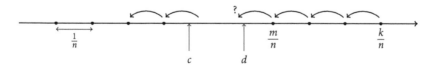

Hunting for rationals between c and d

Proof

Case 1: assume that $d > 1$.

Choose a positive integer n that is greater than $1/(d - c)$. Then $\frac{1}{n} < d - c$.

Choose next a positive integer k that is greater than nd. This ensures that $\frac{k}{n} > d$, and therefore that the set

$$M = \left\{ k \in \mathbb{N} : \frac{k}{n} \geq d \right\}$$

of positive integers is non-empty. Appealing to the above proposition, let m be the smallest element of M. (*Now the intuition is that $\frac{m}{n}$ is the smallest fraction with denominator n that lies at or to the right of d, so one further step to the left will drop us into the gap between c and d.*)

It cannot be the case that $m = 1$: because $\frac{m}{n} \geq d$, and we assumed that $d > 1 \geq \frac{1}{n}$. So $m - 1$ is still a positive integer, but is not in M. That means $\frac{m-1}{n} < d$. In addition,

$$\frac{m-1}{n} = \frac{m}{n} - \frac{1}{n} > \frac{m}{n} - (d-c) \geq d - (d-c) = c$$

that is, the rational $\frac{m-1}{n}$ lies strictly between c and d.

Case 2: now suppose that d is any real number at all.

Choose a positive integer p that is greater than $1 - d$. Then $c + p < d + p$ and $d + p > 1$; so, by Case 1, there is a rational number q such that $c + p < q < d + p$. It follows that $c < q - p < d$ where $q - p$, a rational minus an integer, is still rational. Hence the result.

3.3.3 **Note**   It is now easy to see that, between any two distinct real numbers c and d, there are actually infinitely many rationals: because if not, then the *finite* set $\mathbb{Q} \cap (c, d)$ would have a smallest element q', and then there would be *no* further rationals between c and q', contradicting denseness. (Alternatively, once we have one rational q between c and d, then denseness says there is another q_1 between c and q, and another q_2 between c and q_1, and another q_3 between c and q_2, and so on endlessly.)

3.3.4 **EXERCISE**

1. If a and b are rational, $b \neq 0$, and x is irrational, show that $a + bx$ must be irrational.

2. Given distinct real numbers $c < d$, show that there is an irrational number lying between them.

Partial solution

1. Use proof by contradiction: assume that $a + bx$ equals a rational number, rearrange to obtain a formula for x and conclude that x is actually rational (in contradiction to what was given).

2. Use denseness of $\mathbb{Q}$ twice to find rationals a and b such that $c < a < b < d$ and then consider the number $a + \frac{1}{2}(b - a)\sqrt{2}$.

3.3.5 **Note**   Now it follows (by just the same style of argument as in the previous Note) that between any two distinct real numbers there are actually infinitely many irrational numbers, as well as infinitely many rational numbers: both $\mathbb{Q}$ and $\mathbb{R} \setminus \mathbb{Q}$ have this denseness property. The informal mental picture we should now be building is that, no matter how small a segment of $\mathbb{R}$ we look at and no matter how high a magnification we use, we shall always see an interleaved mix of rationals and irrationals – indeed, infinitely many of each of them. This has important consequences for limits of sequences and for sups and infs:

3.3.6 **Proposition**

1. Every real number is the limit of a sequence of rationals.
2. Every real number is the limit of a sequence of irrationals.

Proof

For any $x \in \mathbb{R}$ and each $n \in \mathbb{N}$ in turn, we can use denseness to find a rational number between $x - \frac{1}{n}$ and $x + \frac{1}{n}$: call this rational number q_n since it may well depend on n. So

$$x - \frac{1}{n} < q_n < x + \frac{1}{n},$$

that is, $|x - q_n| < \frac{1}{n}$.[12] Given $\varepsilon > 0$, if we choose an integer $n_0 > \frac{1}{\varepsilon}$, it follows that $n \geq n_0$ will guarantee that $|x - q_n| < \varepsilon$, so $q_n \to x$ as $n \to \infty$. The proof of the second part is almost identical.

3.3.7 **EXERCISE**   Given any real number x, show that

1. x is the limit of a sequence of rationals each of which is less than x;
2. x is the limit of a sequence of irrationals each of which is greater than x.

3.3.8 **Proposition**

1. Every real number x is the supremum of the set of rationals that are less than x.
2. Every real number x is the infimum of the set of rationals that are greater than x.
3. Every real number x is the supremum of the set of irrationals that are less than x.
4. Every real number x is the infimum of the set of irrationals that are greater than x.

---

[12] The next chapter will provide us with a slick and tidy way to finish the argument from that point: see 4.1.18.

Proof

Let $A = \mathbb{Q} \cap (-\infty, x)$ comprise all the rationals less than x. Certainly x is an upper bound of that set. Also, for any $\varepsilon > 0$, denseness says that there is a rational q between $x - \varepsilon$ and x. This q belongs to A, and $q > x - \varepsilon$. Hence x is the supremum of A. The other three parts are proved in just the same way.

3.3.9 Note The following statement is untrue: 'every non-empty subset of $\mathbb{Q}$ that is bounded above in $\mathbb{Q}$ has a least upper bound in $\mathbb{Q}$'. For instance, consider the set A of rationals whose squares are less than 2. It is clearly non-empty, and bounded above in $\mathbb{Q}$ by, for example, $\frac{3}{2}$. Now suppose it did have a least upper bound λ in $\mathbb{Q}$.

- If $\lambda < \sqrt{2}$, then we can (due to denseness) find a rational q such that $\lambda < q < \sqrt{2}$ which gives $q^2 < 2$. This shows that q belongs to A and yet exceeds the alleged upper bound λ for A.
- If $\lambda > \sqrt{2}$, then we can find rational r such that $\sqrt{2} < r < \lambda$. Any rational a that is $\geq r$ will have $a^2 \geq r^2 > 2$ and therefore cannot belong to A; in other words, every element of A must be less than r. Consequently r is an upper bound in $\mathbb{Q}$ for A and yet is less than the least such upper bound.
- If $\lambda = \sqrt{2}$ then $\sqrt{2}$ has to be rational.

All three cases have now run into contradiction, and the demonstration[13] is complete.

This outcome contrasts strongly with the completeness principle for real numbers, which can be expressed as 'every non-empty subset of $\mathbb{R}$ that is bounded above in $\mathbb{R}$ has a least upper bound in $\mathbb{R}$': in other words, $\mathbb{Q}$ is not complete but $\mathbb{R}$ is. In many ways, this is the most important difference between how $\mathbb{R}$ works as a number system and how $\mathbb{Q}$ does.

3.3.10 Example To verify that the infimum of the set $B = \{q : q$ is rational and $q^2 \leq 2\}$ is $-\sqrt{2}$.

Solution

(a) $-\sqrt{2}$ is not rational so it cannot belong to B.
   Any rational $q < -\sqrt{2}$ has $|q| > \sqrt{2}$ and $q^2 = |q|^2 > 2$, so it cannot belong to B either.
   Hence every element of B must be greater than $-\sqrt{2}$, that is, $-\sqrt{2}$ is a lower bound for B.

---

[13] It is possible to revamp this argument in a way that avoids all mention of $\sqrt{2}$, and that therefore takes place *entirely* within the family of rational numbers.

(b) If $\varepsilon > 0$ then $\varepsilon' = \min\{\varepsilon, \sqrt{2}\}$ is also[14] greater than 0 so, by denseness, there is a rational r between $-\sqrt{2}$ and $-\sqrt{2} + \varepsilon'$.

Our choice of ε' ensures that this r will be negative, and

$$-\sqrt{2} < r < 0 \Rightarrow |r| < \sqrt{2} \Rightarrow r^2 = |r|^2 < 2,$$

that is, $r \in B$.

Hence $-\sqrt{2}$ is inf B.

3.3.11 EXERCISE

- Verify that the supremum of $\{q : q \text{ is rational and } q^2 \leq 2\}$ is $\sqrt{2}$.

---

[14] The step that most often puzzles the reader is the replacement of ε by ε' at this point. Why is something like this necessary at all? Because if ε were too big, the next step might go wrong. If, for instance, ε were 3 then, when we chose a rational r between $-\sqrt{2}$ and $-\sqrt{2}+\varepsilon$, our otherwise random choice of rational lies in the interval $(-1.414, +1.586)$ (to three decimal places) and, for all we know, r could be $+1.5$. This number has a square **greater** than 2 and therefore fails to lie within B, destroying the punch-line of our demonstration. Making sure that the 'new epsilon' is no bigger than $\sqrt{2}$ guarantees that this will not happen.

4 Up and down – increasing and decreasing sequences

If a sequence converges to a limit, then it is perfectly possible that it does not do so *steadily*. For instance, the sequence $(\frac{1}{1}, \frac{1}{3}, \frac{1}{2}, \frac{1}{4}, \frac{1}{3}, \frac{1}{5}, \frac{1}{4}, \frac{1}{6}, \cdots)$ does tend to 0 as limit, but in a 'big step forward, small step back' fashion: as you scan along the list of terms, you find that you are moving *away* from 0 half the time. On the other hand, the sequence $(\frac{1}{1}, -\frac{1}{2}, \frac{1}{3}, -\frac{1}{4}, \frac{1}{5}, -\frac{1}{6}, \frac{1}{7}, -\frac{1}{8}, \cdots)$ converges to zero 'in both directions at once', alternately over- and underestimating the eventual limit. However, many important sequences move *steadily* in one direction (up or down) and, for those that do, it is often easier to determine whether they converge or not. The opening section of this chapter will focus on sequences of this kind.

4.1 Monotonic bounded sequences must converge

4.1.1 Definition   A sequence $(x_n)_{n \geq 1}$ is called

- an *increasing* sequence if $x_n \leq x_{n+1}$ for every n;
- a *decreasing* sequence if $x_n \geq x_{n+1}$ for every n;
- a *monotonic* (or *monotone*) sequence if it is either increasing or decreasing.[1]

So these are the sequences that 'move steadily in one direction – up or down' as you scan along the list of terms. There are some very obvious examples, such as the decreasing sequences $\left(\frac{1}{n}\right)$, $\left(\frac{1}{n^2}\right)$, $\left(\frac{1}{n^3}\right)$, $(-n)$ and $(-\sqrt{n})$, and the increasing sequences (n), (n^2) and $\left(1 - \frac{1}{n}\right)$. For less transparent examples, it is often useful to calculate the difference[2] $x_n - x_{n+1}$ to see whether it is always positive (in which case the sequence is decreasing) or always negative (in which case it is increasing) or sometimes positive and sometimes negative (in which case it cannot be monotonic).

---

[1] Notice that, since the inequalities are non-strict (that is, they allow equality), a constant sequence is both increasing and decreasing.

[2] or sometimes – provided that the terms are positive – the ratio x_{n+1}/x_n if that might simplify through a lot of cancelling: the ratio will be always greater than or equal to 1 if the sequence is increasing, but less than or equal to 1 if it is decreasing.

Undergraduate Analysis: A Working Textbook, Aisling McCluskey and Brian McMaster 2018.

4.1.2 **Example** With

$$x_n = \frac{6n-1}{4n+3}$$

decide whether (x_n) is or is not monotonic.

Solution

In this case,

$$x_n - x_{n+1} = \frac{6n-1}{4n+3} - \frac{6(n+1)-1}{4(n+1)+3} = \frac{6n-1}{4n+3} - \frac{6n+5}{4n+7}$$

which, when you bring it to a common denominator, comes to

$$\frac{-22}{(4n+3)(4n+7)}.$$

This is obviously negative for all n, so the sequence is increasing (and thus monotonic).

4.1.3 **Example** With

$$x_n = 3 + \frac{4}{n} - \frac{2}{n^2}$$

show that (x_n) is a decreasing sequence.

Solution

In the present case,

$$x_n - x_{n+1} = 3 + \frac{4}{n} - \frac{2}{n^2} - \left(3 + \frac{4}{n+1} - \frac{2}{(n+1)^2}\right)$$

which, when it has all been brought to a common denominator, simplifies to

$$\frac{4n^2 - 2}{n^2(n+1)^2}.$$

Since n is at least 1, this expression is always positive, so (x_n) is decreasing, as predicted.

A sequence is not, of course, the same thing as a set. For instance, sets are not allowed to include repeated elements but sequences may; also, sequences present their elements in a particular order but sets do not. However, every sequence $(x_n)_{n \in \mathbb{N}}$ gives rise to a set, namely the set $\{x_n : n \in \mathbb{N}\}$ of all its terms, and it is sometimes helpful to look at the two together. For instance, this gives us an easy way to talk about the *boundedness* of a sequence:

4.1.4 Definition A sequence is said to be *bounded* if the set of all its terms is bounded.

Likewise, a sequence is said to be *bounded above* if the set of all its terms is bounded above, and *bounded below* if the set of all its terms is bounded below.

In view of our earlier discussions on boundedness (paragraph 3.2.5 in particular), there are several slightly different ways to recognise when this happens:

4.1.5 Lemma A sequence $(x_n)_{n\in\mathbb{N}}$ is bounded if and only if one of the following (equivalent) conditions holds:

1. Some bounded closed interval $[a, b]$ includes x_n for every n,
2. Some bounded interval of the form $[-K, K]$ includes x_n for every n,
3. There is some constant $K > 0$ such that $|x_n| < K$ for every n,
4. All terms of the sequence lie within some fixed distance from 0,
5. All terms of the sequence lie within some fixed distance from a number $c \in \mathbb{R}$.

There are important connections between convergence and boundedness of a sequence, one of which is:

4.1.6 Lemma Every convergent sequence is bounded.

Proof

Suppose that $(x_n)_{n\in\mathbb{N}}$ converges and that its limit is ℓ. By the definition (and choosing $\varepsilon = 1$ for convenience), there is a positive integer n_0 such that all the terms of the sequence after the n_0^{th} lie between $\ell - 1$ and $\ell + 1$, that is, less than 1 unit distant from ℓ. The earlier terms $x_1, x_2, x_3, \cdots, x_{n_0-1}$ may well be further away from ℓ, but there are only a finite number of them: so we can find the *biggest* distance from one of them to ℓ...call it M. If we now let $M' = \max\{M, 1\}$, then *every* x_n lies within the distance M' from ℓ, and so $(x_n)_{n\in\mathbb{N}}$ is bounded.

4.1.7 Alert The converse of this lemma is certainly not true! For instance, the 'alternating sequence' $((-1)^n)_{n\in\mathbb{N}}$ is bounded (it lies entirely inside $[-1, 1]$, for example) but not convergent. However, we'll now demonstrate a correct *partial* converse, namely the result embedded in the name of this section:

4.1.8 Theorem

1. A sequence that is increasing and bounded above must converge. (The limit is the supremum of the set of all its terms.)

2. A sequence that is decreasing and bounded below must converge. (The limit is the infimum of the set of all its terms.)

Proof

(1) Suppose that $(x_n)_{n\in\mathbb{N}}$ is increasing and bounded above. The set $X = \{x_n : n \in \mathbb{N}\}$ is non-empty and bounded above so, by the completeness principle, its supremum (let's denote it by ℓ) does exist. The definition of supremum then tells us that, if ε is any positive number:

- $x_n \leq \ell$ for every $n \geq 1$, and
- $\ell - \varepsilon < x_m$ for at least one positive integer m.

Now use the fact that the sequence is increasing, and we get:

$$\ell - \varepsilon < x_m \leq x_{m+1} \leq x_{m+2} \leq x_{m+3} \leq x_{m+4} \leq \cdots \leq \ell,$$

in other words, every term of the sequence from number m onwards lies between $\ell - \varepsilon$ and ℓ, whence $|x_n - \ell| < \varepsilon$ for every $n \geq m$. Thus, $(x_n)_{n\in\mathbb{N}}$ converges to ℓ.

Increasing plus bounded ...

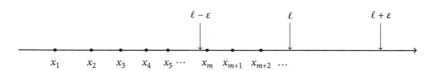

Increasing plus bounded implies convergent

Exercise

Prove part (2) of this theorem: the proof will closely resemble what we have just set out, but a lot of the inequalities will be the other way around.

4.1.9 Example   To use the 'bounded + monotonic' theorem to show that the following sequence (x_n) is convergent: where, for each positive integer n, x_n is defined to be the product of fractions

$$x_n = \frac{3}{4} \times \frac{8}{9} \times \frac{15}{16} \times \cdots \times \frac{n^2 - 1}{n^2}.$$

Solution

Since all the fractions involved here are positive, it is clear that $x_n > 0$ for all n, and so the sequence (x_n) is bounded below. Also, comparing the formulae for x_n and for x_{n+1}, we see that $x_{n+1} = x_n \times \dfrac{(n+1)^2 - 1}{(n+1)^2} = x_n \times \left(1 - \dfrac{1}{(n+1)^2}\right)$.

Since the extra multiplier $\left(1 - \dfrac{1}{(n+1)^2}\right)$ is positive and less than 1, it follows that $x_{n+1} < x_n$, that is, that the sequence (x_n) is decreasing. According to the last theorem (part (2)) it must converge.

4.1.10 Example   To find the infimum of the set

$$X = \left\{3 + \frac{4}{n} - \frac{2}{n^2} : n \in \mathbb{N}\right\}.$$

Solution

As pointed out already, the *sequence* whose n^{th} term is the typical number in this set is decreasing and tends to 3. By the small print of the 'bounded + monotonic' theorem, 3 has to be the infimum of the set.

4.1.11 EXERCISE   If

$$a_n = \frac{2n^3 - 5n^2 - 4n - 2}{2n^3 + n^2}$$

find the supremum of the set $\{a_n : n \in \mathbb{N}\}$.

Hint

Most of the work consists in checking that the *sequence* (a_n) is increasing. Verify first that

$$1 - \frac{2}{n^2} - \frac{6}{2n+1} = a_n.$$

4.1.12 EXERCISE   Show that the sequence $(n - \sqrt{n})$ is increasing but not bounded.

4.1.13 Example   Let $t \in (0,1)$ be a constant. We show that the sequence $(t^n)_{n \in \mathbb{N}}$ converges to 0.

Solution

Since $0 < t < 1$, all the powers of t are positive, and $t^n t < t^n 1 = t^n$, that is, $t^{n+1} < t^n$ for every n. So the given sequence is decreasing and bounded below by zero, and therefore must converge to *some* limit ℓ. Now we need to identify ℓ.

The sequence $(t^{n+1})_{n \in \mathbb{N}} = (t^2, t^3, t^4, t^5, \cdots)$ is the original sequence with its first term removed so (see earlier comment) it also converges to (the same) ℓ. Yet also (by part (4) of the algebra of limits) $t^{n+1} = t \times t^n$ has to converge to $t\ell$. That gives $\ell = t\ell$ (since limits are unique when they exist) which, since $t \neq 1$, forces $\ell = 0$ as predicted.

4.1.14 **EXERCISE** Let $t \in (1, \infty)$ be a constant. Show that the sequence $(t^n)_{n \in \mathbb{N}}$ diverges.

Partial solution

Use proof by contradiction. If this sequence *did not diverge*, it would have to converge to a limit ℓ which (as in the last example) would have to satisfy $\ell = t\ell$. Check that the sequence is increasing, and consider the consequences for the number ℓ.

We have delayed proving part (3) of the algebra of limits theorem until now, because we wanted the *convergent sequences are bounded* theorem to help in the demonstration. This proof is harder than most we have done so far, so we shall first roughwork our way through what needs to be shown and how we might show it, and then crystallise out a proper proof from that discussion.

4.1.15 **Limit of a product** If $a_n \to \ell$ and $b_n \to m$ then $a_n b_n \to \ell m$.

Roughwork

Knowing only that the separate 'error' terms $|a_n - \ell|$ and $|b_n - m|$ can be made as small as we wish just by taking n sufficiently large, we must show the same kind of smallness for $|a_n b_n - \ell m|$. Unfortunately for us, $a_n b_n - \ell m$ will not factorise or simplify at all. We might, however, get somewhere if we un-simplify it by bringing in an extra term that has something in common with each half, such as $a_n m$ or $b_n \ell$. Try that out:

$$a_n b_n - \ell m = a_n b_n - a_n m + a_n m - \ell m = a_n(b_n - m) + (a_n - \ell)m.$$

Because both $|a_n - \ell|$ and $|b_n - m|$ can be made really small, we could now force each half of this to be smaller than $\frac{\varepsilon}{2}$. For $(a_n - \ell)m$ this will be easy: just insist that $a_n - \ell$ is smaller in modulus than $\frac{\varepsilon}{2m}$ (but notice the danger of dividing by m in the case where $m = 0$). For $a_n(b_n - m)$, however, it is not so straightforward: insisting that $b_n - m$ be smaller (in modulus) than $\frac{\varepsilon}{2a_n}$ will not work because that quantity varies with n. We need to replace that a_n by a constant somehow, and it needs to be a constant big enough to deal with all of the a_ns at once. That is exactly what we wanted boundedness of (a_n) for: find a positive constant K so that $|a_n|$ is always $\leq K$, then insist that $b_n - m$ shall be smaller (in modulus) than $\frac{\varepsilon}{2K}$, and everything ought to work. Now let's see if it does:

Proof

Given $\varepsilon > 0$, use first the fact that $a_n \to \ell$ to find n_0 such that[3] $|a_n - \ell| < \dfrac{\varepsilon}{2|m| + 1}$ whenever $n \geq n_0$. Next, use the fact that (a_n), being convergent, must also be

---

[3] The +1 on the bottom line has been put in purely to make sure that we are at no risk of dividing by zero, and to avoid having to treat $m = 0$ as a special case.

bounded, so we can find a constant $K > 0$ such that $|a_n| < K$ for every value of n. Now use the convergence of b_n to m to find another integer n_1 such that $|b_n - m| < \dfrac{\varepsilon}{2K}$ whenever $n \geq n_1$.

For each $n \geq \max\{n_0, n_1\}$ we now have

$$|a_n b_n - \ell m| = |a_n b_n - a_n m + a_n m - \ell m| = |a_n(b_n - m) + (a_n - \ell)m|$$
$$\leq |a_n(b_n - m)| + |(a_n - \ell)m|$$

(using the triangle inequality there)

$$= |a_n||b_n - m| + |a_n - \ell||m| \leq K\left(\frac{\varepsilon}{2K}\right) + \left(\frac{\varepsilon}{2|m| + 1}\right)|m|$$
$$< \frac{\varepsilon}{2} + \frac{\varepsilon}{2} = \varepsilon.$$

The proof is finished. (Notice that we were more careful with the modulus signs in the final proof than we had been in the opening roughwork.)

4.1.16 **EXERCISE** Fill in the details in the following outline proof of part (6) of the algebra of limits theorem. If $b_n \to m$, and neither m nor any of the b_n is zero, then

$$\frac{1}{b_n} \to \frac{1}{m}.$$

Outline proof

Notice first that

$$\left|\frac{1}{b_n} - \frac{1}{m}\right| = \frac{|b_n - m|}{|m||b_n|}.$$

That last expression could get into deep trouble[4] if b_n were to become *close* to zero, so we need to prevent that.[5] Find n_1 so that $\dfrac{|m|}{2} < |b_n| < \dfrac{3|m|}{2}$ whenever $n \geq n_1$. For such values of n, check that

$$\frac{|b_n - m|}{|m||b_n|} < \frac{2|b_n - m|}{m^2}.$$

Next, find n_2 for which $|b_n - m| < \dfrac{m^2 \varepsilon}{2}$ whenever $n \geq n_2$.

Put all the pieces together.

Lastly, it is now quite easy to prove part (6) from part (3) and the above: begin by writing $\frac{a_n}{b_n}$ as $a_n \times \frac{1}{b_n}$ and using part (3) on that product.

---

[4] In order to make a fraction small (in modulus), we need to make its top line small *but* prevent its bottom line from becoming too small.

[5] Since $|b_n| \to |m|$, we can keep $|b_n|$ close enough to $|m|$ – say, between one half of $|m|$ and three halves of $|m|$ – to keep it well away from zero.

We'll conclude this section with two more results that connect limits with inequalities where, this time, the inequalities are between the terms of two or more sequences rather than, as above, between the terms of a single sequence.

4.1.17 Theorem: limits across an inequality   If (a_n) and (b_n) are two convergent sequences such that $a_n \leq b_n$ for every $n \in \mathbb{N}$, then $\lim a_n \leq \lim b_n$.

Proof

For brevity, put $\ell_1 = \lim a_n, \ell_2 = \lim b_n$. If ℓ_1 were not $\leq \ell_2$ then the number $\varepsilon = \dfrac{\ell_1 - \ell_2}{2}$ would be strictly greater than zero. Convergence tells us that, for sufficiently large n, both $|a_n - \ell_1|$ and $|b_n - \ell_2|$ will be smaller than ε, that is,

$$\ell_1 - \varepsilon < a_n < \ell_1 + \varepsilon, \quad \ell_2 - \varepsilon < b_n < \ell_2 + \varepsilon.$$

Our choice of ε, however, arranges that $\ell_1 - \varepsilon$ and $\ell_2 + \varepsilon$ are the same number – indeed, that is precisely why we chose it so. Therefore (for large values of n like this) $b_n < a_n$, and this contradiction establishes the result.

Remarks

1. Be careful not to use this result on sequences about whose convergence you are unsure. For instance, $(-1)^n$ is certainly less than 2 for every value of n ... but does this tell us that $\lim(-1)^n \leq 2$? No, because $\lim(-1)^n$ does not exist.

2. Also be aware that the *strict* inequality $<$ is *not* preserved under limits in this way: that is, $a_n < b_n$ for all n does *not* guarantee that $\lim a_n < \lim b_n$. A simple illustration of this is that $-\frac{1}{n}$ is certainly strictly less than $+\frac{1}{n}$ for every n, and each tends to 0, but it would be foolish to claim that $0 < 0$ as a consequence.

4.1.18 Theorem: the 'sandwich', or the 'squeeze'   Of three sequences (a_n), (b_n), (c_n) suppose we know that $a_n \leq b_n \leq c_n$ for all n, and also that (a_n), (c_n) converge to the same limit ℓ. Then also $b_n \to \ell$.

Proof

Given any $\varepsilon > 0$ we can first use the given convergence to find positive integers n_a, n_c such that $|a_n - \ell| < \varepsilon$ for $n \geq n_a$ and $|c_n - \ell| < \varepsilon$ for $n \geq n_c$. Then for each $n \geq \max\{n_a, n_c\}$ we shall have both of these inequalities true at once, and so

$$\ell - \varepsilon < a_n < \ell + \varepsilon, \quad \ell - \varepsilon < c_n < \ell + \varepsilon.$$

Combining pieces of this display with the given inequality $a_n \leq b_n \leq c_n$ we see that:

$$\ell - \varepsilon < a_n \leq b_n \leq c_n < \ell + \varepsilon$$

(for all $n \geq \max\{n_a, n_c\}$), which places b_n between $\ell \pm \varepsilon$, as required.

Examples

1. To find the limit of the sequence whose n^{th} term is

$$\frac{6n - 5 \sin(n^2 + \pi \sqrt{n})}{2n + 3}.$$

Solution

The awkward-looking trigonometric term must lie between -5 and $+5$, so the nth term here lies between $\dfrac{6n - 5}{2n + 3}$ and $\dfrac{6n + 5}{2n + 3}$. Since each of these converges to 3 via the algebra of limits, so must the given sequence.

2. To find the limit of the sequence whose n^{th} term is

$$\frac{n^2}{n^2 + 3n \cos(n^3 + n + 1) - 2 \sin(\pi \ln 5n - 16)}.$$

Solution

Take care with the inequalities when estimating bottom lines of fractions. We know $-1 \leq \cos \theta \leq +1$ and $-1 \leq \sin \theta \leq +1$ no matter what (real) number θ may be, so

$$n^2 - 3n - 2 \leq n^2 + 3n \cos(n^3 + n + 1) - 2 \sin(\pi \ln 5n - 16) \leq n^2 + 3n + 2$$

is guaranteed. Taking reciprocals, **and assuming $n \geq 4$ to avoid problems with $n^2 - 3n - 2$ and other terms being possibly negative**, we now get

$$\frac{1}{n^2 - 3n - 2} \geq \frac{1}{n^2 + 3n \cos(n^3 + n + 1) - 2 \sin(\pi \ln 5n - 16)} \geq \frac{1}{n^2 + 3n + 2}$$

(note the reversal of the inequalities) and therefore

$$\frac{n^2}{n^2 - 3n - 2} \geq \frac{n^2}{n^2 + 3n \cos(n^3 + n + 1) - 2 \sin(\pi \ln 5n - 16)} \geq \frac{n^2}{n^2 + 3n + 2}.$$

Since (*via* the algebra of limits) the first and third of these expressions converge to 1, so must the given sequence that is squeezed between them. The circumstance, that **we ignored the first three terms**, has no effect on limiting behaviour, of course.

EXERCISE   For each positive integer n, let:

$$t_n = \frac{4n^2}{7n^3} + \frac{4n^2 - 1}{7n^3 + 3} + \frac{4n^2 - 2}{7n^3 + 6} + \frac{4n^2 - 3}{7n^3 + 9} + \cdots + \frac{4n^2 - n}{7n^3 + 3n}.$$

Investigate the limiting behaviour of the sequence $(t_n)_{n \geq 1}$.

Partial solution

As we scan along the list of $n + 1$ separate fractions whose sum defines t_n, the numerators decrease and the denominators increase; consequently the largest of these fractions is the first and the smallest is the last. Therefore

$$\frac{4n^2 - n}{7n^3 + 3n}(n + 1) < t_n < \frac{4n^2}{7n^3}(n + 1).$$

4.2 Induction: infinite returns for finite effort

Mathematics – to make a terribly obvious point – is peculiarly full of 'universal' statements: statements that claim to be true not just for particular values of the unknowns that they contain, nor even for an overwhelming majority of values, but for *all* of them that, in context, make sense. Some such statements will be extremely familiar to you; to give a few examples:

- **Every** positive integer greater than 1 can be expressed as a product of prime numbers.
- **Every** right-angled triangle has the square on its longest side equal in area to the total of the squares on the other two sides.
- **Every** quadratic equation has two (real or complex) solutions (counting by multiplicity).
- $(x - 1)(x^4 + x^3 + x^2 + x + 1) = x^5 - 1$ for **every** real (or complex) value of x.
- For **any** real number x we can find a positive integer n that is bigger than x.
- For **each** real number t between -1 and $+1$ we can find $\theta \in [-\pi/2, \pi/2]$ such that $\sin(\theta) = t$.
- **Each** bijective mapping has a unique inverse.

Universal statements are (relatively) difficult to prove, because you must provide a proof that works for *every* scenario that the statement claims to work for. (*Untrue* universals are, on the other hand, (relatively) easy to disprove because you only need to find *one* instance that they claim to work for, but in which they give a false result.) Sometimes we may be lucky enough to find a single demonstration that deals with all values of the variables at once: for instance, the fourth statement above can be verified just by multiplying out the brackets on the left-hand side, cancelling everything you can, and looking at what is left over (noticing incidentally that the actual value of x makes no difference); the seventh statement can be confirmed by merely constructing what the desired inverse has to be, and then checking that it works (and noticing that *which* bijective map we started with does not affect the construction in any way). On other occasions, though, we find that we need to break up the demonstration into a number of cases, depending on the value(s) of the variable(s). For example, the proof we gave of the result

$$\lim(kx_n) = k \lim x_n$$

divided into the two cases $k = 0$ and $k \neq 0$ and, later, the discussion of whether (x^n) converges or not usually splits into cases such as $0 < x < 1, -1 < x < 0, x = 0$, $x = 1, x = -1, |x| > 1$ because either the result or the argument (or both) will run differently depending on the variable's value.

In this area, the 'worst conceivable situation' is that in which we appear to be forced to consider an infinite number of cases. At the time of writing, the notorious '$3n+1$' problem seems to be stuck in this nightmare zone. The '$3n+1$' problem is this: given a positive integer n, *either* divide it by 2 (if it is even) *or* multiply it by 3 and add 1 (if it is odd); now repeat that process on your answer, and on the answer to that, and so on. **Question: do you always get to 1 in a finite number of moves?** Here is an illustration, starting on 58:

$$58 \to 29 \to 88 \to 44 \to 22 \to 11 \to 34 \to 17 \to 52 \to 26 \to 13 \to 40 \to$$
$$\to 20 \to 10 \to 5 \to 16 \to 8 \to 4 \to 2 \to 1$$

Well, we reached 1 *that* time. Does that always happen, no matter what n we start with? *Nobody knows*, perhaps in part because although an enormous number of individual initial n's have been checked out, and many special cases have been successfully handled, nobody has yet devised a *finite* list of special-case arguments that comprehensively covers *all* positive integers. *Warning:* do not invest a disproportionate amount of your time into exploring this problem; you have plenty of other things to do.

(**Mathematical**) **induction** is a pattern of proof that is highly successful in establishing universal statements that are controlled by a positive integer. Its strength lies in the fact that a (usually quite routine) demonstration along generally predictable lines will cover all positive integer cases at once: so that when it works (which is not always, but very frequently), it proves the truth of an infinite number of statements all at once. Here in outline is what the pattern of proof by induction is:

- Step 0: express the result that you are trying to prove as a sequence $(S(n))_{n \in \mathbb{N}}$ of statements, where $S(n)$ involves the typical positive integer n.
- Step 1: check that the first statement $S(1)$ is actually true.
- Step 2: assume the truth of a particular (but unspecified) $S(k)$.
- Step 3: deduce from this that the next statement $S(k + 1)$ is also true.

That's all you need to do. At that point, induction says that *all* of the statements $S(n)$ are true statements.

4.2.1 **Example**   To show using induction that $7^{2n-1} + 5^{2n+1} + 12$ is exactly divisible by 24, for every positive integer n.

Solution

We'll follow slavishly the pattern of proof set out above; once you are familiar with induction, you can take some shortcuts.

- Step 0: For each $n \in \mathbb{N}$ let $S(n)$ be the statement: $7^{2n-1} + 5^{2n+1} + 12$ is exactly divisible by 24.
- Step 1: $S(1)$ says that $7 + 125 + 12$ is divisible by 24. Since the total is 144, this is indeed true.
- Step 2: Assume the truth of a particular $S(k)$; that is, that $7^{2k-1} + 5^{2k+1} + 12$ really is divisible by 24.
- Step 3: Now $7^{2(k+1)-1} + 5^{2(k+1)+1} + 12 - (7^{2k-1} + 5^{2k+1} + 12)$ simplifies to $7^{2k-1}(49 - 1) + 5^{2k+1}(25 - 1)$ which certainly is a multiple of 24 (write it as $24m$, say) because $49 - 1$ and $25 - 1$ are. Therefore

$$7^{2(k+1)-1} + 5^{2(k+1)+1} + 12 = (7^{2k-1} + 5^{2k+1} + 12) + 24m$$

which, using Step 2, is the total of two multiples of 24, and therefore itself a multiple of 24. In other words, $S(k + 1)$ is also true.

By induction, all of the statements are true: that is, $7^{2n-1} + 5^{2n+1} + 12$ is exactly divisible by 24 for every positive integer n.

4.2.2 Example: the sum of the first n perfect squares   We show that, for every positive integer n,

$$1^2 + 2^2 + 3^2 + 4^2 + \cdots + n^2 = \frac{n(n + 1)(2n + 1)}{6}.$$

Solution

- Step 0: For each n in turn let $S(n)$ be the statement:
$$1^2 + 2^2 + 3^2 + 4^2 + \cdots + n^2 = \frac{n(n + 1)(2n + 1)}{6}.$$
- Step 1: $S(1)$ says that $1^2 = \dfrac{(1)(1 + 1)(2 + 1)}{6}$ which is certainly true.
- Step 2: Assume the truth of a particular $S(k)$; that is, that
$$1^2 + 2^2 + 3^2 + 4^2 + \cdots + k^2 = \frac{k(k + 1)(2k + 1)}{6}.$$
- Step 3: Adding the next perfect square to each side will not damage the equation, so

$$1^2 + 2^2 + 3^2 + 4^2 + \cdots + k^2 + (k + 1)^2 = \frac{k(k + 1)(2k + 1)}{6} + (k + 1)^2$$

$$= \frac{k + 1}{6}(k(2k + 1) + 6(k + 1)) = \frac{k + 1}{6}\left(2k^2 + k + 6k + 6\right)$$

$$= \frac{k+1}{6}(2k^2 + 7k + 6) = \frac{k+1}{6}((k+2)(2k+3))$$
$$= \frac{(k+1)(k+1+1)(2(k+1)+1)}{6}.$$

In other words, $S(k+1)$ is also true.

By induction, all of the statements are true.

4.2.3 **Example: Bernoulli's inequality**   Let x be a real number with $x \geq -1$. We show that, for every positive integer n, $(1+x)^n \geq 1 + nx$.

Solution

- Step 0: For each n in turn let $S(n)$ be the statement: $(1+x)^n \geq 1 + nx$.
- Step 1: $S(1)$ says that $(1+x)^1 \geq 1 + x$ which is not very interesting but certainly true.
- Step 2: Assume the truth of a particular $S(k)$; that is, that $(1+x)^k \geq 1 + kx$.
- Step 3: Now because $x \geq -1$, we know that $(1+x)$ is positive or zero, so it is safe to multiply a non-strict inequality by it. Thus:

$$(1+x)(1+x)^k \geq (1+x)(1+kx) = 1 + x + kx + kx^2$$

and therefore, since kx^2 cannot be negative,

$$(1+x)^{k+1} \geq 1 + x + kx = 1 + (k+1)x.$$

In other words, $S(k+1)$ is also true.

By induction, all of the statements are true.

4.2.4 **Example**   If $n+1$ distinct straight lines are drawn on a plane surface, we show that they cannot have more than $n(n+1)/2$ crossing points.

Solution

- Step 0: For each value of n in turn let $S(n)$ be the statement: $n+1$ distinct straight lines can't cross at more than $n(n+1)/2$ points.
- Step 1: $S(1)$ says that 2 distinct straight lines cannot cross at more than $1(2)/2 = 1$ point, which is a simple geometric truth.
- Step 2: Assume the truth of a particular $S(k)$; that is, that $k+1$ such lines have at most $k(k+1)/2$ crossing points.
- Step 3: If we are now given $(k+1)+1 = k+2$ such lines, imagine looking at the first $k+1$ of them. By step 2, those lines have at most $k(k+1)/2$ crossing points. Imagine we now draw in the last (the $(k+2)^{th}$) line: it hits each of the

previous $k + 1$ lines at most once, so the total number of crossing points now is, at the most, $k(k + 1)/2 + (k + 1)$. This rearranges easily as $(k + 1)(k + 2)/2$, which is the same as $(k+1)((k+1)+1)/2$. In other words, $S(k+1)$ is also true.

By induction, all of the statements are true.

4.2.5 EXERCISES

1. Show by induction that, for every positive integer n:

$$1^3 + 2^3 + 3^3 + \cdots + n^3 = \frac{n^2(n + 1)^2}{4}.$$

2. Use induction to verify that, whenever n is a positive integer:

$$1 \times 2 \times 3 + 2 \times 3 \times 4 + 3 \times 4 \times 5 + \cdots + n(n+1)(n+2) = \frac{n(n + 1)(n + 2)(n + 3)}{4}.$$

3. Verify that $12^n - 23^n + 34^n$ is divisible by 22 for all positive integer values of n.

4.2.6 Comments

- It's a very common experience, when you first meet induction, to feel that it is cheating in some sense! It can seem that, instead of proving the wished-for result properly, you are just assuming (at step 2) that it is true already. This is not, however, what is going on. The desired result is something that claims to work for all positive integers and, in that phrase, the most important word is *all*. At step 2, what we are assuming is definitely not that the relevant statement is true *for all positive integers* but merely *for one particular positive integer*. This is perfectly reasonable: in fact, at step 1 we already confirmed that it actually is true for $n = 1$, so there is nothing outrageous or illogical about supposing that it might be true for some (other) values. This is all that we are doing at step 2.

- Step 3 is the only part of the argument at which you usually have to pause and think a bit. The question to be pondered is: how am I going to turn statement number k into statement number $k + 1$? – and there is no all-purpose answer: it will depend on what these statements are trying to say. Looking back at our four little case studies above, in the first one the two lumps of algebra were very similar in appearance, and it was a reasonable guess that if we subtracted them we might see in a convenient form what the difference was. For the second, adding in the next perfect square was the natural way to trade up from the sum of k squares to the sum of $k + 1$ of them. In the third, powers of $(1 + x)$ were the essential ingredient, and we had to think how to turn $(1 + x)^k$ into $(1 + x)^{k+1}$ . . . to which the simple answer, once the question is posed, is: multiply by another $(1 + x)$, and first ask yourself whether it is actually safe to do that to an inequality. In the fourth example, how can we predict the behaviour of $k + 2$ straight lines when we already know only how $k + 1$ lines behave? How else, apart from keeping one line aside for the moment, letting the remaining $k + 1$ lines do what we know they can, and then bringing in the last line to see how it might interact with the others? In many cases, then, there is a kind of inevitability about how you *trade up* from statement k to statement $k + 1$, but you may need to look quite carefully at those statements before you see what it is.

- As to *why* induction is a valid method of proof, it may help if you imagine the various component statements $S(1), S(2), S(3), \ldots$ stacked one above the previous one, like the rungs of an (endless) ladder 'heading off to infinity'. By checking the truth of $S(1)$ you are, almost literally, getting your foot on the bottom rung of the ladder – testing that it is strong enough to take the weight of careful inspection. The main part of the induction process, then, the demonstration that $S(k + 1)$ follows as a logical deduction from $S(k)$, says that you can always climb from any 'sound' rung to the one above it. So, start climbing: the first rung is strong/valid/true, therefore the one above (that is, the second rung) is also. From that observation, it follows that the one above that (the third rung) is equally sound. From that, so is the fourth. From that, so is the fifth . . . when is this process going to stop? Never! The way in which the positive integers are naturally ordered is that *any* particular one of them can be reached in a finite number of steps starting at 1 and increasing by 1 each time. For that reason, *any* particular $S(n)$ is accessible by the process set out in the induction template, and must therefore be a true statement.

- If you really want to understand why induction works, think back to paragraph 3.3.1 (every non-empty set of positive integers possesses a least element). Once we know that $S(1)$ is true and that $S(k)$ implies $S(k + 1)$ for each $k \geq 1$, then *suppose* that some of the $S(n)$'s are not true: that is, that the set

$$W = \{n \in \mathbb{N} : S(n) \text{ is false}\}$$

is not empty. By 3.3.1, W has to possess a least element w. Now w cannot be 1 since $S(1)$ is known to be true, so $w - 1$ is still a *positive* integer and it is strictly smaller than the smallest element of W: therefore $w - 1$ is not in W, which tells us that $S(w - 1)$ must be a true statement. Yet since $S(w - 1)$ implies $S(w)$, this guarantees that $S(w)$ is also true, which *contradicts* w being an element of W. In consequence of that contradiction, none of the $S(n)$s can have been false.

- Occasionally, $n = 1$ is not the best place at which to start an induction argument. Just as in the case of (other) sequences, it can sometimes be convenient to begin at $n = 0$, or at $n = 2$, or at some other initial value. In fact, our third case study above would have been slightly easier to read if we had expressed it as: 'If n distinct straight lines are drawn on a plane surface, where $n \geq 2$, show that they cannot have more than $n(n - 1)/2$ crossing points'. The only change in procedure that such a twist requires is that Step 1 ought to alter to become: 'Step 1: check that the statement $S(n)$ with the lowest possible value of n is actually true'. (Plus, of course, that we keep in mind that the core argument $S(k)$ implies $S(k + 1)$ only needs to work for the realistic values of k.)

We shall finish the section by doing two further examples: one that illustrates how to carry out these minor changes when '$n = 1$' is not the right starting point, and one that observes that what we called Step 0 can on occasions be a little tricky to get right.

4.2.7 Example For each integer $n \geq 9$, we show that $n! > 4^n$.

Solution

- Step 0: For each $n = 9, 10, 11, 12, \cdots$ let $S(n)$ be the statement: $n! > 4^n$.
- Step 1: $S(9)$ says that $9! > 4^9$. A little calculation shows that the left-hand side is 362880 and the right-hand side is 262144, so this statement is correct.
- Step 2: Assume the truth of a particular $S(k)$: that is, that $k! > 4^k$.
- Step 3: (In order to turn $k!$ into the expected left-hand side $(k+1)!$ of statement $k+1$, we need to multiply by $k+1$, which is of course, positive . . . indeed, it is at least 10 since $k \geq 9$.)

$$k! > 4^k \Rightarrow (k+1)k! > (k+1)4^k \geq (10)4^k > (4)4^k = 4^{k+1}$$

in other words, $S(k+1)$ is also true.

By induction, all of the statements are true from $n = 9$ onwards.

4.2.8 Example To verify that every integer ≥ 2 can be expressed as a product of prime factors.

Attempted solution

- Step 0: For each integer $n \geq 2$ let $S(n)$ be the statement 'n can be expressed as a product of primes'. (That's the obvious way to break the claimed result into layers, isn't it?)
- Step 1: $S(2)$ says that 2 is a product of primes; but 2 is itself a prime, so this statement is vacuously true: $2 = 2$ gives the prime factorisation of 2.
- Step 2: Assume the truth of a particular $S(k)$; that is, k can be expressed as a product of primes.
- Step 3: Suddenly we hit a snag. There is no evident way to get from the prime factors of k to the prime factors of $k+1$. Indeed, no prime factor of k can possibly divide into $k+1$ since they differ by 1.

  However, we can re-word Step 0 and try again:

Reattempted solution

- Step 0: For each integer $n \geq 2$ let $S(n)$ be the statement 'each integer from 2 up to n can be expressed as a product of primes'. (If we can prove all of those true, we shall have what we want.)

- Step 1: $S(2)$ says just that 2 is a product of primes; but, as before, this statement is vacuously true: $2 = 2$ gives the prime factorisation of 2.
- Step 2: Assume the truth of a particular $S(k)$; that is, that each integer from 2 up to k can be expressed as a product of primes.
- Step 3: Now with a view to $S(k+1)$, we know from Step 2 that each integer from 2 to k can be prime-factorised, and we only still need to look at $k+1$. If $k+1$ happens to be prime, there is no need to do anything: $k+1 = k+1$ is a trivial prime factorisation. Otherwise, $k+1$ is not prime and (by definition of *prime*) can be written as the product of two *smaller* numbers, say, $k+1 = a.b$ where a, b are at least 2 but less than $k+1$. Yet then Step 2 tells us that each of a and b can be written as the product of a list of primes and, putting the two lists together, we have a prime factorisation of $a.b = k+1$. So $S(k+1)$ is confirmed.

By induction, all the statements $S(n)$ are true, and so all integers from 2 upwards can be prime-factorised.

4.2.9 EXERCISES

1. Verify that $n < 2^n$ for every positive integer n.
2. Show that, for each integer $n \geq 3$:

$$3(2n)! < 2^{2n}(n!)^2.$$

3. Provided that $n \geq 4$, show that $3^n > n^3$.

Partial draft solution

In roughwork for the last of these three problems, the step from *truth for k* to *truth for k+1* amounts to showing that $3k^3 > (k+1)^3$, that is, that $3 > (1 + \frac{1}{k})^3$. Since k is at least 4, the right-hand side here is $\leq (1.25)^3$ which satisfactorily calculates out at a little under 2.

4.2.10 **Note: binomial coefficients**   It may be useful to round off this section by revising the binomial theorem and the coefficients that appear in connection with it. Whenever n is a positive integer and k is an integer such that $0 \leq k \leq n$, the symbol

$$\binom{n}{k} = \frac{n!}{k!(n-k)!}$$

is called a *binomial coefficient*. It is, amongst other interpretations, the number of different possible selections of k objects that can be chosen[6] from n distinct objects. Straightforward calculations confirm that (for all relevant n and k) $\binom{n}{0} = \binom{n}{n} = 1$, $\binom{n}{1} = n$, $\binom{n}{n-k} = \binom{n}{k}$ and, most importantly:

---

[6] This is why it is usually pronounced as 'n choose k'.

$$\binom{n}{k-1} + \binom{n}{k} = \binom{n+1}{k}.$$

We shall, on several occasions, make use of the theorem itself:

4.2.11 The binomial theorem For each $n \in \mathbb{N}$,

$$(1+x)^n = \sum_{k=0}^{n} \binom{n}{k} x^k.$$

Proof

- Step 0: For each $n \in \mathbb{N}$ in turn, let $S(n)$ denote the statement:
  $(1+x)^n = \sum_0^n \binom{n}{k} x^k$.
- Step 1: $S(1)$ says that $(1+x)^1 = \binom{1}{0} + \binom{1}{1}x$ which is trivially correct.
- Step 2: Assume the truth of a particular $S(j)$; that is, that $(1+x)^j = \sum_0^j \binom{j}{k} x^k$.
  (Notice that we are, as usual, taking care not to use the same symbol with more than one meaning.)
- Step 3: In order to turn $(1+x)^j$ into the expected left-hand side $(1+x)^{j+1}$ of
  statement number $j+1$, we need to multiply by another $(1+x)$ and *carefully*
  gather up[7] each power of x that appears:

$$(1+x)^{j+1} = (1+x) \sum_{k=0}^{j} \binom{j}{k} x^k$$

$$= (1+x)\left\{ 1 + \binom{j}{1}x^1 + \binom{j}{2}x^2 + \binom{j}{3}x^3 + \cdots + \binom{j}{j-1}x^{j-1} + x^j \right\}$$

$$= 1 + \left\{ \overline{\binom{j}{1}+1}(x^1) + \overline{\binom{j}{2}+\binom{j}{1}}(x^2) + \overline{\binom{j}{3}+\binom{j}{2}}(x^3) + \cdots + \overline{1+\binom{j}{j-1}}(x^j) \right\} + x^{j+1}$$

$$= 1 + \left\{ \overline{\binom{j}{1}+\binom{j}{0}}(x^1) + \overline{\binom{j}{2}+\binom{j}{1}}(x^2) + \overline{\binom{j}{3}+\binom{j}{2}}(x^3) + \cdots + \overline{\binom{j}{j}+\binom{j}{j-1}}(x^j) \right\} + x^{j+1}$$

$$= 1 + \left\{ \binom{j+1}{1}x^1 + \binom{j+1}{2}x^2 + \binom{j+1}{3}x^3 + \cdots + \binom{j+1}{j}x^j \right\} + x^{j+1}$$

$$= \sum_{k=0}^{j+1} \binom{j+1}{k} x^k$$

  – using the identity from 4.2.10 at the last-but-one line. In other words,
  $S(k+1)$ is also true.

By induction, all of the statements are true.

---

[7] We are using overscoring as another form of bracketing, to try to improve readability in these few lines of algebra.

4.3 Recursively defined sequences

Up to this point, every individual sequence that we have worked on has been specified by writing down either a formula for its nth term or a list whose pattern was obvious enough that we could find such a formula easily. Not all important sequences work like this: for instance, although the pattern in the Fibonacci sequence $1, 1, 2, 3, 5, 8, 13, 21, 34, \cdots$ is clear enough, it is far from obvious how to obtain an explicit formula for its nth term; rather, the pattern is expressed by noting that each term from the third onwards is the sum of the preceding two:

$$f_n = f_{n-1} + f_{n-2} \quad provided \; that \; n \geq 3$$

and, to complete the definition, $f_1 = 1, f_2 = 1$. This is an instance of what is called *recursive definition*.

Another instance that you are likely to have met is the Newton – Raphson approximation process. In this, faced with an equation of the form $f(x) = 0$ that we cannot solve precisely, but in which the function f can be differentiated (see Chapter 12 if this idea is unfamiliar), we make an initial rough guess $x = x_1$ at the solution, perhaps via a sketch graph, and then improve that guess to a second one:

$x_2 = x_1 - \dfrac{f(x_1)}{f'(x_1)}$ (where f' denotes the derivative).

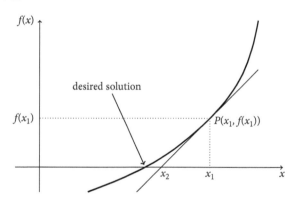

Tangent to curve at P crosses x-axis at x_2

In general, x_2 will be significantly closer to the true solution than x_1 was, and we can now repeat the improvement process

$$x_3 = x_2 - \frac{f(x_2)}{f'(x_2)},$$

$$x_4 = x_3 - \frac{f(x_3)}{f'(x_3)},$$

$$x_5 = x_4 - \frac{f(x_4)}{f'(x_4)}$$

and so on. In most cases this sequence of improving approximations converges to a limit whose exact value is a solution to the original equation. The definition of this sequence

$$x_{n+1} = x_n - \frac{f(x_n)}{f'(x_n)},$$

together with a random (but not grossly inaccurate) guess value for x_1, is another example of recursive definition.

The essential characteristic of this style of definition is that, instead of telling us explicitly what the nth term is, it provides us with an algorithm for determining it from the value(s) of one or more previous terms in the list. In time, that means that we could in principle work out any particular term that was of interest to us ...but how could we hope to find the limit of such a sequence if, indeed, it had one?

4.3.1 Example   To identify, if it exists, the limit of the sequence $(x_n)_{n \geq 1}$ defined recursively by:

$$x_1 = 12; \quad x_{n+1} = \sqrt{3x_n + 28} \ (n \geq 1).$$

Draft solution

With so little given information, we need to start by calculating the first few terms. They work out (to four decimal places) as follows:

$$12, 8, 7.2111, 7.0451, 7.0097, 7.0021, 7.0004, \cdots$$

It must be stressed that this is very little evidence as to what happens as n goes to infinity! Nevertheless, it is enough to let us make a clutch of informed guesses: we guess that all the terms lie between 12 and 7, that the sequence is decreasing, and that the limit is 7. Now we have a definite proposal to try to establish:

Solution

- Step 0: For each $n \geq 1$ let $S(n)$ be the statement: $7 < x_n \leq 12$.
- Step 1: $S(1)$ says that $7 < 12 \leq 12$, which is true.
- Step 2: Assume the truth of a particular $S(k)$; that is, that $7 < x_k \leq 12$.
- Step 3: (How are we to get from x_k to the next term? As the recursive definition told us, we multiply by 3, add 28, and take the square root. So:)

$$21 < 3x_k \leq 36, \quad 49 < 3x_k + 28 \leq 64, \quad 7 < \sqrt{3x_k + 28} = x_{k+1} \leq 8 \leq 12.$$

In other words, $S(k + 1)$ is also true.

By induction, all of the statements $S(n)$ are true, that is, all terms of the sequence do lie between 7 exclusive and 12 inclusive.

(Next, to compare x_n with x_{n+1}, the square root makes it less easy to see what is going on, so let us instead square both sides and actually compare x_n^2 with x_{n+1}^2.

Keep in mind that $0 < a < b$ implies that $\sqrt{a} < \sqrt{b}$, as we pointed out in paragraph 1.2; we shall use this several times in the next few pages.)

Notice that

$$x_n^2 - x_{n+1}^2 = x_n^2 - 3x_n - 28 = (x_n - 7)(x_n + 4)$$

and because we already know that all terms lie between 7 and 12, that is a product of two *positive* numbers, therefore positive itself. So $x_n^2 > x_{n+1}^2$ and, taking (positive) square roots, we get $x_n > x_{n+1}$ for all values of n, that is, the sequence is decreasing.

We now know that (x_n) is both bounded and decreasing, and must therefore converge to *some* limit ℓ. Also (x_{n+1}), being merely the sequence (x_n) without its first term, converges to the same ℓ, and (x_n^2) converges to ℓ^2 by algebra of limits. Take limits across the equation

$$x_{n+1}^2 - 3x_n - 28 = 0 \quad \text{for all } n$$

and we get the quadratic $\ell^2 - 3\ell - 28 = 0$ which factorises predictably into $(\ell - 7)(\ell + 4) = 0$. Thus, the only conceivable values of ℓ are 7 and -4. Yet $7 < x_n \leq 12$ (for all n) informs us[8] that $7 \leq \ell \leq 12$, so -4 is really not a possible value, and only 7 remains. We conclude that (x_n) must indeed converge to the limit 7.

We'll do a second example, but this time cutting the 'running commentary' back to the extent that is normally expected in a solution.

4.3.2 Example For the sequence (a_n) specified by the formulae $a_1 = 0, a_{n+1} = \sqrt{4a_n + 77}$:

1. Show that $0 \leq a_n < 11$ for every $n \geq 1$,
2. Show that the sequence is increasing,
3. Explain why it must possess a limit, and evaluate that limit.

Solution

- Step 0: For each n let $S(n)$ be the statement: $0 \leq a_n < 11$.
- Step 1: $S(1)$ says $0 \leq 0 < 11$ which is true.
- Step 2: Assume that $0 \leq a_k < 11$ for some particular k.
- Step 3: Then $4(0) + 77 \leq 4a_k + 77 < 4(11) + 77$, and so $0 \leq \sqrt{77} \leq \sqrt{4a_k + 77} = a_{k+1} < \sqrt{121} = 11$. Therefore $S(k + 1)$ is also true.

By induction, all the statements $S(n)$ are true. This proves (1).

---

[8] See taking limits across an inequality: Theorem 4.1.17.

Using (1), $a_n^2 - a_{n+1}^2 = a_n^2 - 4a_n - 77 = (a_n + 7)(a_n - 11)$ is the product of a positive and a negative, therefore negative. That is, $a_n^2 < a_{n+1}^2$ and so $a_n < a_{n+1}$ for each n. This proves (2).

Since (a_n) is now bounded and increasing, it must converge. Let ℓ be its limit. Also $a_{n+1} \to \ell$. Taking limits across $a_{n+1}^2 - 4a_n - 77 = 0$ (for all n) gives $\ell^2 - 4\ell - 77 = (\ell + 7)(\ell - 11) = 0$, so ℓ can only be -7 or 11. Yet $\ell = -7$ is impossible because every term is at least 0. Therefore $a_n \to 11$.

4.3.3 EXERCISE   For the sequence (a_n) specified by the formulae $a_1 = 16$, $a_{n+1} = \sqrt{20 + a_n}$:

1. Show that $5 < a_n \le 16$ for every $n \ge 1$,

2. Show that the sequence is decreasing,

3. Explain why it must possess a limit, and evaluate that limit.

4.3.4 EXERCISE   Find (if it exists) the limit of the indicated sequence:

$$\sqrt{5}, \sqrt{5 + \sqrt{5}}, \sqrt{5 + \sqrt{5 + \sqrt{5}}}, \sqrt{5 + \sqrt{5 + \sqrt{5 + \sqrt{5}}}}, \cdots .$$

Draft solution

We first need a clearer idea of what this sequence is. The pattern is that, for each term in turn, the next one is created by adding 5 and taking the square root. In other words, $a_{n+1} = \sqrt{5 + a_n}$. Also $a_1 = \sqrt{5}$ to get the process started. This flags up the recursive nature of the problem. Calculate the first few terms and they seem to be increasing towards a limit of approximately 2.791. What number could that be?

If there is indeed a limit ℓ then (a_{n+1}) will also converge to ℓ so, from $a_{n+1}^2 = 5 + a_n$ we get $\ell^2 = 5 + \ell$. This quadratic has solutions $(1 \pm \sqrt{21})/2 = 2.7913, -1.7913$ approximately, so now we can see exactly what number that limit is going to be.

Break the problem into the same three sections as before:

1. Show that $\sqrt{5} \le a_n < (1 + \sqrt{21})/2$ for every $n \ge 1$,

2. Show that the sequence is increasing,

3. Explain why it must possess a limit, and evaluate that limit.

4.3.5 EXERCISE   Investigate the following sequence:

$$4, \sqrt{3(4) - 2}, \sqrt{3(\sqrt{3(4) - 2}) - 2}, \sqrt{3(\sqrt{3(\sqrt{3(4) - 2}) - 2}) - 2},$$

$$\sqrt{3(\sqrt{3(\sqrt{3(\sqrt{3(4) - 2}) - 2}) - 2}) - 2}, \cdots .$$

4.3.6 **EXERCISE** Investigate the following sequence:

$$1.5, \sqrt{3(1.5) - 2}, \sqrt{3(\sqrt{3(1.5) - 2}) - 2}, \sqrt{3(\sqrt{3(\sqrt{3(1.5) - 2}) - 2}) - 2},$$

$$\sqrt{3(\sqrt{3(\sqrt{3(\sqrt{3(1.5) - 2}) - 2}) - 2}) - 2}, \cdots.$$

4.3.7 **EXERCISE**

1. The sequence $(c_n)_{n \in \mathbb{N}}$ is defined recursively by the two formulae

$$c_1 = 2, \quad c_{n+1} = \frac{2 + 2c_n}{3 + c_n} \text{ (while } n \in \mathbb{N}).$$

   Show that
   - $1 < c_n \leq 2$ for all $n \in \mathbb{N}$,
   - $(c_n)_{n \in \mathbb{N}}$ is a decreasing sequence,
   - $(c_n)_{n \in \mathbb{N}}$ converges: and determine what its limit is.

2. Alter part 1 by changing only the first term, to $c_1 = 2/3$, and re-work the problem (changing whatever has to be changed).

4.4 POSTSCRIPT: The epsilontics game – the 'fifth factor of difficulty'

You have now seen two versions of the 'challenge-response game' that is endemic in analytic arguments and that, for many students, contributes to the perception that analysis is difficult. Look again at the definition (2.7.1) of limit of a sequence, and the standard way (3.2.7) of identifying the supremum of a set:

For each $\varepsilon > 0$ there is some positive integer n_ε such that $\cdots$

For each $\varepsilon > 0$ there exists $x_\varepsilon \in A$ such that $\cdots$

In most cases, the actual *mathematical calculation* of finding or estimating n_ε or $x_\varepsilon \in A$ is pretty routine. The difficulty, such as it is, lies in the English words 'For each', 'there is', 'there exists' or in the logic that is represented by them.

It is important to bear in mind that each argument such as these is, in some sense, a game – a competition. Our imaginary opponent puts forward a value of ε as a challenge. In order to win the point, we have to come up with a response (n_ε, x_ε, ...) that satisfies the requirements of the game. A complete argument (a 'proof', a 'solution') is then simply a winning strategy: a procedure that will find a winning response for every legitimate challenge – 'for every positive epsilon'.

Having worked through the examples and exercises in the first few chapters, you should by now be reasonably acclimatised to the interactive nature of arguments of this kind, but they do pose an initial barrier to the casual reader.

We shall encounter several more variations of this 'challenge-response' epsilon-tics game as we proceed.

5 Sampling a sequence – subsequences

5.1 Introduction

If we sample an (infinite) sequence by picking out a finite number of its terms, it would be unreasonable to expect such a sample to be at all 'representative' in the sense of telling us anything useful about the sequence as a whole. Certainly it will not tell us anything about its possible limit: for we have pointed out several times that changing or deleting any finite number of terms does not affect limiting behaviour in any way. What if, instead, we sample an infinite selection of terms? Then at least our selection (in the original order) will constitute a sequence in its own right, and we might reasonably expect that its behaviour will tell us something, but not everything, about the original sequence; on the other hand, knowledge about the whole sequence is likely to give us all the information we might need about the newly selected one. This short chapter gives a more precise description of the idea that we are sketching here, and develops and applies a few rather predictable results plus one that is less obvious and much more powerful (*the Bolzano-Weierstrass theorem*) and which will play a key role later in the text.

5.2 Subsequences

Informally, a subsequence of a sequence is an endless list of its terms in their original order. So if $(x_n)_{n \geq 1}$ is any sequence, and we imagine its terms strung out in an unending list

$$x_1, x_2, x_3, x_4, x_5, x_6, x_7, x_8, x_9, x_{10}, \cdots$$

then a subsequence will be created by scanning along that list and lifting out an unending selection of what we find; for instance,

$$x_7, x_{16}, x_{21}, x_{22}, x_{39}, x_{122}, \cdots$$

or

$$x_4, x_8, x_{13}, x_{400}, x_{401}, x_{605}, x_{677}, x_{759}, \cdots .$$

Undergraduate Analysis: A Working Textbook, Aisling McCluskey and Brian McMaster 2018.
© Aisling McCluskey and Brian McMaster 2018. Published 2018 by Oxford University Press

How should we best write a general subsequence when we don't know in advance *which* terms are to be picked out? We *could* relabel the first item chosen as y_1 and the second chosen as y_2 and the third as y_3 and so on, and the resulting symbol $(y_n)_{n \in \mathbb{N}}$ is certainly a perfectly good notation for a sequence, but it has lost any visible connection with the original sequence that we started with. A better method is to give labels to the places in the original sequence where we find the subsequence's items: if our first choice occurred at the n_1^{th} place in the original list, and our second at the n_2^{th} place, and our third at the n_3^{th} place and so on, then the chosen numbers that build up the subsequence are

$$x_{n_1}, x_{n_2}, x_{n_3}, x_{n_4}, \cdots$$

and the entire subsequence can now be written as $(x_{n_k})_{k \geq 1}$. This is a slightly cluttered symbol, but it succeeds in capturing two of the important aspects of a subsequence: that it actually is an infinite sequence, and that its terms are *some* of the terms from the original (x_n). The third important aspect – that the order has to be the same as was in the original (no back-tracking allowed) – is captured by insisting that $n_1 < n_2 < n_3 < n_4 < \cdots$, in other words, that the sequence of labels $(n_1, n_2, n_3, n_4, \cdots)$ has to be strictly increasing.

With that discussion behind us, we can now write down concisely what a subsequence is (and how to denote it):

5.2.1 **Definition**   Suppose we are given a sequence $(x_n)_{n \geq 1}$.
For any strictly increasing sequence $(n_k)_{k \geq 1}$ of positive integers –
that is, $n_1 < n_2 < n_3 < \cdots < n_k < \cdots$ –
the sequence $(x_{n_k})_{k \geq 1} = (x_{n_1}, x_{n_2}, x_{n_3}, \cdots, x_{n_k}, \cdots)$ is a subsequence of it.

Notice how we used different symbols, in the sequence and in the subsequence, for 'the thing that is going to infinity'. Indeed, we must be careful not to use a symbol with two meanings at once anywhere in mathematics, and especially not when working with subsequences: if we had denoted our subsequence by such a notation as $(x_{n_n})_{n \geq 1}$ then confusion would be practically guaranteed. Of course there is no need to use the particular letter k that we chose here: $(x_{n_i})_{i \geq 1}$ or $(x_{n_p})_{p \geq 1}$ would have done equally well.

5.2.2 **Examples**

1. For any sequence $(x_n)_{n \geq 1}$, the following are a few of its subsequences: the sequence of even-numbered terms $(x_{2k})_{k \geq 1}$, the sequence of odd-numbered terms $(x_{2k-1})_{k \geq 1}$, the sequence $(x_{k^2})_{k \geq 1} = (x_1, x_4, x_9, x_{16}, x_{25}, \cdots)$, the sequence $(x_{k!})_{k \geq 1} = (x_1, x_2, x_6, x_{24}, x_{120}, \cdots)$, the sequence $(x_{2^k})_{k \geq 1} = (x_2, x_4, x_8, x_{16}, x_{32}, \cdots)$.

2. In particular, $(5k-1)$ is a subsequence of (n), (k^{-2}) is a subsequence of (n^{-1}), $(\frac{1}{k!})$ is a subsequence of $(\frac{1}{n})$, $(\sin(7k^2 + 3k + 6))$ is a subsequence of $(\sin n)$, $(\sqrt[3k]{3k})$ is a subsequence of $(\sqrt[n]{n})$.

3. The sequence

$$\left((-1)^n\left(2+\frac{1}{n}\right)\right)_{n\in\mathbb{N}} = \left(-2-\frac{1}{1},2+\frac{1}{2},-2-\frac{1}{3},2+\frac{1}{4},\cdots\right) \text{ does not}$$

converge, but two of its subsequences (those consisting of the odd terms and the even terms) are

$$\left(-2-\frac{1}{1},-2-\frac{1}{3},-2-\frac{1}{5},-2-\frac{1}{7},\cdots\right)$$

$$=\left((-1)^{2k-1}\left(2+\frac{1}{2k-1}\right)\right)_{k\in\mathbb{N}} = \left(-\left(2+\frac{1}{2k-1}\right)\right)_{k\in\mathbb{N}}$$

and

$$\left(2+\frac{1}{2},2+\frac{1}{4},2+\frac{1}{6},2+\frac{1}{8},\cdots\right)$$

$$=\left((-1)^{2k}\left(2+\frac{1}{2k}\right)\right)_{k\in\mathbb{N}} = \left(2+\frac{1}{2k}\right)_{k\in\mathbb{N}}$$

both of which do converge (to -2 and to 2 respectively).

4. The sequence $(n(1+(-1)^n))_{n\in\mathbb{N}} = (0,4,0,8,0,12,0,16,\cdots)$ is unbounded, but at least one of its subsequences (that consisting of the odd terms) is not only bounded but convergent; indeed, it is constant at 0. Another of its subsequences is $(2k(1+(-1)^{2k})) = (4k) = (4,8,12,16,\cdots)$ which is unbounded.

5. The sequence $(\sin(n\pi/2))$ is the 'endlessly recycling' sequence

$$(1,0,-1,0,1,0,-1,0,1,\cdots)$$

from which we could evidently extract a subsequence converging to 1, another converging to -1 and a third converging to 0 (as well as many others that do not converge).

6. Any subsequence of a subsequence of $(x_n)_{n\geq 1}$ is itself a subsequence of $(x_n)_{n\geq 1}$.

Let us formalise the insight that convergence of the whole sequence implies convergence of every subsequence:

5.2.3 Theorem Each subsequence of a convergent sequence converges, and to the same limit as does the original sequence.

Proof

Suppose that $x_n \to \ell$ and that $(x_{n_k})_{k\geq 1}$ is a subsequence of $(x_n)_{n\geq 1}$. We need to show that $x_{n_k} \to \ell$ (as $k \to \infty$).

Notice that, because $n_1 < n_2 < n_3 < \cdots < n_k < \cdots$, we get

$$n_1 \geq 1, n_2 \geq 2, n_3 \geq 3, \cdots$$

and, in general,

$$n_k \geq k.$$

If $\varepsilon > 0$ is given, $x_n \to \ell$ tells us that there is a positive integer n_0 such that $n \geq n_0$ forces $|x_n - \ell|$ to be $< \varepsilon$. Now notice that $k \geq n_0$ forces $n_k \geq k \geq n_0$, that is, $n_k \geq n_0$, from which we get $|x_{n_k} - \ell| < \varepsilon$. (Less formally, big enough values of k guarantee that the error $|x_{n_k} - \ell|$ will be smaller than ε.) Hence $x_{n_k} \to \ell$, as required.

5.2.4 Example To show that the sequence $\left((-1)^n \left(2 + \frac{1}{n}\right)\right)_{n \in \mathbb{N}}$ is divergent.

Solution

In 5.2.2 (3) we noticed that this sequence has a subsequence whose limit is -2 and another whose limit is 2. By the theorem, this could not happen if the full sequence converged. Hence the result.

Here is a kind of weak converse to Theorem 5.2.3:

5.2.5 Example Suppose we are given a sequence $(x_n)_{n \geq 1}$ and a number ℓ such that the subsequence of odd-numbered terms $(x_{2k-1})_{k \geq 1}$ and the subsequence of even-numbered terms $(x_{2k})_{k \geq 1}$ both converge to ℓ. To show that $(x_n)_{n \geq 1}$ itself converges to ℓ.

Solution

Given $\varepsilon > 0$, the convergence of the two subsequences tells us that there are positive integers k_o and k_e such that

$$|x_{2k-1} - \ell| < \varepsilon \text{ whenever } k \geq k_o, \text{ and}$$
$$|x_{2k} - \ell| < \varepsilon \text{ whenever } k \geq k_e.$$

Put $n_0 = \max\{2k_o - 1, 2k_e\}$. It follows that when $n \geq n_0$ (whether n is odd or even) we have $|x_n - \ell| < \varepsilon$. Hence $x_n \to \ell$.

5.2.6 HARDER EXERCISE It is not difficult to modify 5.2.5 to show that, if we break up a sequence into three, or four, or indeed any finite number of subsequences (using all of the terms) and find that all these subsequences converge to the same limit, then so must the original sequence. (*Suggestion:* induction.) However, this does not work if we break it into an infinite number of subsequences: see if you can devise a sequence $(y_n)_{n \geq 1}$ such that

- for each prime number p, the subsequence $(y_{(p^k)})_{k \geq 1}$ converges to zero, and
- all of the terms y_n for which n is *not* a power of a prime equal zero, and yet
- the whole sequence $(y_n)_{n \geq 1}$ does not converge to zero.

5.2.7 **EXERCISE**   Prove that

- each subsequence of an increasing sequence is increasing,
- each subsequence of a decreasing sequence is decreasing.

Partial proof

If $(a_n)_{n\in\mathbb{N}}$ is increasing and $(a_{n_k})_{k\in\mathbb{N}}$ is one of its subsequences then, for each k, we have $n_k < n_{k+1}$. Fill in the integers that lie between them, and we see

$$n_k < n_k + 1 < n_k + 2 < \cdots < n_{k+1} - 1 < n_{k+1}.$$

Since the original sequence was increasing, that yields

$$a_{n_k} \leq a_{n_k+1} \leq a_{n_k+2} \leq \cdots \leq a_{n_{k+1}-1} \leq a_{n_{k+1}},$$

so $a_{n_k} \leq a_{n_{k+1}}$.

Expressing the last exercise briefly as 'sequence is monotonic implies subsequence is monotonic', it would be foolish to expect anything like a full converse (along the lines of 'subsequence is monotonic implies sequence is monotonic') since a single subsequence simply does not contain enough information about the parent sequence for such a conclusion to be at all plausible. All the same, it would not be unreasonable of us to expect some kind of *partial* converse, in which the monotonicity of a subsequence told us *something* about the ordering of the terms of the entire sequence. It is therefore a little surprising to see, from the next result, that possession of a monotonic subsequence tells us *absolutely nothing* about the parent sequence

5.2.8 **Theorem**   *Every* sequence has a monotonic subsequence.

Proof

Let $(x_n)_{n\in\mathbb{N}}$ be any sequence. To help us (almost literally) see through this curious proof, let us call a positive integer m *farsighted* if x_m is greater than all the later terms in the sequence, that is, if $x_m > x_q$ for every $q > m$. (Imagine that you are standing on x_m and trying to see off to infinity over the heads of all the later x_q's in the sequence; if you can do that, then the m that you are using is farsighted.) Now *either* there are infinitely many farsighted integers, *or else* there are only finitely many (perhaps even none).

In the first case, there is an (endless) succession $m_1 < m_2 < m_3 < \cdots$ of farsighted integers, and by definition of farsighted we get

$$x_{m_1} > x_{m_2} > x_{m_3} > \cdots$$

that is, a decreasing subsequence of $(x_n)_{n\in\mathbb{N}}$.

In the second case we can count through to an integer r that is bigger than *every* farsighted integer and know that, from r onwards, no integer is farsighted, that is, for every integer s we encounter, there will be a greater integer s' for which $x_s \leq x_{s'}$. (Every x_s has its view of infinity obstructed by some later $x_{s'}$, so to speak.) This yields firstly $x_r \leq x_{r_1}$ for some $r_1 > r$, and then in turn $x_{r_1} \leq x_{r_2}$ for some $r_2 > r_1$, $x_{r_2} \leq x_{r_3}$ for some $r_3 > r_2$ and so on without end. Look: we are forming an increasing subsequence

$$x_r \leq x_{r_1} \leq x_{r_2} \leq x_{r_3} \leq \cdots$$

this time. The demonstration is complete.

5.2.9 EXERCISES Find the limit of each of the following sequences:

1.
$$\left(0.9^{n^2+5n+17}\right),$$

2.
$$\left(\frac{1}{n^3}\right),$$

3.
$$\left(\frac{(5n^4 + 2n^2 - 1)^3 - 1}{3(5n^4 + 2n^2 - 1)^3 + 2}\right).$$

Draft solution

In each case, it will be enough to recognise the given sequence as a subsequence of some (simpler) sequence whose limit you already know or can easily calculate.

5.2.10 EXERCISES Show that the following sequences are divergent:

1.
$$\left((-1)^n \left(\frac{3+4n}{2+n}\right)\right)_{n \in \mathbb{N}}.$$

2.
$$\left(5 \sin\left(\frac{n\pi}{7}\right) + 3\cos\left(\frac{n\pi}{11}\right)\right)_{n \in \mathbb{N}}.$$

3. The sequence $(1/pmax(n))_{n \geq 2}$ given by $pmax(n) =$ the largest prime factor of n (assuming $n \geq 2$).

Fragment of solution

In the second of these, you may save yourself a good deal of trigonometry by noticing the value of the nth term when n is an odd number times 77, and again when n is an even number times 77. (The idea of focusing on 77 is driven simply by a wish to avoid fractions if we can reasonably do so without losing too much information.)

5.3 Bolzano-Weierstrass: the overcrowded interval

There is an ancient mathematical insight sometimes called the pigeonhole principle: if you put $n + 1$ pigeons into n pigeonholes, then at least one pigeonhole will contain more than one pigeon. (More broadly, if you put $kn + 1$ or more letters into n mailboxes, where k and n are positive integers, then at least one mailbox must contain $k + 1$ or more letters.) The Bolzano-Weierstrass result starts out as an infinite version of this: if you distribute an infinite number of pigeons across two pigeonholes, then at least one pigeonhole will end up containing an infinite number of pigeons.

Let's re-express that in terms of sequences. If all the terms of a sequence lie within the union $A \cup B$ of two sets A, B of real numbers, then at least one of the two sets must include an infinite number of those terms.[1] Formally: if $(x_n)_{n \in \mathbb{N}}$ is a sequence of elements of $A \cup B$, then either A or B (or both) will include the number x_n for infinitely many values of n.

Hopefully, that sounds like little more than common sense ...and yet it has an un-obvious and powerful consequence:

5.3.1 **The Bolzano-Weierstrass theorem**   Every bounded sequence has a convergent subsequence.

Proof

Let $(x_n)_{n \in \mathbb{N}}$ be any bounded sequence. By its boundedness, we can find a closed interval $I_0 = [-M, M]$ (for some positive M) that includes *every* term x_n.

Now I_0 is the union of its left half and its right half: indeed, we can write $I_0 = [-M, 0] \cup [0, M]$. So we can pick one of the two halves that includes x_n for infinitely many values of n. Call whichever half we pick I_1, and select also n_1 for which $x_{n_1} \in I_1$.

Next, repeat this argument upon I_1: for I_1 is the union of its left half and its right half, and we can pick one of the two halves that includes x_n for infinitely many values of n. Call whichever half we pick this time I_2, and select also n_2 **greater than** n_1 for which $x_{n_2} \in I_2$. The phrase **greater than** n_1 is legitimate because we have an infinite number of possible n_2s to pick from, and can therefore surely arrange to make that choice larger than our previous n_1, whatever it was.

Next, repeat this argument upon I_2: I_2 is the union of its left half and its right half, and we can pick one of the two halves that includes x_n for infinitely many values of n. Call whichever half we pick this time I_3, and select also n_3 **greater than** n_2 for which $x_{n_3} \in I_3$.

---

[1] Unlike real pigeonholes, the sets A and B do not have to be disjoint; unlike real pigeons, the terms of the sequence do not have to be all distinct from one another.

And so on without end (and you can make an appeal to induction if you feel that it's necessary).

This generates a sequence $(I_1, I_2, I_3, \cdots, I_k, \cdots)$ of closed intervals, each contained in the previous one and exactly half of its length, and also a subsequence $(x_{n_1}, x_{n_2}, x_{n_3}, \cdots, x_{n_k}, \cdots)$ of the original sequence, such that (for all k) x_{n_k} belongs to I_k. All we still need to do is to use the 'shrinking' nature of the intervals to show that the subsequence converges.

Write the typical I_k as $[a_k, b_k]$ and notice that, since the length of I_1 was M and the (half) length of I_2 was $M/2$ and the length of I_3 was $M/4$ and so on, b_k is actually $a_k + M/2^{k-1}$. Since I_{k+1} is either the left or the right half of I_k, we also have

$$a_k \le a_{k+1} \le b_{k+1} \le M$$

for all k, so (a_k) is an increasing bounded sequence, and therefore converges to some limit ℓ. Finally,

$$a_k \le x_{n_k} \le b_k = a_k + M/2^{k-1}$$

lets us appeal to the squeeze because, since $2^{-k} \to 0$ as $k \to \infty$ (see 4.1.13), a_k and $a_k + M/2^{k-1}$ both converge to (the same) ℓ. Thus we have (x_{n_k}) converging to ℓ.

An alternative approach to proving Bolzano-Weierstrass is outlined in paragraph 5.3.7.

For the purposes of this text, the really important applications of Bolzano-Weierstrass happen only after we have defined *continuous functions*. Until we reach that point, 5.3.2 and 5.3.3 will provide a little insight into how it may be used. In any case, please take note of 5.3.5 and 5.3.6 which have a direct bearing on our ongoing study of *sequences*.

5.3.2 **Example**   Let us (temporarily) call a sequence $(x_n)_{n \in \mathbb{N}}$ *channelled* if

$$|x_1 - x_2| < 1/2, |x_2 - x_3| < 1/3, |x_3 - x_4| < 1/4, \cdots$$

and, in general, $|x_n - x_{n+1}| < 1/(n+1)$ for every positive integer n. To prove that every bounded sequence has a channelled subsequence.

Solution

Thanks to what Bolzano-Weierstrass tells us, it will be enough just to show that every *convergent* sequence has a channelled subsequence; so let us tackle that.

Suppose $(y_n)_{n \in \mathbb{N}}$ converges to ℓ. Then there is $n_1 \in \mathbb{N}$ such that $|y_n - \ell| < \frac{1}{4}$ whenever $n \ge n_1$. There is also $n_2 \in \mathbb{N}$ such that $|y_n - \ell| < \frac{1}{6}$ whenever $n \ge n_2$, and we can make sure that $n_2 > n_1$ (just by increasing n_2 if necessary). Then there is also $n_3 \in \mathbb{N}$ such that $|y_n - \ell| < \frac{1}{8}$ whenever $n \ge n_3$, and we can make sure that $n_3 > n_2$ (just by increasing n_3 if necessary).

We are now into an induction process, and it will generate more and more positive integers $n_3 < n_4 < n_5 < n_6 < \cdots$ such that (in the sequence (y_n)):

- each two terms from number n_4 onwards will be less than $\frac{1}{10}$ away from ℓ,
- each two terms from number n_5 onwards will be less than $\frac{1}{12}$ away from ℓ,
- each two terms from number n_6 onwards will be less than $\frac{1}{14}$ away from ℓ,

– and so on. Using these distance estimates (and the triangle inequality), we see that

- $|y_{n_1} - y_{n_2}| \leq |y_{n_1} - \ell| + |\ell - y_{n_2}| < 1/4 + 1/4 = 1/2$,
- $|y_{n_2} - y_{n_3}| \leq |y_{n_2} - \ell| + |\ell - y_{n_3}| < 1/6 + 1/6 = 1/3$,
- $|y_{n_3} - y_{n_4}| \leq |y_{n_3} - \ell| + |\ell - y_{n_4}| < 1/8 + 1/8 = 1/4$,
- $|y_{n_4} - y_{n_5}| \leq |y_{n_4} - \ell| + |\ell - y_{n_5}| < 1/10 + 1/10 = 1/5$

— and so on. In other words, the subsequence $(y_{n_1}, y_{n_2}, y_{n_3}, y_{n_4} \cdots)$ is channelled.

5.3.3 **EXERCISE** Let us (equally temporarily) call a sequence $(x_n)_{n \in \mathbb{N}}$ *superchannelled* if

$$|x_1 - x_2| < 10^{-1}, |x_2 - x_3| < 10^{-2}, |x_3 - x_4| < 10^{-3}, \cdots$$

and, in general, $|x_n - x_{n+1}| < 10^{-n}$ for every positive integer n. Prove that every bounded sequence has a superchannelled subsequence.

5.3.4 **HARDER EXERCISE** Suppose we are given a sequence of points $(P_1, P_2, P_3, \cdots)$ in the coordinate plane, each lying inside the square whose corners have coordinates $(0,0), (1,0), (1,1)$ and $(0,1)$, and suppose also that for every positive ε there is at least one of these points that lies below the horizontal line $y = \varepsilon$. Show that there is a point on the x-axis such that every open disc centred on that point includes infinitely many of the points P_n.

Draft solution

Think of the coordinates (x_n, y_n) of the typical point P_n in the sequence. For each positive integer k (thinking $\varepsilon = \frac{1}{k}$) we get a positive integer n_k such that $0 < y_{n_k} < \frac{1}{k}$. Can we use Bolzano-Weierstrass on the sequence (x_{n_k}) of all the x-coordinates of the associated points? What does it tell us if we do?

5.3.5 **Example** Let $(a_n)_{n \in \mathbb{N}}$ be a bounded sequence that is *not* convergent. We show that it must have two convergent subsequences possessing different limits.

Solution

Find $M > 0$ such that $[-M, M]$ contains the entire sequence. By Bolzano-Weierstrass, there is a subsequence $(a_{n_k})_{k \in \mathbb{N}}$ that converges to a limit ℓ. Yet $(a_n)_{n \in \mathbb{N}}$

itself does not converge to ℓ: which means that there is some positive number ε so small that a_n never settles permanently between $\ell - \varepsilon$ and $\ell + \varepsilon$. This in turn means that, whichever n_0 in $\mathbb{N}$ we think of, there is some greater $n > n_0$ for which x_n is outside those borderlines ...put more tidily, there is a whole subsequence of $(a_n)_{n \in \mathbb{N}}$ lying in $[-M, \ell - \varepsilon] \cup [\ell + \varepsilon, M]$. By pigeonholing, one of the two intervals here contains a subsequence. Use Bolzano-Weierstrass *on this* and it gives us another sub-subsequence (still a subsequence of the original, of course) converging to a limit ℓ' which, since it has to lie in $[-M, \ell - \varepsilon] \cup [\ell + \varepsilon, M]$ (by the theorem on taking limits across an inequality – see 4.1.17), cannot be the same number as ℓ.

That is actually a more useful result than it initially looks. Establishing that a sequence converges *by the original definition* is, of course, seldom easy ...but using that same definition to show that a sequence *diverges* can be very awkward indeed. What the theorems on 'convergent implies bounded' and 'convergence of subsequences' plus the last example, put together, tell us is that *it is never necessary*. Of course a sequence that is unbounded or that possesses two subsequences with different limits cannot be convergent (by earlier results), but now we see that the converse is also valid:

5.3.6 Proposition   A sequence is divergent *if and only if* it is unbounded or has two subsequences with different limits.

Proof

A sequence that is unbounded, or possesses subsequences with differing limits, must be divergent by Lemma 4.1.6 and Theorem 5.2.3. The converse – that a divergent sequence must have one of these characteristics – is what Example 5.3.5 demonstrated.

5.3.7 EXERCISE   Use the *every sequence has a monotonic subsequence* theorem (5.2.8) to give an alternative proof of the Bolzano-Weierstrass result.

(This is quite a simple exercise to carry out, and most people regard this alternative proof as both shorter and easier to follow than the one we presented earlier, and yet somehow less informative, more like 'rabbit out of the hat' show-off maths. Of course, you are free to use whichever works better for you.)

5.3.8 A look forward   By this point, we have built up a good range of techniques for establishing convergence and evaluating limits that are capable of working across a wide variety of sequences. A wide variety, but by no means all – for there are many important and useful sequences that need some additional (and sometimes quite individual) attention. The business of our next chapter is to gather together and explore many of these 'routine-procedure-resistant' examples.

6 Special (or specially awkward) examples

6.1 Introduction

We apologise in advance for the rather fragmentary character of this chapter but, as we pointed out at the end of Chapter 5, there are many convergent sequences that do not readily give up their secrets under routine uses of the techniques we have so far developed, and you will need to become acquainted with their limits at some point. It might as well be now. Keep in mind the squeeze, which will often turn out to play a role here.

6.2 Important examples of convergence

6.2.1 Geometric sequences For which real values of x does the geometric sequence

$$(x^n)_{n\in\mathbb{N}}$$

converge?

Solution

We have already seen (see 4.1.13 and 4.1.14) that for $0 < x < 1$ we get $x^n \to 0$ and that for $x > 1$ we get $(x^n)_{n\in\mathbb{N}}$ divergent, and an appeal to 2.7.13 shows that for $-1 < x < 0$ we get $x^n \to 0$ again. Now we only need to address the few missing cases.

If $x = 1$ then it is obvious that $x^n \to 1$, and if $x = 0$ then it is obvious that $x^n \to 0$.

If $x = -1$ then the odd and even powers of x are (respectively) -1 and 1, so evidently $(x^n)_{n\in\mathbb{N}}$ is again divergent.

In the final case $x < -1$, if $(x^n)_{n\in\mathbb{N}}$ were to converge to some limit ℓ then (algebra of limits) $|x^n| = |x|^n \to |\ell|$ which is impossible since $|x| > 1$.

We conclude that $(x^n)_{n\in\mathbb{N}}$ converges only when $-1 < x \le 1$.

6.2.2 Negative powers of n For any $t > 0$ we have $n^{-t} \to 0$.

Undergraduate Analysis: A Working Textbook, Aisling McCluskey and Brian McMaster 2018.
© Aisling McCluskey and Brian McMaster 2018. Published 2018 by Oxford University Press

Solution

Since n^{-t} is certainly positive, we need only ensure that (given $\varepsilon > 0$) we can find $n_0 \in \mathbb{N}$ so large that $n^{-t} < \varepsilon$ will always happen once $n \geq n_0$. Straightforward roughwork will show how big n_0 needs to be:

$$n^{-t} < \varepsilon \Leftrightarrow n^t > 1/\varepsilon$$
$$\Leftrightarrow t \ln n > \ln(1/\varepsilon)$$
$$\Leftrightarrow \ln n > \ln(1/\varepsilon)/t$$
$$\Leftrightarrow n > e^{\ln(1/\varepsilon)/t} (= 1/\sqrt[t]{\varepsilon}).$$

A formal proof is now easy to construct, beginning with: 'Given $\varepsilon > 0$, choose an integer n_0 greater than $e^{\ln(1/\varepsilon)/t}$. Then for any $n \geq n_0$, we have ...'.

6.2.3 n^{th} roots of a constant   If $a > 0$, how does the sequence $(\sqrt[n]{a})_{n \in \mathbb{N}}$ behave?

Solution

In the special case $a = 1$ it is clear that this sequence converges to 1.

Now if $a > 1$ we shall have $\sqrt[n]{a} > 1$ for every n (for the nth power of a positive number less than 1 will still be less than 1), so let us write

$$\sqrt[n]{a} = 1 + h_n$$

where h_n is positive, and its subscript n just serves to remind us that its actual value may vary with n. Raise both sides to the power of n and think what the binomial theorem tells us:

$$a = (\sqrt[n]{a})^n = (1 + h_n)^n = 1 + n(h_n) + \frac{n(n-1)}{2!}(h_n)^2 + \cdots$$

where there are several more terms to come, but all we need to know about them is that *they are all positive*. Each term on the right-hand side is therefore smaller than a, and we focus on the second one:

$$n(h_n) < a.$$

This rearranges to give

$$0 < h_n < \frac{a}{n}$$

and an easy application of the squeeze gives us $h_n \to 0$, and $\sqrt[n]{a} = 1 + h_n \to 1$.

In the third case $0 < a < 1$, notice that a^{-1} is greater than 1 so, by the second case, we already know that $\sqrt[n]{a^{-1}} \to 1$. By the algebra of limits it follows that $\sqrt[n]{a} \to 1^{-1} = 1$ so the desired limit is once again 1.

6.2.4 Example To find the limit of the sequence whose n^{th} term is

$$x_n = \sqrt[3n^2+n-2]{123}.$$

Solution

This is a subsequence of the sequence $(\sqrt[n]{a})$ (for $a = 123$) whose limit we know to be 1. Therefore $x_n \to 1$ also.

6.2.5 EXERCISE Determine the limit of $(a_n)_{n \in \mathbb{N}}$ where $a_n = \sqrt[n]{3^n + 7^n + 9^n}$.

Draft solution

Begin with the obvious remark that

$$9^n < 3^n + 7^n + 9^n < 9^n + 9^n + 9^n = (3)9^n.$$

Take n^{th} roots right across and think squeeze.

Somewhat surprisingly, almost exactly the same argument works on what appears at first sight to be a much harder problem:

6.2.6 n^{th} root of n The sequence $(\sqrt[n]{n})$ converges to 1.

Solution

Ignoring the case $n = 1$ (which, as usual, cannot affect the limiting behaviour) we see that $\sqrt[n]{n} > 1$ for every other n, so let us write

$$\sqrt[n]{n} = 1 + h_n$$

where h_n is positive and will vary with n. Raise both sides to the power of n and use the binomial theorem:

$$n = (\sqrt[n]{n})^n = (1 + h_n)^n = 1 + n(h_n) + \frac{n(n-1)}{2!}(h_n)^2 + \cdots$$

where there are several more terms to come and *they are all positive*. Each term on the right-hand side is therefore smaller than n, but this time we focus on the third one:

$$\frac{n(n-1)}{2!}(h_n)^2 < n.$$

Dividing both sides by n and remembering that n is at least 2, this rearranges to give

$$(h_n)^2 < \frac{2}{n-1}$$

and so

$$0 < h_n < \sqrt{\frac{2}{n-1}}.$$

Now it is fairly obvious (or see paragraph 2.7.12) that $\sqrt{\frac{2}{n-1}} \to 0$, the squeeze gives us $h_n \to 0$, and $\sqrt[n]{n} = 1 + h_n \to 1$.

6.2.7 Example To determine the limit of (a_n) where $a_n = \sqrt[5n+2]{7n+10}$.

Solution

The nth term broadly resembles $\sqrt[n]{n}$ but, of course, we have to do better than *broad resemblance*. One approach (assuming $n > 2$) is:

$$\sqrt[5n+2]{5n+2} < \sqrt[5n+2]{7n+10} < \sqrt[3.5n+5]{7n+10} = (\sqrt[7n+10]{7n+10})^2$$

(think carefully about the use of the index laws[1] in that calculation) at which point we see that the first and last items involve subsequences of $(\sqrt[n]{n})$ and therefore converge to 1, after which the squeeze tells us that $a_n \to 1$ also.

6.2.8 EXERCISE

1. Determine the limit of $\sqrt[n]{20n - 15}$. (It may help to begin by observing that $5n \le 20n - 15 < 20n$ for each n.)

2. Determine the limit of $(a_n)_{n \in \mathbb{N}}$ where

$$a_n = (n^{-1}) \sqrt[n]{1^n + 2^n + 3^n + \cdots + n^n}.$$

6.2.9 Factorials grow faster than powers For any constant t, the sequence

$$\left(\frac{t^n}{n!} \right)_{n \in \mathbb{N}}$$

converges to zero. This may seem contrary to common sense at first sight; for example, try $t = 10$, calculate the first few terms

$$10, 50, 166.666\ldots, 416.666\ldots, 833.333\ldots\cdots$$

---

[1] In particular, you should take note that if $x > 1$ and $a < b$ (where a and b are positive) then

$$x^{\frac{1}{a}} > x^{\frac{1}{b}}.$$

This is another of those plausible-looking arithmetical results whose full understanding will have to wait until we have properly defined the elementary functions, including general powers, in Chapter 18.

– and there is certainly no sign yet that they are drifting closer to zero. But we need to look at what happens for seriously big values of n, and the early, small values may be misleading.

Solution

We can assume that t is positive. (Because, firstly, in the case where $t = 0$ the result is immediate; and, secondly, once we have proved it in the case $t > 0$, if we are then challenged with a sequence

$$(a_n)_{n \in \mathbb{N}} = \left(\frac{u^n}{n!}\right)_{n \in \mathbb{N}}$$

where u is negative, we shall already know that

$$\frac{|u|^n}{n!} \to 0.$$

That is, $|a_n| \to 0$. Now an earlier exercise (2.7.13) tells us that $a_n \to 0$ also.)

Comparing a_{n+1} with a_n, we find that

$$a_{n+1} = \frac{t}{n+1}\, a_n,$$

and that fraction multiplier $t/(n+1)$ will be small once n becomes substantially bigger than t. More precisely, once n exceeds $2t$, the fraction multiplier will be less than a half and, therefore, each term will be more than 50% smaller than the one before it. This lets us set up a clear proof that the terms tend to zero:

As soon as the integer n exceeds $2t$, we have

$$\frac{t^{n+1}}{(n+1)!} = \frac{t}{n+1}\frac{t^n}{n!} < \frac{t}{2t}\frac{t^n}{n!} = \frac{1}{2}\frac{t^n}{n!}$$

– that is, $a_{n+1} < 0.5 a_n$. So, picking an integer $n_0 > 2t$, we have

$$a_{n_0+1} < 0.5 a_{n_0},\, a_{n_0+2} < (0.5)^2 a_{n_0},\, a_{n_0+3} < (0.5)^3 a_{n_0}, \cdots$$
$$\cdots a_{n_0+k} < (0.5)^k a_{n_0}\, (k \geq 1).$$

Since $(0.5)^k \to 0$ as $k \to \infty$ and all the terms are positive, the sequence (a_n) converges to zero (by the squeeze) as claimed. (Notice that we entirely ignored the first n_0 terms but, as usual, this does not affect limiting behaviour.)

6.2.10 How does the nth root of $n!$ behave?   It gets seriously big. More precisely, the sequence $(x_n)_{n \geq 1}$ given by

$$x_n = \frac{1}{\sqrt[n]{n!}}$$

converges to 0.

Solution

For any positive constant ε put $t = 1/\varepsilon > 0$ and call in the previous result, that

$$\frac{t^n}{n!} \to 0.$$

In particular we can find a positive integer n_0 such that $n \geq n_0$ guarantees that

$$\frac{t^n}{n!} < 1$$

which in turn gives

$$\frac{1}{n!} < t^{-n} = \varepsilon^n$$

and

$$\frac{1}{\sqrt[n]{n!}} < \varepsilon$$

therefore establishing the claim.

6.2.11 **EXERCISE** Investigate the limiting behaviour of the sequence whose n^{th} term, for $n > 10$, is

$$\frac{1}{\sqrt[n]{(10)(11)(12)(13) \cdots (n-2)(n-1)n}}.$$

6.2.12 **EXERCISE** Determine the limit, as $n \to \infty$, of:

$$(i) \; \frac{10^{6n^2}}{(n^2)!}, \quad (ii) \; \frac{2^n}{n! + \sqrt{n!}}, \quad (iii) \; \frac{1}{\sqrt[n!]{(n!)!}}.$$

Remark

Watch out for subsequences, and for the squeeze.

6.2.13 **Difference between large square roots** Determine the limiting behaviour of $\sqrt{n+a} - \sqrt{n+b}$ for positive constants a and b.

Solution

Arrange the labelling so that $a > b$ (for if a and b are equal, the sequence is constantly zero). Notice first (difference of two squares) that

$$(\sqrt{n+a} - \sqrt{n+b})(\sqrt{n+a} + \sqrt{n+b}) = (\sqrt{n+a})^2 - (\sqrt{n+b})^2$$
$$= (n+a) - (n+b) = a - b$$

from which we find that

$$0 < \sqrt{n+a} - \sqrt{n+b} = \frac{a-b}{\sqrt{n+a} + \sqrt{n+b}}.$$

The bottom line of that last quotient is more than $\sqrt{n} + \sqrt{n} = 2\sqrt{n}$ so it can be made arbitrarily large. It follows that the quotient itself tends to zero, and the squeeze gives us

$$\sqrt{n+a} - \sqrt{n+b} \to 0.$$

6.2.14 EXERCISE

1. Use the same kind of algebraic reorganisation to show that, if $x_n \to \ell$ where each x_n and ℓ itself are positive, then $\sqrt{x_n} \to \sqrt{\ell}$.

2. (Using the algebra of limits when it becomes appropriate) determine the limit (as $n \to \infty$) of
$$\sqrt{4n^2 + 2n + 1} - \sqrt{4n^2 - n + 5}.$$

6.2.15 How does the sequence $\left(\left(1 + \frac{1}{n}\right)^n \right)_{n \geq 1}$ behave? This is a classic instance of common sense misleading us. Thinking informally (and sloppily) leads us to expect that, since $\left(1 + \frac{1}{n}\right)$ converges to 1, this is essentially a large power of 1 and therefore, going to the limit, is also 1. The weakest point in that draft argument is that we have to allow n to go off to infinity *before* regarding $\left(1 + \frac{1}{n}\right)$ as being 1, and therefore we cannot talk about its nth power at all after we have done that.

What's actually going on in this sequence is a contest between $\left(1 + \frac{1}{n}\right)$ itself getting smaller as n increases while, on the other hand, the fact that we are simultaneously raising it to higher and higher powers is attempting to make it bigger. The first step towards a proper understanding of this example is to admit that it is far from obvious which of these two tendencies is going to win.

The second step is to swallow our pride and calculate the first few terms:

$$2, \; 2.25, \; 2.370370, \; 2.441406, \; 2.488320, \; 2.521626, \; 2.546500, \; 2.565785, \cdots$$

(where the answers have been reported to six decimal places). It appears that this is an increasing sequence so far, but the numerical value of the limit (if it even exists) is not yet obvious. Still, we now have a possible strategy: to try to show that the sequence is, indeed, always increasing, and perhaps bounded as well?

At a first or second reading you might choose to ignore the proofs of the next three paragraphs, and you will be none the worse for it. Do not, however, ignore the result that emerges.

6.2.16 Recall, from an earlier example (4.2.3): – that if $x \geq -1$ and n is any positive integer, then

$$(1+x)^n \geq 1 + nx.$$

This is the vital ingredient in the demonstration of the following unexpected and unattractive lemma:

6.2.17 Lemma For each $n \in \mathbb{N}$ we have $\left(1 - \dfrac{1}{(n+1)^2}\right)^n \geq \dfrac{n+1}{n+2}.$

Proof

Take $x = -\dfrac{1}{(n+1)^2}$ in the preceding Recall, and simplify the resulting algebra:

$$\left(1 - \frac{1}{(n+1)^2}\right)^n \geq 1 - \frac{n}{(n+1)^2}$$

$$= 1 - \frac{n}{n^2+2n+1} > 1 - \frac{n}{n^2+2n} = 1 - \frac{1}{n+2}$$

$$= \frac{(n+2)-1}{n+2} = \frac{n+1}{n+2}.$$

Carefully note the disturbance of inequality caused by ditching the $+1$ from the bottom line. This shrinks the bottom line, and therefore *increases* the fraction; but the fraction has an overall minus attached, so the nett result is to *decrease* the total.

Now it turns out that this ugly duckling of a lemma is precisely what is needed to prove the result that we want:

6.2.18 Proposition The sequence $\left(\left(1 + \frac{1}{n}\right)^n\right)_{n\geq 1}$ is increasing.

Proof

Denoting the typical term by x_n, we carefully simplify the ratio[2] $\frac{x_{n+1}}{x_n}$ and seek evidence that the answer is greater than 1.

$$\frac{x_{n+1}}{x_n} = \frac{\left(1+\frac{1}{n+1}\right)^{n+1}}{\left(1+\frac{1}{n}\right)^n}$$

$$= \frac{\left(\frac{n+2}{n+1}\right)^{n+1}}{\left(\frac{n+1}{n}\right)^n}$$

$$= \left(\frac{n+2}{n+1}\right)\left(\frac{n+2}{n+1} \times \frac{n}{n+1}\right)^n$$

$$= \left(\frac{n+2}{n+1}\right)\left(\frac{n^2+2n}{(n+1)^2}\right)^n$$

[2] This is a case where *subtracting* x_n from x_{n+1} would not readily have helped us.

$$= \left(\frac{n+2}{n+1}\right)\left(\frac{n^2 + 2n + 1 - 1}{(n+1)^2}\right)^n$$

$$= \left(\frac{n+2}{n+1}\right)\left(\frac{(n^2 + 2n + 1) - 1}{(n+1)^2}\right)^n$$

$$= \left(\frac{n+2}{n+1}\right)\left(1 - \frac{1}{(n+1)^2}\right)^n$$

$$\geq \left(\frac{n+2}{n+1}\right)\left(\frac{n+1}{n+2}\right) = 1$$

by the lemma. Hence the result, that $x_{n+1} \geq x_n$ for all $n \geq 1$.

6.2.19 Proposition The sequence $\left(\left(1 + \frac{1}{n}\right)^n\right)_{n \geq 1}$ converges to a limit that is somewhere between 2.5 and 3.

Proof

Since we know that it is increasing and that the sixth term already exceeds 2.5, we need only confirm that it is bounded above by 3. That is disarmingly simple, using the binomial theorem again:

$$\left(1 + \frac{1}{n}\right)^n = 1 + \frac{n}{1}\left(\frac{1}{n}\right)^1 + \frac{n(n-1)}{2!}\left(\frac{1}{n}\right)^2$$

$$+ \frac{n(n-1)(n-2)}{3!}\left(\frac{1}{n}\right)^3 + \cdots + \left(\frac{1}{n}\right)^n$$

$$< 1 + \frac{n}{1}\left(\frac{1}{n}\right)^1 + \frac{n(n)}{2!}\left(\frac{1}{n}\right)^2 + \frac{n(n)(n)}{3!}\left(\frac{1}{n}\right)^3 + \cdots + \frac{1}{n!}$$

$$= 1 + 1 + \frac{1}{2!} + \frac{1}{3!} + \frac{1}{4!} + \cdots + \frac{1}{n!}$$

$$< 1 + 1 + \frac{1}{2} + \frac{1}{2 \cdot 2} + \frac{1}{2 \cdot 2 \cdot 2} + \cdots + \frac{1}{2^{n-1}}$$

$$= 1 + \frac{1 - \left(\frac{1}{2}\right)^n}{1 - \frac{1}{2}} < 1 + 2 = 3$$

and the proof is complete. Notice that we assumed n to be big enough that all the terms of the binomial expansion that we listed actually came into play: but that is harmless since an upper bound for later terms in this (increasing) sequence is certainly an upper bound also for the earlier, smaller ones.

6.2.20 Important notes

1. Taking more care in our estimations will let us find much more accurate values for the limit in question. On the lower side, the limit of this *increasing* sequence has to exceed each term that we choose to calculate exactly, such as

term number 10 (which is 2.593742 to six decimal places), term number 20 at 2.653298, term number 100 at 2.704814. On the upper side, we could sharpen the above argument along the following lines:

$$\left(1+\frac{1}{n}\right)^n < 1+1+\frac{1}{2!}+\frac{1}{3!}+\frac{1}{4!}+\cdots+\frac{1}{n!}$$

$$< 1+1+\frac{1}{2}+\frac{1}{6}+\frac{1}{24}+\frac{1}{120}\left(1+\frac{1}{6}+\frac{1}{6\times7}+\frac{1}{6\times7\times8}+\cdots+\frac{1}{6\times7\times8\times\cdots\times n}\right)$$

$$< \left(1+1+\frac{1}{2}+\frac{1}{6}+\frac{1}{24}\right)+\frac{1}{120}\left(1+\frac{1}{6}+\frac{1}{6\times6}+\frac{1}{6\times6\times6}+\cdots+\frac{1}{6^{n-5}}\right)$$

$$= \frac{65}{24}+\frac{1}{120}\left(\frac{1-\left(\frac{1}{6}\right)^{n-4}}{1-\frac{1}{6}}\right) < \frac{65}{24}+\frac{1}{120}\left(\frac{1}{1-\frac{1}{6}}\right) = 2.718333 \text{ (to 6 decimal places)}.$$

2. The limit is actually the irrational number written as e, as in $\log_e$ (that is, $\ln$) and e^x and exponential growth. Its numerical value is approximately 2.71828.

3. Summary:

$$\left(1+\frac{1}{n}\right)^n \to e.$$

4. This turns out to be a special case (the case $x = 1$) of a highly important limit that you need to know, but whose detailed proof will have to wait until much later (Chapter 18, in fact; paragraph 18.2.16) in this account:

$$\left(1+\frac{x}{n}\right)^n \to e^x.$$

6.2.21 Examples   To find the limits of the sequences whose n^{th} terms are as follows:

$$(1+2n^{-1})^n, \quad \left(1-\frac{\pi}{n}\right)^n, \quad \left(1+\frac{1}{n!+1}\right)^{n!+1},$$

$$\left(1+\frac{1}{n^2}\right)^{n^2+n}, \quad \left(\frac{n}{n+3}\right)^{n+2}.$$

Solution

The first three are immediate from the above theorem and its subsequent notes: they are e^2, $e^{-\pi}$ and e (since the third is a subsequence of $\left(\left(1+\frac{1}{n}\right)^n\right)_{n\geq1}$).

The fourth one – let us denote it by x_n – needs a little more attention. We can express x_n as

$$\left(1+\frac{1}{n^2}\right)^{n^2}\left(1+\frac{1}{n^2}\right)^n$$

– and the first of these factors is easy enough to deal with: it converges to e because it represents a subsequence of $\left(\left(1 + \frac{1}{n} \right)^n \right)$ (which, of course, tends to e). The remaining problem is to estimate the second factor which (check the index laws) is the nth root of the first factor. Now the (convergent) first factor is bounded: it lies always between two (evidently positive) constants a and b:

$$a \le \left(1 + \frac{1}{n^2} \right)^{n^2} \le b$$

for all n and so, now taking n^{th} roots:

$$\sqrt[n]{a} \le \left(1 + \frac{1}{n^2} \right)^{n} \le \sqrt[n]{b}$$

for all n. At this point, the squeeze gives us $\left(1 + \frac{1}{n^2} \right)^{n} \to 1$ and so

$$x_n = \left(1 + \frac{1}{n^2} \right)^{n^2} \left(1 + \frac{1}{n^2} \right)^{n} \to e \times 1 = e.$$

Number five also needs some rearrangement of its algebra:

$$\left(\frac{n}{n+3} \right)^{n+2} = \left(\frac{n+3}{n} \right)^{-n-2}$$

$$= \left(\left(1 + \frac{3}{n} \right)^n \right)^{-1} \left(1 + \frac{3}{n} \right)^{-2} \to (e^3)^{-1} \times (1)^{-2} = e^{-3}.$$

using, incidentally, parts of the algebra of limits.

6.2.22 **EXERCISE** Determine the limits of:

$$\left(\frac{n+3}{n+6} \right)^{5n+2}, \quad \left(\frac{n^2+5}{n^2} \right)^{n^2}, \quad \left(1 + \frac{1}{n^2+3n} \right)^{n^2+4n-1}, \quad \left(\frac{n^2+2n+8}{n^2+n+7} \right)^{n},$$

$$\left(1 - \frac{0.5}{n} \right)^{n^2+n+10}, \quad \left(1 - \frac{4}{n^2} \right)^{n}.$$

(The fourth one is not at all easy.)

6.2.23 **EXERCISE** Use induction to show that, for each positive integer n:

$$n! \ge 3 \left(\frac{n}{3} \right)^n.$$

(You may find that the result established in the *proof* of 6.2.19 is useful here.)

The final case study in this set – unlike the one we have just done – is included merely for interest and, if you are short of time, you may safely leave it out. It concerns the Fibonacci sequence $(f_n)_{n \in \mathbb{N}}$ that we mentioned briefly in our work on recursively defined sequences, the sequence

$$1, 1, 2, 3, 5, 8, 13, 21, 34, \cdots$$

that is created by choosing the first and second terms to be 1 and, from then on, letting $f_{n+2} = f_{n+1} + f_n$ ($n \geq 1$), that is, creating each term by adding the two that are immediately before it. It seems highly unlikely at first sight that the sequence of numbers (f_n) is settling towards a limit (indeed, it would be fairly easy to show that it is unbounded, and therefore divergent) but its *rate of growth* is quite a different matter.

Even a casual look at the sequence (f_n) will show that the terms are growing at quite a steady rate, increasing by about 60% each time once the pattern is securely established. This is rather curious ...why should a sequence formed by *adding* turn out to be one that is propagated by *multiplying* by about 1.6, and what is the limiting value of this multiplier if, indeed, it has a limiting value? That is, can we find the limit of the ratio, the 'growth rate' $\frac{f_{n+1}}{f_n}$? Note that our usual trick of trying to show that the sequence (of ratios) is monotonic and bounded will not work this time: a glance at the first few ratios

$$1, 2, 1.5, 1.666666, 1.6, 1.625, 1.615385, 1.619048, \cdots$$

shows that it is neither increasing nor decreasing, but is apparently 'homing in from both sides' towards a presumed (but still unproven) limit.

Let us rough-work our way blindly towards a possible solution. *If* (and that might be a big *if* because we are still only guessing) $\frac{f_{n+1}}{f_n}$ converges to some limit ℓ, then certainly ℓ is no less than 1 since the Fibonacci sequence is increasing, so $\ell \neq 0$, and $1/\ell$ makes sense. Look at the Fibonacci formation law

$$f_{n+2} = f_{n+1} + f_n,$$

divide across by f_{n+1} to get

$$\frac{f_{n+2}}{f_{n+1}} = 1 + \frac{f_n}{f_{n+1}},$$

and observe that the left-hand side also converges to ℓ (for it is just the ratio sequence with its first term left out). Taking limits on both sides, we deduce that $\ell = 1 + \frac{1}{\ell}$ or, more readably:

$$\ell^2 - \ell - 1 = 0.$$

This quadratic will not factorise, so we solve it instead by the quadratic formula to obtain $\frac{1+\sqrt{5}}{2}$ and $\frac{1-\sqrt{5}}{2}$, that is, approximately 1.618034 and -0.618034. At this point we can be pretty confident that we know what number the limit is bound to be, but bear in mind that we have not yet proved that any limit exists for the growth-rate sequence.

Let's try a different approach. A sequence whose growth-rate is constant (say, permanently equal to x), as opposed to merely settling towards a constant as limit, can only take the form

$$a, ax, ax^2, ax^3, \cdots ax^{n-1}, \cdots .$$

Is it at all possible for such a sequence to satisfy the Fibonacci recurrence relation? Let's see:

$$f_{n+2} = f_{n+1} + f_n \text{ now says } ax^{n+1} = ax^n + ax^{n-1}$$

which cancels down to $x^2 = x + 1$, that is, to the same quadratic equation $x^2 - x - 1 = 0$ that we met a moment ago, and whose two possible solutions we calculated. To save some writing, let us use temporary symbols for those solutions, say

$$\alpha = \frac{1 + \sqrt{5}}{2} \text{ and } \beta = \frac{1 - \sqrt{5}}{2}.$$

(Keep in mind that, because of the equation that they satisfy, $\alpha + 1 = \alpha^2$ and $\beta + 1 = \beta^2$.) So the only constant-growth-rate sequences that satisfy the Fibonacci recurrence look like $a\alpha^{n-1}$ or $b\beta^{n-1}$.

Wild surmise: is it possible that the *real* Fibonacci sequence is a combination of these? Something like $a\alpha^{n-1} + b\beta^{n-1}$?

If this were true then, to make the first and second terms both equal 1, we'd have to replace a and b by numbers p and q that make the tentative formula yield exactly these values for $n = 1$ and $n = 2$, that is:

$$p + q = 1, \quad p\alpha + q\beta = 1.$$

Yet these are very easy equations to solve for p and q! We get

$$p = \frac{1 - \beta}{\alpha - \beta}, \quad q = \frac{\alpha - 1}{\alpha - \beta}$$

which, when you substitute in the values we have for α and β, simplify to

$$p = \frac{5 + \sqrt{5}}{10}, \quad q = \frac{5 - \sqrt{5}}{10}$$

(approximately 0.7236 and 0.2764 respectively).

After so much conjecture, we at last have a proposal to set up and defend:

6.2.24 Proposition   With the parameters α, β, p and q as evaluated above, the Fibonacci sequence is explicitly described by the formula

$$f_n = p\alpha^{n-1} + q\beta^{n-1} \ (n \geq 1).$$

Proof

The (originally described) Fibonacci sequence is *fully specified* by the conditions $f_1 = 1$, $f_2 = 1$, $f_{n+2} = f_{n+1} + f_n$ ($n \geq 1$). That is, once we have agreed that these conditions are to hold, there is no ambiguity as to what every term in that sequence has to be. (You may take that as obvious, on the grounds that we could calculate from these conditions the value of any particular term that we wanted. If you are not convinced by that, an induction argument upon the statement 'all the terms up to and including the nth term are completely specified by the given conditions' can easily be constructed.)

Yet the (possibly new?) sequence $g_n = p\alpha^{n-1} + q\beta^{n-1}$ ($n \geq 1$) does satisfy $g_1 = 1$ and $g_2 = 1$ since we picked the numbers p and q expressly so as to make that happen, and also

$$g_{n+1} + g_n = p\alpha^n + q\beta^n + p\alpha^{n-1} + q\beta^{n-1} = p\alpha^{n-1}(\alpha + 1) + q\beta^{n-1}(\beta + 1)$$
$$= p\alpha^{n-1}(\alpha^2) + q\beta^{n-1}(\beta^2) = p\alpha^{n+1} + q\beta^{n+1}$$
$$= g_{n+2}.$$

That is, (g_n) satisfies all of the conditions that *completely specified* the Fibonacci sequence. This can only mean that (g_n) *actually is* the Fibonacci sequence, and our proof is complete.

6.2.25 Note   For ease of use, we can slightly simplify the formula just obtained for f_n as follows:

$$f_n = p\alpha^{n-1} + q\beta^{n-1}$$

$$= \frac{5 + \sqrt{5}}{10}\left(\frac{1 + \sqrt{5}}{2}\right)^{n-1} + \frac{5 - \sqrt{5}}{10}\left(\frac{1 - \sqrt{5}}{2}\right)^{n-1}$$

$$= \frac{\sqrt{5}(1 + \sqrt{5})}{5 \times 2}\left(\frac{1 + \sqrt{5}}{2}\right)^{n-1} - \frac{\sqrt{5}(1 - \sqrt{5})}{5 \times 2}\left(\frac{1 - \sqrt{5}}{2}\right)^{n-1}$$

$$= 5^{-\frac{1}{2}}\left(\frac{1 + \sqrt{5}}{2}\right)^n - 5^{-\frac{1}{2}}\left(\frac{1 - \sqrt{5}}{2}\right)^n$$

$$= 5^{-\frac{1}{2}}\alpha^n - 5^{-\frac{1}{2}}\beta^n.$$

6.2.26 **Example**   The limit of the growth-rate in the Fibonacci sequence is $\alpha = \frac{1+\sqrt{5}}{2}$.

Solution

Using the formula for the nth Fibonacci number f_n obtained in the Note, we have:

$$\frac{f_{n+1}}{f_n} = \frac{\alpha^{n+1} - \beta^{n+1}}{\alpha^n - \beta^n}$$

$$= \frac{\alpha - \beta\left(\frac{\beta}{\alpha}\right)^n}{1 - \left(\frac{\beta}{\alpha}\right)^n}$$

which, because the number $\frac{\beta}{\alpha}$ is numerically less than 1 (in fact it is approximately -0.382) converges to

$$\frac{\alpha - 0}{1 - 0} = \alpha = \frac{1 + \sqrt{5}}{2}$$

as we claimed.

6.2.27 **Postscript**   Of the two components $5^{-\frac{1}{2}}\alpha^n$ and $5^{-\frac{1}{2}}\beta^n$ of our formula for the nth Fibonacci number, the first is by far the more important. For example, if we put $n = 15$, the formula yields (to five decimal places) $f_{15} = 609.99967 + 0.00033$. The reason is that β has modulus smaller than 1, and therefore its powers tend to zero rather rapidly. More exactly, when we regard $5^{-\frac{1}{2}}\alpha^n$ as an approximation to f_n, the error term

$$|f_n - 5^{-\frac{1}{2}}\alpha^n| = 5^{-\frac{1}{2}}|\beta|^n$$

which decreases to the limit zero. Notice also that even for $n = 1$, the error term $5^{-\frac{1}{2}}|\beta|^1$ is only about 0.2764, so all of the error terms are much less than 1. It follows that f_n is always the integer closest to $5^{-\frac{1}{2}}\alpha^n$. An additional detail – taking note of the sign of $5^{-\frac{1}{2}}\beta^n$ – is that

- for odd values of n, f_n is the integer just greater than $5^{-\frac{1}{2}}\alpha^n$, whereas
- for even values of n, f_n is the integer just less than $5^{-\frac{1}{2}}\alpha^n$.

6.2.28 **EXERCISES**

1. With f_n continuing to denote the nth Fibonacci number, what is the limiting behaviour of

$$\frac{f_{n+2}}{f_n}?$$

2. If we were to alter the defining conditions of the Fibonacci sequence by changing only the first two terms, say, to

$$g_1 = 7, \ g_2 = 4, \ g_{n+2} = g_{n+1} + g_n \ (n \geq 1)$$

or to

$$g_1 = a, \ g_2 = b, \ g_{n+2} = g_{n+1} + g_n \ (n \geq 1)$$

for any constants a and b, what effect would that have on the limiting behaviour of the growth rate?

3. Investigate the sequence $(h_n)_{n \geq 1}$ defined by

$$h_1 = 1, \ h_2 = 1, \ h_{n+2} = 3h_{n+1} + 4h_n \ (n \geq 1).$$

7 Endless sums – a first look at series

7.1 Introduction

Every ten-year-old school child knows that *zero point endlessly many threes* means one third. This is not in doubt. What does need some critical analysis, however, is whether *zero point endlessly many threes* is a legitimate symbol at all.

We are so over-familiar with the decimal system of representing numbers that it is all too easy to forget what a superb invention it was. Its beauty and power reside in the ability it gives us to write down any whole number whatsoever, and a great many non-integers too, using only twelve symbols: the digits 0, 1, 2, 3, 4, 5, 6, 7, 8 and 9, the decimal point (or, if you prefer, the decimal comma) and the minus sign. The power derives from the fact that it is a *positional* system: each symbol carries information not only from its shape but also from where it occurs in relation to the other symbols (especially the decimal point). Thus, for instance, 12825 actually means $1(10)^4 + 2(10)^3 + 8(10)^2 + 2(10)^1 + 5(10)^0$ and the two 2s have different meanings, different significances, because of where they sit. In the same way, positive numbers less than 1 can be denoted by placing digits of lower and lower significance to the right of the decimal point: 0.4703 means $4(10)^{-1} + 7(10)^{-2} + 0(10)^{-3} + 3(10)^{-4}$.

Seeking to extend that notation to non-terminating decimals[1] raises an issue that virtually no ten-year-old is in a position to handle with full rigour. The phrase *zero point endlessly many threes* suggests that we write down, or at least imagine writing down,

$$0.33333333 \ldots \textit{ and so on for ever,}$$

and that this ought to mean

$$\frac{3}{10} + \frac{3}{100} + \frac{3}{1000} + \frac{3}{10000} + \frac{3}{100000} + \cdots \textit{ and so on for ever.}$$

The first (practical) problem here is that no-one has ever lived long enough to write down an infinite list of threes, nor indeed will the universe last long enough for this to occur; but the deeper (conceptual) difficulty is that this is not how addition works. Adding is essentially a finite procedure: we know what adding two numbers

---

[1] which, at the least, obliges us to admit one more symbol, namely the row of dots $\cdots$

Undergraduate Analysis: A Working Textbook, Aisling McCluskey and Brian McMaster 2018.
© Aisling McCluskey and Brian McMaster 2018. Published 2018 by Oxford University Press

means, and from that we can add three, or four, or a million, just by grouping them together in pairs or by implementing some kind of induction argument; but *adding an infinite list of numbers* does not make sense.

However, we have already faced and overcome this difficulty while talking about where the idea of sequence limits comes from. Instead of grappling with a virtual symbol such as that invoked by the slightly mystical phrase *zero point endlessly many threes*, look instead at the sequence of perfectly ordinary numbers

$$(\, 0.3, 0.33, 0.333, 0.3333, 0.33333, \cdots \,)$$

$$= \Big(\frac{3}{10}, \frac{3}{10} + \frac{3}{100}, \frac{3}{10} + \frac{3}{100} + \frac{3}{1000},$$

$$\frac{3}{10} + \frac{3}{100} + \frac{3}{1000} + \frac{3}{10000}, \frac{3}{10} + \frac{3}{100} + \frac{3}{1000} + \frac{3}{10000} + \frac{3}{100000}, \cdots \Big).$$

If this has a limit, then that limit will be the natural way to interpret *zero point endlessly many threes*. It does, and – to the undoubted satisfaction of many former ten-year-olds – the limit is one third.

More importantly, though, this discussion provides a fruitful suggestion as to how we might seek to make sense of the sum of an arbitrary endless list of numbers, namely:

- don't attempt to add all of them,
- just add the first n
- and then look for a limit of that partial total as $n \to \infty$.

7.2 Definition and easy results

Informally, a series arises whenever we try to add all the terms of a sequence. So, if $(a_k)_{k \in \mathbb{N}}$ is any sequence, the associated series question is: can we make sense of

$$a_1 + a_2 + a_3 + \cdots + a_n + \cdots ?$$

As we indicated in the introduction, the standard way to attempt to do this is to create a second sequence

$$(a_1, \; a_1 + a_2, \; a_1 + a_2 + a_3, \; a_1 + a_2 + a_3 + a_4, \cdots)$$

and seek a limit for it.

Some more notation at this point will help to reduce how much we have to write: the sum of the first n terms of $(a_k)_{k \in \mathbb{N}}$ is called the nth *partial sum* and denoted by $\sum_{k=1}^{k=n} a_k$ rather than $a_1 + a_2 + a_3 + \cdots + a_n$, and we'll often denote it even more briefly by s_n or S_n. The series itself – *the issue that we are trying to resolve* – is denoted by $\sum_{k=1}^{\infty} a_k$ or $\sum_{1}^{\infty} a_k$.

The capital Greek letter $\sum$ is *sigma*, and the attached $k = 1$ and $k = n$ are usually abbreviated to merely 1 and n, or even left out altogether if the context makes it clear what range of values is involved. Just as for sequences, it is perfectly fine to start off at $k = 0$ or at $k = 2$ or elsewhere if $k = 1$ is not the best first step: for instance, $8 + 16 + 32 + 64 + \cdots + 2^n$ can be expressed either as $\sum_{k=3}^{n} 2^k$ or as $\sum_{k=1}^{n-2} 2^{k+2}$.

(There is nothing special about the choice of the letter k here – you may use just about any symbol you please since it is a 'dummy variable' having no effect on the outcome, but don't use n or it will create confusion. Symbols such as $\sum_{n=1}^{n} 2^n$ really don't mean anything unambiguous – if we are told to begin with $n = 1$ and continue until $n = n$, where are we meant to stop? Not until $n = n$? But n is always equal to n, isn't it . . . ? This is the kind of vicious circle that may drag us in if we employ the same symbol with two different meanings at the same time.)

7.2.1 Definition   A *series* is a pair of sequences $(a_k)_{k \in \mathbb{N}}$ and $(s_n)_{n \in \mathbb{N}}$ linked together by the conditions $s_n = \sum_{k=1}^{n} a_k$ for each n. The first is called *the sequence of terms* and the second (as we already said) is called *the sequence of partial sums*. The series itself is denoted by the symbol $\sum_{1}^{\infty} a_k$ or $\sum_{1}^{\infty} a_n$.

It is the *second* of the two sequences whose limiting behaviour we have to focus on when we examine a series – so much so that, for nearly all practical purposes, it is legitimate to shorten the above official definition to:

Shortened definition: a series $\sum_{1}^{\infty} a_k$ is the sequence $(s_n)_{n \in \mathbb{N}}$ of partial sums (where s_n means $\sum_{k=1}^{n} a_k = a_1 + a_2 + a_3 + \cdots + a_n$).

An advantage of the shortened definition is that it makes the rest of this definition paragraph so obvious as to be scarcely worth saying:

7.2.2 Continuing definition   A series is said to be

- *convergent* if the sequence of partial sums is convergent,
- *divergent* if the sequence of partial sums is divergent,
- *bounded* if the sequence of partial sums is bounded,

and so on. One slight disturbance of this pattern is important: when the sequence of partial sums does converge to some limit ℓ, we don't call ℓ the limit of the series; we call it the *sum to infinity* of the series (or, more briefly, the *sum* of the series).

Irritatingly, and despite what we said above about never using the same symbol with two different meanings at the same time, for historical reasons the symbol $\sum_{1}^{\infty} a_k$ is often used to mean *both* the series *and* (provided it does converge) the sum to infinity of that series. We apologise for this, but you may need to know it when you read textbooks. So, for example, the statements

$$\sum_{1}^{\infty} \left(\frac{1}{2}\right)^k \quad \text{converges, and its sum is 1}$$

and

$$\sum_{1}^{\infty}\left(\frac{1}{2}\right)^k = 1$$

should be viewed as saying exactly the same thing, although in the first the sigma symbol means the series itself, while in the second the sigma symbol means the sum (to infinity) of that series.

One more ambiguity alert: when working on a series, be careful not to use too many pronouns! You are dealing with two sequences at once, so try to avoid phrases like '*it* tends to zero' or '*it* is bounded' or '*it* converges' unless the context really does make it clear *which* of the two you mean.

7.2.3 **Example: geometric series** For any $x \in (-1, 1)$, the geometric series

$$\sum_{k=0}^{\infty} x^k$$

converges, and its sum is $\dfrac{1}{1-x}$.

Solution

If we multiply out the product

$$(1 - x)(1 + x + x^2 + x^3 + \cdots + x^{n-1})$$

we see that all of the terms cancel in pairs except for 1 and $-x^n$. Dividing across by (non-zero) $(1 - x)$ shows that the nth partial sum of this series

$$s_n = 1 + x + x^2 + x^3 + \cdots + x^{n-1} = \frac{1 - x^n}{1 - x}.$$

Taking limits (as $n \to \infty$) we find that $s_n \to \dfrac{1}{1-x}$ as predicted.

(Note also the effect of starting such a geometric series at a point other than $k = 0$: for an integer $m \geq 0$ the series

$$x^m + x^{m+1} + x^{m+2} + x^{m+3} + \cdots = \sum_{k=m}^{\infty} x^k$$

can be thought of as

$$x^m \left(\sum_{r=0}^{\infty} x^r \right) = x^m \left(\frac{1}{1-x} \right) = \frac{x^m}{1-x}.$$

This is legitimate because each partial sum can be factorised in just such a fashion, after which we can let the number of terms tend to infinity and obtain the claimed conclusion in the limit.)

7.2.4 Example   We confirm that the recurring decimal indicated by the phrase *zero point endlessly many nines* represents the number 1.

Solution

The phrase actually means the limit – if it has a limit – of the sequence

$$0.9, 0.99, 0.999, 0.9999, \cdots ,$$

in other words, the sum of the series $\sum_{k=0}^{\infty} 0.9(1/10)^k$.
  To investigate this, we must look at the nth partial sum

$$s_n = \sum_{k=0}^{n-1} 0.9(1/10)^k = 0.9 \sum_{k=0}^{n-1} (1/10)^k.$$

By the above example, the limit of this is

$$0.9 \frac{1}{1 - \frac{1}{10}} = \frac{0.9}{0.9} = 1$$

as predicted.

7.2.5 EXERCISE

1. Determine the meaning (and the numerical value) of

$$-3.2283737373737 \cdots .$$

2. Think about how you would prove that *every* recurring decimal represents a *rational* number.

  If $x \geq 1$ or $x \leq -1$ then the geometric series $\sum_{k=0}^{\infty} x^k$ does not converge, but diverges. You can check this out by examining the nth partial sum and using the definition, but it is easier and quicker to engage the following little theorem instead:

7.2.6 Theorem   If a series $\sum_{k=1}^{\infty} x_k$ converges, then $x_k \to 0$.

Proof

That the series converges tells us that the nth partial sum s_n converges to some limit ℓ. Also s_{n+1} converges to the same limit ℓ since loss of its first term has no bearing on the limit. Then $x_{n+1} = s_{n+1} - s_n \to \ell - \ell = 0$. Hence the result. (We lost x_1 in that demonstration but, once again, early terms don't have any impact on the limit of a sequence.)

7.2.7 Alert: The converse of this result is not true. That is, $x_k \to 0$ is not enough to guarantee that $\sum_{k=1}^{\infty} x_k$ converges. We'll do an example next to demonstrate this.

7.2.8 Example: the harmonic series diverges The series
$\sum_1^\infty \frac{1}{k} = 1 + \frac{1}{2} + \frac{1}{3} + \frac{1}{4} + \cdots$ diverges.

Solution

Scan along the list of fractions and you will realise that, since they are steadily decreasing, the biggest in any block is the first in that block and the smallest is the last. So, for instance,

$$\frac{1}{3} + \frac{1}{4} > \frac{1}{4} + \frac{1}{4} = 2\left(\frac{1}{4}\right) = \frac{1}{2},$$

$$\frac{1}{5} + \frac{1}{6} + \frac{1}{7} + \frac{1}{8} > 4\left(\frac{1}{8}\right) = \frac{1}{2},$$

$$\frac{1}{9} + \frac{1}{10} + \cdots + \frac{1}{15} + \frac{1}{16} > 8\left(\frac{1}{16}\right) = \frac{1}{2}$$

and so on. The pattern emerging here (and writing s_n for the nth partial sum as usual) is

$$s_4 > 1 + \frac{1}{2} + \frac{1}{2} = 1 + 2\left(\frac{1}{2}\right), s_8 > 1 + \frac{1}{2} + \frac{1}{2} + \frac{1}{2} = 1 + 3\left(\frac{1}{2}\right),$$

$$s_{16} > 1 + \frac{1}{2} + \frac{1}{2} + \frac{1}{2} + \frac{1}{2} = 1 + 4\left(\frac{1}{2}\right)$$

and, in general,

$$s_{(2^n)} > 1 + n\left(\frac{1}{2}\right)$$

for each $n \geq 2$.

We see from this that the subsequence $\left(s_{(2^n)}\right)$ of (s_n) is unbounded,[2] so the partial-sum sequence (s_n) itself is also unbounded and therefore cannot converge.

7.2.9 EXERCISE

1. Find a positive integer N for which the Nth partial sum of the harmonic series is greater than 2018.

2. Let $(a_k)_{k \in \mathbb{N}}$ be any given sequence of positive numbers converging to a limit $\ell > 0$. Show that the following series diverges:

$$\sum_1^\infty \frac{a_k}{k}$$

[2] If this subsequence were bounded, we could find a constant M such that, for every positive integer n, $s_{(2^n)} < M$. The previous display now gives us $1 + n/2 < M$, that is, $n < 2M - 2$ for every positive integer, which is absurd.

Partial draft solution

For the first part, use an inequality from the solution of the previous example. For the second part, begin by noting that there is a positive integer n_0 such that $a_k > \ell/2$ for each $k \geq n_0$, then observe that there is a power of 2 (say, 2^m for some $m \in \mathbb{N}$) that is greater than n_0, and then revisit the estimation process that established the previous example.

Generally speaking, series whose terms are exclusively positive are easier to work with, as we shall see in the next two sections. There is, however, a class of series that have many negative terms but whose convergence is nevertheless easy to establish:

7.2.10 **The alternating series test** Suppose that $(a_k)_{k \geq 1}$ is a decreasing sequence of positive numbers that converges to zero. Then the 'alternating series'

$$\sum_{1}^{\infty} (-1)^{k-1} a_k = a_1 - a_2 + a_3 - a_4 + a_5 - \cdots$$

converges.

Proof

Since the terms a_k are positive and getting steadily smaller, the typical 'even' partial sum

$$s_{2n} = a_1 + (-a_2 + a_3) + (-a_4 + a_5) + (-a_6 + a_7) + \cdots + (-a_{2n-2} + a_{2n-1}) - a_{2n}$$

is less than a_1 (and is therefore bounded above). Furthermore,

$$s_{2n+2} - s_{2n} = a_{2n+1} - a_{2n+2} > 0$$

so (s_{2n}) is also an increasing sequence. Therefore it converges to *some* limit ℓ.

For very much the same reasons, the 'odd' partial sums form a bounded, decreasing sequence (s_{2n-1}) which also converges to some limit ℓ'.

Lastly, $s_{2n} = s_{2n-1} + a_{2n}$ and, taking limits across that line, we get $\ell = \ell' + 0$, in other words, ℓ and ℓ' are the same number. It follows that the entire sequence (s_n) of partial sums converges to ℓ (see 5.2.5 and 5.2.6 for more discussion).

7.2.11 **Example**

1. The 'alternating harmonic series'

$$\sum_{1}^{\infty} (-1)^{k-1} \frac{1}{k} = 1 - \frac{1}{2} + \frac{1}{3} - \frac{1}{4} + \frac{1}{5} - \frac{1}{6} + \cdots$$

converges.

2. For any $a > 0$ the series

$$\sum_{1}^{\infty}(-1)^{k-1}\frac{1}{k^a} = 1 - \frac{1}{2^a} + \frac{1}{3^a} - \frac{1}{4^a} + \frac{1}{5^a} - \frac{1}{6^a} + \cdots$$

converges. (Recall paragraph 2.9.16.)

Solution

Both of these are immediate from the alternating series test.

7.2.12 **EXERCISE**   Decide whether the following are convergent or divergent:

1.
$$\sum_{1}^{\infty}(-1)^{k-1}\left(\frac{2k+1}{3k^2-1}\right);$$

2.
$$\sum_{1}^{\infty}(-1)^{k-1}\left(1+\frac{1}{3k}\right)^{-2k}.$$

7.2.13 **HARDER EXERCISE**   We define a sequence (b_n) by the formulae:

$$b_{2k-1} = \frac{1}{k}; \quad b_{2k} = \frac{1}{2k+2},$$

noting that the odd- and even-numbered terms have different descriptions.

- Is it legitimate to apply the alternating series test to $\sum_{1}^{\infty}(-1)^{k-1}b_k$?
- Does $\sum_{1}^{\infty}(-1)^{k-1}b_k$ converge or diverge?

7.2.14 **Note**   Some parts of the algebra of limits transfer immediately from sequences to series. If two series $\sum a_k$ and $\sum b_k$ converge, with sums s_a and s_b, say, then the series $\sum(a_k + b_k)$ converges to the sum $s_a + s_b$ simply because its nth partial sum $\sum_{k=1}^{n}(a_k + b_k)$ can be rearranged as $\sum_{k=1}^{n}a_k + \sum_{k=1}^{n}b_k$ and therefore does converge to $s_a + s_b$. We write this briefly as

$$\sum(a_k + b_k) = \sum a_k + \sum b_k$$

(provided, of course, that the two sums on the right-hand side do exist). In the same way, and subject to similar provisos:

$$\sum(a_k - b_k) = \sum a_k - \sum b_k,$$
$$\sum(Ca_k) = C\sum a_k$$

where C is any constant. On the other hand, no such result is available for multiplication: we do not obtain a partial sum for $\sum a_k b_k$ by multiplying partial sums for $\sum a_k$ and $\sum b_k$!

7.3 Big series, small series: comparison tests

We suggested earlier that series whose terms are all positive are easier to deal with. Here is the basic and simple reason why:

7.3.1 Theorem   A series of non-negative terms is convergent if and only if its sequence of partial sums is bounded.

Proof

If a series $\sum_1^\infty a_k$ has $a_k \geq 0$ for all values of k then, considering its partial sums s_n:

$$s_{n+1} = s_n + a_{n+1} \geq s_n$$

for all n, that is, (s_n) is an increasing sequence. Therefore (s_n) will converge if it is bounded, and *vice versa*.

In less formal language, this result converts the relatively difficult idea of convergence (for this class of series only) into the relatively easy one of boundedness: you will get convergence precisely when the terms are so small that, no matter how many of them you add together, there is some absolute upper ceiling to how big a total you accumulate. 'Small series converge, big series diverge'. This insight, in turn, we can sharpen up into a group of results called in general *series comparison tests*:

7.3.2 The direct comparison test   Suppose $(a_k)_{k \in \mathbb{N}}$ and $(b_k)_{k \in \mathbb{N}}$ are two sequences of non-negative terms and $a_k \leq b_k$ for every $k \in \mathbb{N}$. Then if $\sum_1^\infty b_k$ converges, $\sum_1^\infty a_k$ must also converge.
    (Equivalently, if $\sum_1^\infty a_k$ diverges, $\sum_1^\infty b_k$ must also diverge.)

Proof

If $\sum_1^\infty b_k$ converges, the previous theorem tells us that there is some upper bound (call it M) for all of its partial sums: that is, for every $n \in \mathbb{N}$ we have

$$\sum_1^n b_k \leq M.$$

Yet the fact that $a_k \leq b_k$ for every $k \in \mathbb{N}$ assures us that $\sum_1^n a_k \leq \sum_1^n b_k$ by simple addition, so

$$\sum_1^n a_k \leq M$$

also. Now the same theorem gives us the convergence of $\sum_1^\infty a_k$.

7.3.3 EXERCISE: the direct comparison test with scaling   Suppose $(a_k)_{k \in \mathbb{N}}$ and $(b_k)_{k \in \mathbb{N}}$ are two sequences of non-negative terms, and we can find a positive constant C such that $a_k \leq Cb_k$ for every $k \in \mathbb{N}$. Then if $\sum_1^\infty b_k$ converges, $\sum_1^\infty a_k$ must also converge.

Comment

You will find that the proof is virtually identical to the previous one: at the line before the last you'll get $\sum_1^n a_k \leq CM$, but CM is also just a constant.

7.3.4 Remark   To increase further the usefulness of comparison between two series, we need to think about the effect of changing or ignoring a few terms at the start. By way of example, consider a geometric series whose sum we already know:

$$1 + 0.1 + (0.1)^2 + (0.1)^3 + (0.1)^4 + \cdots + (0.1)^n + \cdots = \frac{1}{1 - 0.1} = \frac{10}{9}.$$

What does that tell us about

$$(10 + 1) + 0.1 + (300 + (0.1)^2) + (0.1)^3 + (0.1)^4 + \cdots + (0.1)^n + \cdots ?$$

Well, as long as we are sure that only the first and third terms have been altered, the answer ought to be obvious: the altered series converges to a sum of $10 + 300 + \frac{10}{9}$ simply because *every* partial sum from the third one onwards has been increased by 310, and that will feed through to the limit. So we see that – unlike the similar scenario in sequences – changing a finite number of terms in a series *does* affect its convergence behaviour, but only in quite a predictable fashion: the total of the changes that you added to individual terms gets added onto the sum-to-infinity. In particular, if the original series did converge, then so must the altered one ... and *vice versa*, because additions can be cancelled out by adding their negatives. We conclude that:

If a series converges, then after alteration or omission of a finite number of terms, the new series also converges, and vice versa.

More simply, it is always safe to alter or delete a finite number of terms from a series provided that we only want to know whether or not it converges (and do not care what particular sum it converges to). This allows us to modify the previous two results as follows:

7.3.5 The direct comparison test with alterations/omissions   Suppose $(a_k)_{k\in\mathbb{N}}$ and $(b_k)_{k\in\mathbb{N}}$ are two sequences of non-negative terms, and that we can find a positive integer k_0 such that $a_k \leq b_k$ for every $k \geq k_0$. Then if $\sum_1^\infty b_k$ converges, $\sum_1^\infty a_k$ must also converge.

7.3.6 The direct comparison test with scaling and alterations/omissions   Suppose $(a_k)_{k\in\mathbb{N}}$ and $(b_k)_{k\in\mathbb{N}}$ are two sequences of non-negative terms and we can find a positive constant C and a positive integer k_0 such that $a_k \leq Cb_k$ for every $k \geq k_0$. Then if $\sum_1^\infty b_k$ converges, $\sum_1^\infty a_k$ must also converge.

7.3.7 **Examples**   For each of the following definitions of a_k, decide whether $\sum_1^\infty a_k$ converges or diverges.

1.
$$a_k = \frac{5 + 3\sin(k\sqrt{k\pi})}{2^k};$$

2.
$$a_k = \frac{k^2 - 30k\cos(k + 2\sqrt{k})}{4k^3 + 7}.$$

Solution

1. The top line of the fraction that defines a_k always lies between 2 and 8 so, firstly, all the terms are positive (and therefore we can use the theory developed so far) and, secondly, $a_k \leq 8(\frac{1}{2})^k$.
   Since (the geometric series) $\sum(\frac{1}{2})^k$ converges, so does $\sum a_k$ (by the direct comparison test with scaling).

2. There is a risk that several terms here may be negative because of the minus on the top line. However, if $k \geq 31$ then

$$k^2 = k \times k > 30k \geq 30k\cos(anything)$$

   so, from that point on, the top line and a_k itself are definitely positive: so we shall just ignore the first 30 terms. Furthermore, if $k \geq 60$ then

$$\left(\frac{1}{2}\right)k^2 = \left(\frac{1}{2}\right)k \times k > 30k \geq 30k\cos(anything)$$

   and therefore $k^2 - 30k\cos(k + 2\sqrt{k})$ is at least $(\frac{1}{2})k^2$ which, in turn, gives us

$$a_k \geq \frac{k^2}{2(4k^3 + 7)} \geq \frac{k^2}{2(4k^3 + 7k^3)} = \left(\frac{1}{22}\right)\left(\frac{1}{k}\right) \quad provided\ k \geq 60.$$

   Since the harmonic series $\sum \frac{1}{k}$ diverges, so must the 'bigger' series $\sum a_k$ by 'comparison', along with scaling by $\frac{1}{22}$ and omission of terms.

7.3.8 **EXERCISE**

1. Given a sequence $(t_n)_{n \in \mathbb{N}}$ of real numbers, not all of them positive, that converges to a limit $\ell > 0$, show that $\sum_1^\infty t_k(0.99)^k$ converges and that $\sum_1^\infty t_k\left(\frac{1}{k}\right)$ diverges.

2. Provided with the extra information (which we shall confirm soon – see 7.3.13) that $\sum k^{-2}$ converges, show that both

$$\sum \frac{6k - 5}{k(k^2 + 17)}$$

and

$$\sum \frac{6k + 2}{k(8k^2 - 5)}$$

also converge.

There is another variety of comparison test that, in some cases at least, saves us the bother of ignoring initial terms and guessing about scaling constants:

7.3.9 The limit comparison test (sometimes denoted by LCT) Suppose that $(a_k)_{k \in \mathbb{N}}$ and $(b_k)_{k \in \mathbb{N}}$ are two sequences of positive terms and that $\dfrac{a_k}{b_k}$ tends to a non-zero limit. Then either the two series both converge, or they both diverge.

Proof

Let ℓ denote the (non-zero) limit of the ratio of a_k to b_k. Using $\varepsilon = \ell/2$ in the definition of limit, there is a positive integer k_0 such that, for $k \geq k_0$, we have

$$\frac{\ell}{2} < \frac{a_k}{b_k} < \frac{3\ell}{2}.$$

The right-hand portion of that inequality rearranges to give $a_k < (3\ell/2)b_k$ for large values of k, so the direct comparison test with scaling $3\ell/2$ (and ignoring terms up to the k_0^{th}) tells us that if $\sum b_k$ converges then so must $\sum a_k$.

Now the left-hand portion of the displayed line rearranges to produce $b_k < (2/\ell)a_k$ for large values of k, so the same argument establishes that if $\sum a_k$ converges then so must $\sum b_k$.

7.3.10 Examples To decide whether the following converge or diverge:

1.
$$\sum a_k \text{ where } a_k = \frac{k^4 - 3k^3 + 8k^2 - 5}{1 + k + k^2 + k^3 + k^4 + 2k^5} ;$$

2.
$$\sum a_k \text{ where } a_k = \left(\frac{2k - 1}{3k}\right)^k.$$

Solution

1. (Roughly speaking, the biggest power of k will dominate each line of the fraction, that is, the top line will be dominated by k^4 and the bottom line dominated by $2k^5$. So a_k resembles $\dfrac{k^4}{2k^5} = \dfrac{1}{2k}$. Therefore ...)
   Let us consider $b_k = \frac{1}{k}$. We see that

$$\frac{a_k}{b_k} = \frac{k^5 - 3k^4 + 8k^3 - 5k}{1 + k + k^2 + k^3 + k^4 + 2k^5} = \frac{1 - 3k^{-1} + 8k^{-2} - 5k^{-4}}{k^{-5} + k^{-4} + k^{-3} + k^{-2} + k^{-1} + 2}$$

whose limit is 0.5 which is **not** zero. By the LCT, both series must converge or else both series must diverge. Yet the harmonic series $\sum b_k$ diverges, and therefore so does the given series $\sum a_k$.

2. In the second example, try $b_k = \left(\frac{2}{3}\right)^k$. Then

$$\frac{a_k}{b_k} = \left(\frac{k - 1/2}{k}\right)^k = \left(1 + \frac{-0.5}{k}\right)^k$$

whose limit is $e^{-0.5}$ which is **not zero**. Since the geometric series $\sum b_k$ converges, the LCT tells us that $\sum a_k$ must do so also.

7.3.11 EXERCISE   Revisit Exercise 7.3.8 and use the limit comparison test to get quicker, easier solutions of each of the problems that it posed.

7.3.12 Example: a telescoping series   To show, by directly calculating the typical partial sum, that the series $\sum_1^\infty \frac{1}{k(k+1)}$ converges.

Solution

A quick answer depends on noticing[3] that

$$\frac{1}{k} - \frac{1}{k+1} = \frac{k + 1 - k}{k(k+1)} = \frac{1}{k(k+1)},$$

so the nth partial sum of this series

$$\frac{1}{1(1+1)} + \frac{1}{2(2+1)} + \frac{1}{3(3+1)} + \frac{1}{4(4+1)} + \cdots + \frac{1}{n(n+1)}$$

$$= \frac{1}{1} - \frac{1}{2} + \frac{1}{2} - \frac{1}{3} + \frac{1}{3} - \frac{1}{4} + \frac{1}{4} - \frac{1}{5} + \cdots + \frac{1}{n} - \frac{1}{n+1}$$

cancels down almost completely (*telescopes*) to $\frac{1}{1} - \frac{1}{n+1}$ whose limit is obviously 1.

7.3.13 Example   To show that the series $\sum k^{-2}$ converges.

7.3.14 Solution   Use the preceding example and the limit comparison test: if we put

$$a_k = \frac{1}{k^2}, \quad b_k = \frac{1}{k(k+1)}$$

then it is immediate that $\frac{a_k}{b_k} \to 1 \neq 0$, so the convergence of $\sum b_k$ proves the convergence of $\sum a_k$.

---

[3] The so-called *theory of partial fractions*, which you may have come across, helps one to notice such things (especially in more complicated examples).

7.3.15 **EXERCISE**   Find constants a and b such that

$$\frac{1}{k(k+2)} = \frac{a}{k} + \frac{b}{k+2}$$

and hence show that the series $\sum_1^\infty \dfrac{1}{k(k+2)}$ converges (because its partial sums collapse 'telescopically').

7.3.16 **HARDER EXERCISE**   Let $t > 1$ be a constant. Prove that the series

$$\sum_{k=1}^{\infty} k^{-t} = 1 + \frac{1}{2^t} + \frac{1}{3^t} + \frac{1}{4^t} + \frac{1}{5^t} + \frac{1}{6^t} + \frac{1}{7^t} \cdots$$

converges.

Partial solution

We can use the same kind of estimation of partial sums that we employed for the harmonic series (in paragraph 7.2.8) but, instead of grouping the terms in blocks that *end* with a negative power of 2, this time we make them *start* with a negative power of 2, thus:

$$\frac{1}{2^t} + \frac{1}{3^t} < \frac{1}{2^t} + \frac{1}{2^t} = 2\left(\frac{1}{2^t}\right) = \frac{1}{2^{t-1}},$$

$$\frac{1}{4^t} + \frac{1}{5^t} + \frac{1}{6^t} + \frac{1}{7^t} < 4\left(\frac{1}{4^t}\right) = \frac{1}{4^{t-1}},$$

$$\frac{1}{8^t} + \frac{1}{9^t} + \cdots + \frac{1}{14^t} + \frac{1}{15^t} < 8\left(\frac{1}{8^t}\right) = \frac{1}{8^{t-1}}$$

and so on. The pattern emerging this time concerning the partial sums is:

$$s_{(2^n-1)} < 1 + \frac{1}{2^{t-1}} + \frac{1}{4^{t-1}} + \cdots + \frac{1}{(2^{n-1})^{t-1}}$$

for each $n \geq 2$. Now, recognise the right-hand side of the last display as being a partial sum of a convergent geometric series, and therefore bounded above by some constant M (the sum-to-infinity of that geometric series). Lastly, argue that *each* partial sum of the original series is less than $s_{(2^n-1)}$ for a suitably chosen value of n, and is therefore less than M. An appeal to Theorem 7.3.1 (convergent equals bounded for series of positive terms) will now complete the argument.

7.3.17 **Note**   Another consequence of the (very limited) effect of deleting a finite number of terms from a series concerns its so-called *tails*. For a series $\sum_{k=1}^{\infty} x_k$ and a positive integer n, the series

$$\sum_{k=n+1}^{\infty} x_k = x_{n+1} + x_{n+2} + x_{n+3} + x_{n+4} + \cdots$$

is called a *tail* of $\sum_1^{\infty} x_k$ – more precisely, *its nth tail*. Since this is formed by omitting the first n terms, part at least of the following example should be quite obvious.

7.3.18 Example Given a series $\sum_1^{\infty} x_k$, to show that the following three statements are equivalent:

1. $\sum_1^{\infty} x_k$ converges,
2. every tail of $\sum_1^{\infty} x_k$ converges,
3. at least one tail of $\sum_1^{\infty} x_k$ converges.

Solution

- Suppose that statement 1 is true. For each $n \geq 1$, the nth tail is formed by omitting the first n terms of the series and therefore, by Remark 7.3.4, it is itself a convergent series, that is, 2 is true.
- Suppose that statement 2 is true. Then the truth of 3 is immediate.
- Suppose that statement 3 is true. So we can find a positive integer n such that $\sum_{k=n+1}^{\infty} x_k$ converges to some limit s. (For $N > n$) the Nth partial sum of the given series can now be written as

$$\sum_{1}^{N} x_k = (x_1 + x_2 + x_3 + \cdots + x_n) + \sum_{k=n+1}^{N} x_k$$

and, just by the algebra of limits for sequences,

$$\sum_{1}^{N} x_k \to (x_1 + x_2 + x_3 + \cdots + x_n) + s \quad (\text{as } N \to \infty).$$

Thus, $\sum_1^{\infty} x_k$ is a convergent series.

7.3.19 Note It may be helpful to draw attention to the structure of that last demonstration. To claim that a number of statements are equivalent is to say that if any one of them is true then so are all the others. So when we said that statements 1, 2 and 3 were equivalent, we were asserting that 1 implies 2, 2 implies 1, 2 implies 3, 3 implies 2, 1 implies 3 and 3 implies 1. Fortunately there is no need to give a separate demonstration of each of these six implications: for instance, once one has established both 1 implies 2 and 2 implies 3, then 1 implies 3 follows immediately. The usual way to set up an efficient proof of equivalence for three statements is to confirm either that

$$1 \Rightarrow 2 \Rightarrow 3 \Rightarrow 1$$

or that

$$1 \Rightarrow 3 \Rightarrow 2 \Rightarrow 1$$

that is, to set out a *cyclical* proof. This is what we did in 7.3.18 above. As another illustration, if we wished to write out in full detail why the five 'equivalent conditions' in 4.1.5 actually are equivalent, a cyclical proof along the lines of $1 \Rightarrow 2 \Rightarrow 3 \Rightarrow 4 \Rightarrow 5 \Rightarrow 1$ or $1 \Rightarrow 4 \Rightarrow 3 \Rightarrow 5 \Rightarrow 2 \Rightarrow 1$ would be an efficient strategy.

7.3.20 EXERCISE Given a convergent series $\sum_1^\infty x_k$, let s_n denote the sum of the nth tail (which, by 7.3.18, is necessarily convergent). Show that $\lim_{n\to\infty} s_n = 0$.

7.4 The root test and the ratio test

The two most powerful tests to be discussed in this chapter are called the root test (or the nth root test) and the ratio test (or d'Alembert's ratio test). They both require the terms of a series to be non-negative before you can apply them to it. They are quite similar in effectiveness and in the details of their proofs, but in most applications one of them is likely to work much more easily than the other, so it is important to make a sensible choice of which one to try. We'll return later to how you should make that choice.

7.4.1 The nth root test Suppose that $\sum_{n=1}^\infty a_n$ is a series of non-negative terms and that $\sqrt[n]{a_n}$ converges to a limit ℓ (as $n \to \infty$). Then:

1. if $\ell < 1$ then the series converges,

2. if $\ell > 1$ then a_n does not tend to zero, and therefore the series diverges.

Proof

1. Assuming $\ell < 1$, we consider the number half-way between ℓ and 1; we can write this either as $\ell + \varepsilon$ or as $1 - \varepsilon$ if we choose ε to be half the length of the gap, that is, $\varepsilon = \dfrac{1-\ell}{2}$.

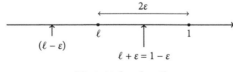

(Limit ℓ is less than 1)

Because $\sqrt[n]{a_n} \to \ell$ we can find n_0 such that, for every $n \geq n_0$:

$$\sqrt[n]{a_n} < \ell + \varepsilon = 1 - \varepsilon, \quad \text{and therefore} \quad a_n < (1 - \varepsilon)^n.$$

Since the geometric series $\sum (1 - \varepsilon)^n$ converges, so does the 'smaller' series $\sum a_n$ according to the comparison test (with alteration/omissions, since we 'lost' the first n_0 terms here).

2. Now assuming $\ell > 1$ (and thinking $\varepsilon = \ell - 1 > 0$) we can find n_1 such that, for every $n \geq n_1$:

$$\sqrt[n]{a_n} > \ell - \varepsilon = 1, \quad \text{and therefore} \ \ a_n > 1.$$

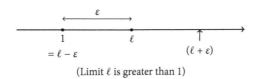

(Limit ℓ is greater than 1)

That certainly guarantees that $a_n \to 0$ cannot be true, so by an earlier theorem (7.2.6) we get $\sum a_n$ to be divergent.

7.4.2 **Remark**   You may well have noticed that we have gone back to using n instead of k for the label on the typical term. This is safe because, at the moment, we are not looking both at terms and at partial sums in the same paragraph, so there are not two different labels to keep separate in our minds. Of course, it is perfectly acceptable to use k or another letter instead of n.

7.4.3 **Warning**   If the limit of the nth root of the nth term is exactly 1, this test tells us nothing at all: for all it knows, the series could be convergent or divergent. So we will need to look for a different test when such a borderline case arises. As illustration of this point, notice that $\sum n^{-1}$ is divergent and $\sum n^{-2}$ is convergent, but in both cases you get a limit of 1 for the nth root of the nth term. (Note that $\sqrt[n]{n^2} = n^{2/n} = \left(\sqrt[n]{n}\right)^2$, and recall the result of 6.2.6.)

7.4.4 **Example**   Does the series

$$\sum \frac{(1 + \frac{3}{n})^{n^2}}{(1 + \frac{\pi}{n})^{n^2}}$$

converge or diverge?

Solution

The typical term is positive so we can try the root test. The nth root of the nth term is

$$\frac{(1 + \frac{3}{n})^{n}}{(1 + \frac{\pi}{n})^{n}}$$

which converges to $\frac{e^3}{e^\pi} = e^{3-\pi}$ which is less than 1, so the given series converges.

7.4.5 Example For precisely which positive values of t does the series

$$\sum \left(\frac{3n^2 - 1}{2n^2 - 1}\right)^n t^n$$

converge?

Solution

All terms are positive, and the nth root of the nth term is

$$\left(\frac{3n^2 - 1}{2n^2 - 1}\right) t$$

which has a limit of $\dfrac{3t}{2}$ so, by the root test,

1. for $0 < t < 2/3$ the limit is < 1 and so the series converges,
2. for $t > 2/3$ the limit is > 1 and the nth term does not tend to zero and the series must diverge.

It remains to ponder what happens when t is exactly $2/3$. Luckily, in that borderline case the nth term itself is

$$\left(\frac{3n^2 - 1}{2n^2 - 1}\right)^n \left(\frac{2}{3}\right)^n = \left(\frac{6n^2 - 2}{6n^2 - 3}\right)^n$$

which is (just) greater than 1 (in the final fraction, the top line exceeds the bottom line by 1). That shows once again that the nth term cannot tend to zero and the series diverges.

We conclude that the given series converges precisely when $0 < t < 2/3$.

7.4.6 EXERCISE

1. Determine whether or not the series

$$\sum \left(1 - \frac{1}{3n - 1}\right)^{n^2 + 2}$$

converges.

2. For which positive values of t does the series

$$\sum \frac{(3n + 1)^n}{n^{n-1}} t^n$$

converge, and for which does it diverge?

7.4.7 D'Alembert's ratio test Suppose that $\sum_{n=1}^{\infty} a_n$ is a series of positive terms and that the growth rate $\dfrac{a_{n+1}}{a_n}$ converges to a limit ℓ (as $n \to \infty$). Then:

1. if $\ell < 1$ then the series converges,

2. if $\ell > 1$ then a_n does not tend to zero, and therefore the series diverges.

Proof

1. Assuming $\ell < 1$, we again consider the number $\ell + \varepsilon = 1 - \varepsilon$ half-way between ℓ and 1, where $\varepsilon = \dfrac{1 - \ell}{2}$.

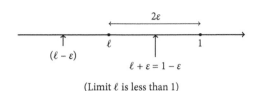

(Limit ℓ is less than 1)

Because $\dfrac{a_{n+1}}{a_n} \to \ell$ we can find m such that, for every $n \geq m$:

$$\frac{a_{n+1}}{a_n} < \ell + \varepsilon = 1 - \varepsilon, \quad \text{and therefore} \quad a_{n+1} < (1 - \varepsilon)a_n.$$

Therefore

$$a_{m+1} < (1 - \varepsilon)a_m, \quad a_{m+2} < (1 - \varepsilon)a_{m+1} < (1 - \varepsilon)^2 a_m,$$
$$a_{m+3} < (1 - \varepsilon)a_{m+2} < (1 - \varepsilon)^3 a_m, \cdots$$

and the pattern emerging is that

$$a_{m+k} < (1 - \varepsilon)^k a_m \text{ for each } k \geq 1.$$

Since the geometric series $\sum (1 - \varepsilon)^k$ converges, so does the 'smaller' series $\sum_{k=1}^{\infty} a_{m+k}$ according to the comparison test with scaling (by a_m). We 'lost' the first m terms this time, but that doesn't prevent the entire series $\sum a_n$ from converging also.

2. Now assuming $\ell > 1$ (and thinking $\varepsilon = \ell - 1 > 0$) we can find p such that, for every $n \geq p$:

$$\frac{a_{n+1}}{a_n} > \ell - \varepsilon = 1, \quad \text{and therefore} \quad a_{n+1} > a_n.$$

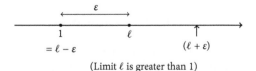

(Limit ℓ is greater than 1)

So (a_n) is an increasing sequence of positive terms if we disregard the first p terms. That guarantees that its limit (if it even has one) cannot be 0, and so once again $\sum a_n$ is divergent.

7.4.8 Warning   If the limit of the growth rate is exactly 1, this test too tells us nothing at all, and we must seek a different test to analyse such a borderline case. For instance, the divergent series $\sum n^{-1}$ and the convergent series $\sum n^{-2}$ both have limits of 1 for their growth rates.

7.4.9 Example   To decide whether the following series converges or diverges:

$$\sum v_n \text{ where } v_n = \frac{(n!)^3\, 2^{5n}}{(3n)!}.$$

Solution

Carefully cancel all you can in the ratio $\dfrac{v_{n+1}}{v_n}$ and you should find that the growth rate is

$$\frac{(n+1)^3\, 2^5}{(3n+1)(3n+2)(3n+3)}\,.$$

Using the algebra of limits, we see that this fraction converges to $32/27$ which is greater than 1, so $\sum v_n$ diverges.

7.4.10 Example   For exactly which positive values of x does the following series converge?

$$\sum w_n \text{ where } w_n = \frac{(n+1)!(2n+2)!\, x^n}{(3n+3)!}$$

Solution

The growth rate $\dfrac{w_{n+1}}{w_n}$ cancels to

$$\frac{(n+2)(2n+3)(2n+4)x}{(3n+4)(3n+5)(3n+6)} = \frac{(2n+3)(2n+4)x}{3(3n+4)(3n+5)}$$

whose limit is $4x/27$. By the ratio test, therefore:

1. for $0 < x < 27/4$ the limit is less than 1 and the series converges, but
2. for $27/4 < x$ the limit exceeds 1, the nth term does not tend to zero and the series diverges.

It remains unclear at first what will happen in the borderline case $x = 27/4$.
   Notice, however, that when $x = 27/4$, the growth rate is actually

$$\frac{(2n+3)(2n+4)27}{12(3n+4)(3n+5)} = \frac{(6n+9)(6n+12)}{(6n+8)(6n+10)}$$

which is greater than 1 (look at the individual factors in the top and bottom lines). Thus $w_{n+1} > w_n$, the terms are increasing, the terms do not converge to zero and the series again diverges.
   We conclude that this series converges only when $0 < x < 27/4$.

7.4.11 EXERCISE Determine whether or not the series $\sum a_n$ converges, where:

$$a_n = \frac{n^n \times n!}{(2n-1)!}.$$

7.4.12 EXERCISE Determine the range of values of the real number x for which the series

$$\sum \frac{(n!)^4 \, x^{2n}}{(4n)!}$$

converges. (Note that the wording of the question allows x itself to be negative.)

7.4.13 Remark Notice that the ratio test is particularly suitable for series whose terms involve several factorials, simply because massive cancellation occurs. For instance,

$$\frac{(n+1)!}{n!} = \frac{1 \times 2 \times 3 \times 4 \times \cdots \times n \times (n+1)}{1 \times 2 \times 3 \times 4 \times \cdots \times n} = n+1,$$

$$\frac{(3n)!}{(3n+3)!} = \frac{1 \times 2 \times 3 \times 4 \times \cdots \times 3n}{1 \times 2 \times 3 \times 4 \times \cdots \times (3n) \times (3n+1) \times (3n+2) \times (3n+3)}$$

$$= \frac{1}{(3n+1)(3n+2)(3n+3)}$$

and so on.

Take care that, when writing down the formula for term number $n+1$, you do so by replacing n by $n+1$ at each of its appearances in term number n so that, for example, $2n+1$ turns into $2n+3$ (that is, $2(n+1)+1$). Don't just 'add 1 to each bracket'.

Series in which the nth term's formula is dominated by n^{th} powers are likely candidates for simplification by the nth root rest, of course. Where the formula contains both factorials and n^{th} powers, the decision is less clear-cut – but unless and until you can access some useful information about the nth root of factorial n, the ratio test is still probably the better bet.

8 Continuous functions – the domain thinks that the graph is unbroken

8.1 Introduction

In order to study continuity, we need to make sure that we understand the ideas of function (mapping, map, transformation), domain, codomain (target), range, one-to-one (injective, 1 – 1), onto (surjective), composite (composition) and inverse function. Let us begin by revising the most basic points now, and promising to review other concepts through this chapter as we come to need them.

There are two styles of definition of the term *function* that you need to be aware of. We'll begin with the *informal* one, which is the one that we shall actually use almost all of the time. A function from a set D to a set C is any rule, however it may be expressed, that for each element of D allows us to determine a single associated element of the set C. If the letter f stands for the function – the rule – then for each $x \in D$ the associated element of C is written as $f(x)$. To describe a particular function, you need to identify both sets D and C and, most importantly, to specify the rule in enough detail to allow the reader to work out what $f(x)$ is for each possible x in D.

If you find that definition rather unsatisfactory, then your unease is justified. Apart from being somewhat vague (which is bad enough already), it suffers from the more serious flaw that it does not spell out the meaning of the words *rule* or *associated*, and these are at least as needful of definition as the word *function* was.[1] The way around that is to ground the ideas entirely in set theory, as in the following *formal* definition.

A *function*, also called a mapping, map or transformation, consists of a list of three sets called (respectively) the *domain*, the *codomain* (or target) and the *graph*, which satisfy a particular condition that we now describe. If the sets are denoted

---

[1] This is more than a little like defining a circle to mean a perfectly round figure, and then going on to define 'round' to mean 'shaped like a circle'. We end up merely noting the equivalence of two ideas – which, of course, is better than nothing – without actually succeeding in defining either of them.

Undergraduate Analysis: A Working Textbook, Aisling McCluskey and Brian McMaster 2018.

by D, C and Γ respectively then, *firstly*, Γ is a subset of the cartesian product set $D \times C$ (that is, Γ consists entirely of ordered pairs of elements (a, b) where each a belongs to D and each b belongs to C) and, *secondly*:

for each $x \in D$ there is a unique $y \in C$ such that $(x, y) \in \Gamma$.

The function is usually represented by a single symbol such as f. Then the entire phrase

f is a function whose domain is D and whose codomain is C

is abbreviated to

$$f : D \to C$$

and usually spoken as

f maps D to C.

For each x in D, the object represented above by y is conventionally written as $f(x)$ and called the *value* of f at x or the *image* of x under f.

Whether, in a particular question, you choose to think of the informal or the formal definition, let us be clear that a function involves three objects: the domain, the codomain, and the process of converting each element of the domain into an element of the codomain. If you change any one of these, then you are looking at a different function.

To convince yourself that the formal and informal definitions are essentially saying the same thing, consider this: if we had drawn the graph of some (formally defined) function f with perfect accuracy then, for each individual x in its domain, we could trace vertically up or down the graph paper until we found a point on the graph whose first coordinate was x, because this was guaranteed by *the condition*. Furthermore, the same condition also guaranteed that only one such point could exist. Then the second coordinate of that point is the (single) value of $f(x)$, and we have uncovered 'the rule' which the informal definition spoke of. Yet, conversely, if we begin with exact knowledge of the rule, and then precisely mark the point $(x, f(x))$ for every single value of x in the domain, we shall have drawn the graph with absolute precision. Thus, perfect knowledge of the rule and perfect knowledge of the graph determine one another. In the ideal world of infinite precision (which is where definitions live), a function is its graph.

A *real function* is a function whose domain and codomain are both subsets of the real line $\mathbb{R}$. Since all of the functions discussed in this text are real functions, we shall drop the qualifier *real* and simply refer to them as *functions*. In most examples,[2] the domain will be either an interval or a union of intervals, and the

---

[2] An important exception is that of a sequence: formally, a sequence is simply a real function $x : \mathbb{N} \to \mathbb{R}$ whose domain is $\mathbb{N}$, although the usual x_n notation for its typical value instead of $x(n)$ rather obscures this.

individual function itself will be spelled out by presenting some sort of formula or algorithm for calculating $f(x)$ for each permissible input value x. When this is so, by way of default the domain will comprise *all* the real numbers x for which the formula for $f(x)$ makes sense, and the default codomain will be $\mathbb{R}$ itself. For example, the phrase

f is the function given by $f(x) = \sqrt{4 - x^2}$

or, in a more abbreviated style,

the function $f(x) = \sqrt{4 - x^2}$

or, in even more tightly compressed style,

the function $\sqrt{4 - x^2}$

will mean that $f : [-2, 2] \to \mathbb{R}$ and that, for each x in the domain, the associated element $f(x)$ of $\mathbb{R}$ is $\sqrt{4 - x^2}$. Likewise,

the function $g(x) = (x - 1)^{-2}(x + 3)^{-1}$

will mean the function $g : (-\infty, -3) \cup (-3, 1) \cup (1, \infty) \to \mathbb{R}$ defined by the formula $g(x) = (x - 1)^{-2}(x + 3)^{-1}$. Note, in each case, how the domain has been defaulted to be as extensive as possible, subject to the formula always returning a *real* number.

You will already be familiar with the graphs of common and important functions such as $\sin x$, $\cos x$ and e^x, of quadratics with formulas of the form $ax^2 + bx + c$ and of 'straight line' functions of the form $mx + c$. Graphs, even rather rough sketch graphs, are a useful way of storing and presenting information about functions and, although they are not in themselves *proofs*, they often enable us to make sensible guesses about how particular functions behave. Indeed, a decent sketch graph frequently helps us to build up a sound, logical proof by supporting and guiding our intuition in a visual way. Accordingly, we shall begin our investigation of continuous functions by a short series of sketch graphs aimed at visually explaining what it is that we are trying to define.

8.2 An informal view of continuity

Take a careful look at these functions and at the back-of-the-envelope graphs we have drawn for them.

$$f : (-1, 5) \to \mathbb{R}, \quad f(x) = x(x-1)(x-2)(x-4); \quad g : (1, 3) \to \mathbb{R}, \quad g(x) = x^2 \lfloor x \rfloor.$$

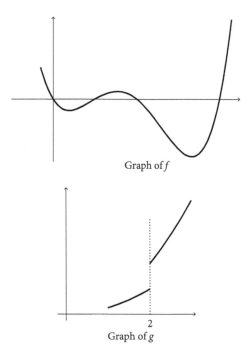

Graph of f

Graph of g

The graph of the first, f, is a classic example of an unbroken 'continuous' curve: we could draw the whole thing without lifting our pen or pencil off the paper. In contrast, g has a broken graph – there is an obvious gap between the part that lies to the left of $x = 2$ and the part that lies to the right. Certainly we could draw pieces of this graph without lifting our pen, but not the part that is close to $x = 2$. Common sense suggests we could say that f is continuous at each point we look at, but that g is not continuous at 2. One of our immediate tasks is to describe the difference between these two situations in mathematically precise terms that do not depend upon our limited artistic skills or on our ability to interpret visual clues, while taking care not to lose the intuition embedded in such diagrams.

$$h : (-3, 0) \rightarrow \mathbb{R}, \quad h(x) = \frac{x^5 + 1}{x^4 - 1} \text{ if } x \neq -1, \quad h(-1) = -1;$$

$$k : (-3, 0) \rightarrow \mathbb{R}, \quad k(x) = \frac{x^5 + 1}{x^4 - 1} \text{ if } x \neq -1, \quad k(-1) = -5/4;$$

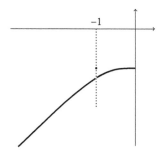

Graph of h

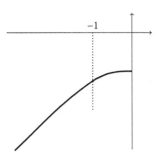

Graph of k

$$m : (-3, -1) \cup (-1, 0) \to \mathbb{R}, \quad m(x) = \frac{x^5 + 1}{x^4 - 1}.$$

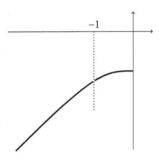

Graph of m

As far as visual intuition goes, k looks to be continuous at $x = -1$ and h does not: the graph of h looks as if one of its points has been placed incorrectly. In any case, it seems impossible to draw the part of the graph of h that lies near to -1 without lifting our pen. It is tempting to say that m also is non-continuous at $x = -1$ because of the (tiny) break in its graph there, but that is misleading: since -1 is not actually in the domain of m, the graph of m cannot properly be said to possess a break at $x = -1$: rather, at the point where $x = -1$, the graph of m does not exist at all. In this sense, the graph of m appears to be unbroken or continuous at every point at which it makes sense.

Here, for future use, is another trio of graphs that collectively make the same points:

$$p : (-7, 0) \cup (0, 7) \to \mathbb{R}, \quad p(x) = \frac{\sin x}{x};$$

$$q : (-7, 7) \to \mathbb{R}, \quad q(x) = \frac{\sin x}{x} \text{ if } x \neq 0, \quad q(0) = 2;$$

$$r : (-7, 7) \to \mathbb{R}, \quad r(x) = \frac{\sin x}{x} \text{ if } x \neq 0, \quad r(0) = 1.$$

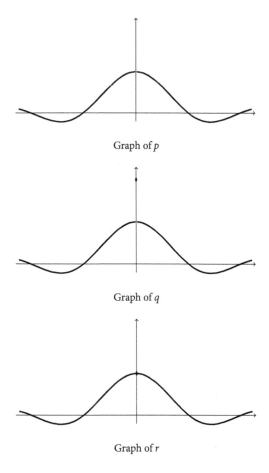

Graph of p

Graph of q

Graph of r

We see that p is not defined at all at $x = 0$ so the question of it being continuous here really does not arise. In the other two cases, a value has been assigned to the function at $x = 0$ and – continuing to think informally – it appears that q is not continuous at 0 because of the break in its graph, but that r (due to a 'wiser' choice of value at 0) is continuous here.

$$s : (-2, 2) \to \mathbb{R}, \quad s(x) = 1 \text{ if } -2 < x \leq 0, \quad s(x) = x^{-1} \text{ if } 0 < x < 2.$$

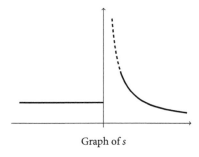

Graph of s

In this case, there is what we might be inclined to call an infinite gap at $x = 0$ and, unlike the previous situation involving p, q and r, there is no way to remedy that by making a judicious choice of a value for the function at $x = 0$.

Now, how can we turn these informal diagrammatic insights into a proper mathematical definition?

To make sure that the graph of a function f does not have any kind of gap or break at a point p in its domain, we need the values of $f(x)$ to approximate $f(p)$ very closely whenever x (in the domain of f) is very close to p. The values of x and of $f(x)$, however, do not constitute sequences – endless lists of one number after another – rather, they range over solid blocks or intervals of numbers, and this is unfortunate for us because we have spent the last hundred or so pages learning how to work with sequences, and would like to be able to capitalise on the skills that we have developed. We could, though, use sequences as probes within these solid blocks to seek out breaks. For instance, in the case of the function, sketched above, that we called $g : (1, 3) \to \mathbb{R}, g(x) = x^2 \lfloor x \rfloor$, if we probe near the point $x = 2$ by using the sequence $(x_n)_{n \geq 1} = \left(2 - \frac{1}{n+1} \right)_{n \geq 1}$ which surely converges to 2, we find that $g(2) = 8$ but that (for every n) $g(x_n) = \left(2 - \frac{1}{n+1} \right)^2$ which converges to 4, and so does not converge to $g(2)$. This bears witness to the presence of some kind of break in the graph at 2, and it does so in a way that is independent of any attempt we might make to draw the graph.

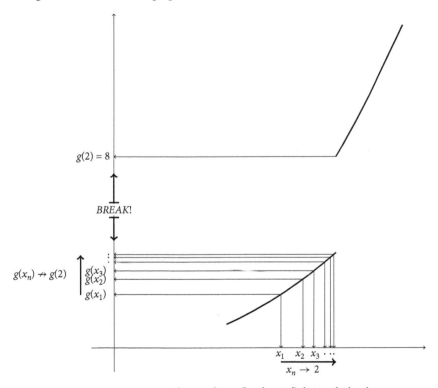

A sequence (approaching a 'break point') detects the break

Of course, if we had used the different probe-sequence $(y_n)_{n\geq1} = \left(2 + \frac{1}{n+1}\right)_{n\geq1}$ which also converges to 2, this time (for all n) $g(y_n) = 2\left(2 + \frac{1}{n+1}\right)^2$ which *does* converge to $8 = g(2)$, so we observe that not every probe near to 2 will succeed in identifying the break.

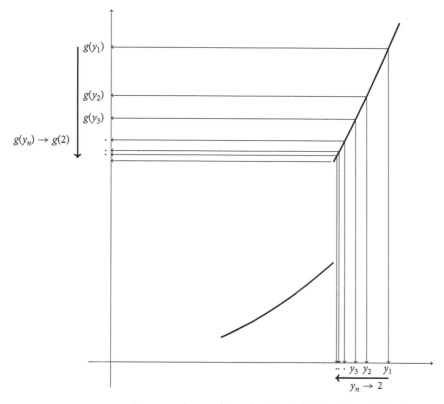

A sequence (approaching a 'break point') fails to detect the break

On the other hand, look again at the first function whose graph we sketched above, the one described by $f : (-1, 5) \to \mathbb{R}$, $f(x) = x(x - 1)(x - 2)(x - 4)$ and which we thought 'looked continuous' at, for instance, the point where $x = 3$. If we send in absolutely any probing sequence (z_n) that converges to 3, we see from the algebra of sequence limits that $f(z_n) = z_n(z_n - 1)(z_n - 2)(z_n - 4) \to 3(3 - 1)(3 - 2)(3 - 4)$ which is precisely the value $f(3)$ of f at 3: so no sequence probe finds any evidence of a break in the graph at $x = 3$.

These examples – and you might usefully try a few more like them just to reinforce the point – allow us to see a way to identify graphs that are continuous at a point $x = p$ by using sequences, and without needing to depend on imprecise and time-consuming sketches: if at least one sequence probe (x_n) (in the domain of f) that converges to p finds that $f(x_n)$ does not converge to $f(p)$, then there is some kind of continuity-fracturing break at p; on the other hand, if all sequences (x_n) that converge to p find that $f(x_n) \to f(p)$, then there is no such break, and we have continuity.

This is, incidentally, not the only way to define and identify continuity for real functions, but in the present context it is the quickest to understand and the easiest to use, so we shall run with it:

8.3 Continuity at a point

8.3.1 Definition   A function $f : D \to C$ is **continuous at a point** p of its domain D when, for every sequence (x_n) in D that converges to p, we have $f(x_n) \to f(p)$.

8.3.2 Example   To show that the (polynomial) function described by $p(x) = x^3 - 6x^2 + 17x + 11$ is continuous at $x = 4$.

Solution

Let $(x_n)_{n \geq 1}$ be any sequence in the domain ($\mathbb{R}$) of p that converges to 4 as limit. Then, using the algebra of limits:

$$p(x_n) = x_n^3 - 6x_n^2 + 17x_n + 11 \to 4^3 - 6(4^2) + 17(4) + 11 = p(4).$$

Hence the result.

8.3.3 EXERCISE   Show that the (rational) function defined by the formula

$$r(x) = \frac{x^4 + 3x^3 + 8x + 24}{x^4 + x^2 - 20}$$

is continuous at $x = -1$. Why can your argument not be modified to show that it is also continuous at $x = -2$?

8.3.4 Example   To show that the 'floor' function $j(x) = \lfloor x \rfloor$ is not continuous at $x = 2$, but is continuous at $x = 2.2$.

Solution

To take the second part of the question first . . . if (x_n) is any sequence whose limit is 2.2 then, for all $n \geq$ some n_0, we shall have $|x_n - 2.2| < 0.2$, that is, $2.0 < x_n < 2.4$ which implies that $j(x_n) = \lfloor x_n \rfloor = 2$. Ignoring the first $n_0 - 1$ terms of the sequence (which, as usual, has no effect on its limiting behaviour) we get $j(x_n) \to 2 = j(2.2)$. Therefore j is continuous at 2.2.

In contrast, if we take the particular sequence (y_n) described by $y_n = 2 - (n+1)^{-1}$ then certainly $y_n \to 2$. However, since each term of this sequence lies between 1.5 and 2 and therefore has floor 1, we find that $j(y_n) \to 1 \neq j(2)$. The discovery of even one sequence convergent to 2 whose convergence is not preserved by j shows that j is not continuous at 2.

8.3.5 EXERCISE   Verify that the function $s(x) = \lfloor x^2 \rfloor - \lfloor x \rfloor^2$ is not continuous at $x = \sqrt{3}$.

8.3.6 Example   To show that the function f defined by

$$f(x) = |x^2 + 3x| \text{ if } x \text{ is rational}; \quad f(x) = -|x^2 + 3x| \text{ if } x \text{ is irrational}$$

is continuous at $x = 0$, but not continuous at $x = 1$.

Solution

Firstly, the domain of f is the whole of $\mathbb{R}$ so 0 is a point of its domain. Also note that, since 0 is rational, $f(0) = |0^2 + 3(0)| = 0$.

For any x we have $-|x^2 + 3x| \leq f(x) \leq |x^2 + 3x|$ so, if (x_n) is any sequence in $\mathbb{R}$ that converges to zero, we get

$$-|x_n^2 + 3x_n| \leq f(x_n) \leq |x_n^2 + 3x_n|.$$

Now (as $n \to \infty$) $|x_n^2 + 3x_n| \to 0$ and $-|x_n^2 + 3x_n| \to 0$. The squeeze then tells us that $f(x_n) \to 0$ also. By our definition, f is continuous at 0.

Secondly, since 1 is rational, $f(1) = |1^2 + 3(1)| = 4$.

We can easily devise a sequence of irrationals that converges to 1: for instance, $y_n = 1 + \frac{\sqrt{2}}{n}$ will do fine. Then $f(y_n) = -|y_n^2 + 3y_n| \to -|1^2 + 3(1)| = -4$. Since 4 and -4 are different, the definition says that f is not continuous at 1.

(It is quite routine to modify that argument to show that f is not continuous anywhere except at $x = 0$ and at $x = -3$.)

8.3.7 EXERCISE   Assuming elementary trigonometric properties, show that the function f that is defined by

$$f(x) = x \sin(x^{-1}) \text{ if } x \neq 0; \quad f(0) = 0$$

is continuous at $x = 0$.

By far the most useful, best behaved functions are those that are continuous not just at individual points, but at every point in their domains. This is where we shall now concentrate our attention.

8.4 Continuity on a set

8.4.1 Definition

- A function $f : D \to C$ is said to be *continuous* (or, for emphasis, *continuous on D*) if it is continuous at every point of its domain D.
- A function $f : D \to C$ is said to be *continuous on a set S* (where S is a subset of D) if the restriction of f to S is continuous.

Recall, at this point, that the **restriction of f to S** is 'the same function as f was' except that it is only defined at the points of S. That is, it is the function $f' : S \to C$ described by $f'(x) = f(x)$ for every $x \in S$. By way of example, the function $f(x) = \lfloor x \rfloor$ is – as we saw earlier – not continuous at $x = 2$, but it is continuous on $[2, 3)$ because it is actually constant there (and, as we shall soon easily verify, constant functions are always continuous). The difference[3] arises because, when we investigate f itself near $x = 2$, we must examine what f does to all sequences that converge to 2 and, since the domain of f is $\mathbb{R}$, that includes sequences that approach 2 from the left as well as sequences that approach 2 from the right; in contrast, when we investigate f on the set $[2, 3)$, we only examine how it behaves on that set, and not therefore what might be happening to the left of 2.

For convenience, we combine definitions 8.3.1 and 8.4.1 into one:

8.4.2 Definition   A function $f : D \to C$ is continuous precisely when:

- for every sequence (x_n) in D that converges to an element p of D, we have $f(x_n) \to f(p)$.

The displayed condition here is often rendered in English as 'f *preserves limits of sequences*'.

Some functions, naturally enough, are extremely simple to prove continuous:

8.4.3 Example   A constant function (say, $f : D \to C$ where $f(x) =$ some constant k for all $x \in D$) is continuous.

Solution

For each $p \in D$ and each sequence (x_n) in D that converges to p, we have $\lim_{n\to\infty} f(x_n) = \lim_{n\to\infty} k = k = f(p)$.

8.4.4 Example   An identity function (that is, a function $f : D \to C$ where $f(x) = x$ for all $x \in D$) is continuous.

Solution

For each $p \in D$ and each sequence (x_n) in D that converges to p, we have $\lim_{n\to\infty} f(x_n) = \lim_{n\to\infty} x_n = p = f(p)$.

---

[3] If it helps you to understand this point, feel free to think in terms of the informal description of continuous functions as those whose domains believe that their graphs are unbroken. The domain of f includes 2 and numbers to the right and to the left of 2, so it is able to 'see' the abrupt change in height of the graph, from 1 immediately to the left of 2, to 2 at and immediately to the right of 2; on the other hand, the domain of f' is only $[2, 3)$ and, consequently, all it 'sees' of the graph is an unbroken horizontal line.

At this point, if we knew that it was safe to add and multiply continuous functions and be certain that the results were continuous, we could begin to build up a useful catalogue of basic continuous functions such as $kx, x^2, kx^2,$ $kx^2 + mx, kx^3 + mx^2 + qx + r$ and many others . . .

8.4.5 Theorem Suppose that the functions f and g are continuous on a set D; then so are

1. $f + g$,
2. $f - g$,
3. their product fg,
4. kf for any constant k,
5. f/g provided that $g(x) \neq 0$ for each $x \in D$, and
6. $|f|$.

Proof

All six parts are proved in the same rather predictable fashion, so we shall only demonstrate the (very slightly trickier) part 5.

Let (x_n) be any sequence in D that converges to an element p of D. Because f and g are continuous on D, we know that $f(x_n) \to f(p)$ and that $g(x_n) \to g(p)$. We additionally know that none of the $g(x_n)$ can be zero. By the algebra of limits for sequences yet again, $f(x_n)/g(x_n) \to f(p)/g(p)$, that is,

$$\left(\frac{f}{g}\right)(x_n) \to \left(\frac{f}{g}\right)(p).$$

Thus f/g is continuous.

8.4.6 Corollary

1. Every polynomial is continuous (on its domain $\mathbb{R}$).
2. Every rational function is continuous (on its domain).

Proof

The typical polynomial

$$p(x) = a_0 + a_1 x + a_2 x^2 + a_3 x^3 + \cdots + a_n x^n$$

can be built up in a finite number of moves from the basic constituents of x (that is, the identity function on $\mathbb{R}$) and constants by adding, multiplying and scaling. Parts 1, 3 and 4 of the theorem assure us that continuity will not be lost in that process.

The typical *rational function* is one polynomial divided by another

$$r(x) = \frac{p_1(x)}{p_2(x)}$$

and its domain comprises all real numbers *except* those at which the denominator $p_2(x)$ takes the value zero. Once we avoid those points (which, of course, are not in the domain of r anyway), the first part of the corollary says that $r(x)$ is one continuous function divided by another that is nonzero, and part 5 of the theorem tells us that $r(x)$ is continuous.

There is a way of combining functions that simply does not apply to sequences and, therefore, was not mentioned in the earlier chapters of this text. If $f : D \to C$ and $g : C \to B$ are two functions such that the codomain of f equals[4] the domain of g then, for each $x \in D$, we see that $f(x)$ lies in the domain of g and therefore $g(f(x))$ makes sense and is an element of B. Thus the correspondence of x to $g(f(x))$ has created a function from D to B. It is called the *composite* or the *composition* of f and g, and denoted by $g \circ f$ in most textbooks. In summary:
If $f : D \to C$ and $g : C \to B$ then their composite is the function

$$g \circ f : D \to B$$

specified by

$$(g \circ f)(x) = g(f(x)) \quad \text{for each} \ \ x \in D.$$

8.4.7 **Warning**   Beware that in some books the composite is written as gf and, in this case, you must take great care not to confuse it with the product g times f. Just to illustrate this: if $f(x) = \sqrt[3]{x}$ and $g(x) = x^2 + 5$ then the ordinary product gf is given by $(gf)(x) = g(x)f(x) = (x^2 + 1)\sqrt[3]{x}$, but the composite $g \circ f$ by $(g \circ f)(x) = g(\sqrt[3]{x}) = (\sqrt[3]{x})^2 + 5$. Notice also that the composite the other way round, $f \circ g$, is given by $(f \circ g)(x) = f(x^2 + 5) = \sqrt[3]{x^2 + 5}$ and is a completely different function from $g \circ f$.

The relevance of the composition idea at this point is:

8.4.8 **Theorem: continuity of composites**   If $f : D \to C$ and $g : C \to B$ are both continuous then so is their composite

$$g \circ f : D \to B.$$

Proof

Let (x_n) be any sequence in D that converges to an element p of D. Because f is continuous, we know that $f(x_n) \to f(p)$. Yet this convergence takes place inside

---

[4] This definition works equally well if the codomain of f is merely a *subset* of the domain of g.

the domain of the continuous function g, and therefore $g(f(x_n)) \to g(f(p))$. In other words, $(g \circ f)(x_n) \to (g \circ f)(p)$. Therefore $g \circ f$ is continuous.

8.4.9 A look forward   Once we get around to checking that $\sqrt[3]{x}$ is continuous, this theorem will assure us that both of the composites in the above Warning paragraph are continuous.

More generally, once we have confirmed continuity for functions such as $\sin, \cos, e^x$ and so on, Theorem 8.4.8 will help to generate a huge and diverse array of functions whose continuity we shall know in advance: expressions such as

$$e^{\sin x}, \sin(e^x), \cos(\sin(e^{\cos x})), e^{\sin x + e^{\cos x}}, e^{\frac{1+x+x^2}{2+3x+4x^2}}, \frac{33 - 7\sin^2 x \cos^4 x}{\sin x + \cos x + 5} \cdots$$

will be immediately seen to be continuous for no more complicated reason than that they have been created out of known basic continuous components by combining them through virtually any algebraic processes (including composition) that took care only to avoid dividing by zero. In short: continuity, at nearly any point, of nearly any formula that you can write down, will be a foregone conclusion *except*, of course, where division by zero raises its ugly head.[5]

8.5 Key theorems on continuity

Next, we set out to justify the claim that continuous functions are easier to work with, and 'behave better' than others.

8.5.1 Lemma   If s is the supremum of a set A of real numbers, then there is a sequence of points of A that converges to s.

Proof

For each positive integer n, the definition of supremum tells us that there is an element a_n of A such that $s - \frac{1}{n} < a_n \le s$. Now the squeeze shows that $a_n \to s$.

8.5.2 The intermediate value theorem ('IVT')   Let f be continuous on (at least) a closed bounded interval $[a, b]$ and suppose that either $f(a) < \lambda < f(b)$ or $f(a) > \lambda > f(b)$. Then there is a number $c \in (a, b)$ such that $f(c) = \lambda$.

(In other words, any number that lies *intermediate* between two values of a continuous function on an interval *actually is* a value of that function.)

---

[5] And provided, of course, that you never try to apply a function to a number that does not lie in its domain: taking square roots or logarithms of a negative number, for instance, will destroy not only the continuity of an expression, but also its very meaning.

Proof

Take the first case[6] $f(a) < \lambda < f(b)$. We put

$$X = \{x \in [a, b) : f(x) < \lambda\},$$

and we see that X is not empty (for a at least belongs to it) and is bounded above by b. Put $s =$ the supremum of X.

By the lemma, there is a sequence (a_n) in X such that $a_n \to s$. Continuity gives us $f(a_n) \to f(s)$ and, since each $f(a_n) < \lambda$, therefore (using 4.1.17)

$$f(s) = \lim_{n\to\infty} f(a_n) \leq \lambda \ldots\ldots(1)$$

In particular, since $f(s) < f(b)$, s must be strictly less than b, and the whole of $(s, b]$ lies outside X: that is, $f(y) \geq \lambda$ at every point y of $(s, b]$.

Now choose any sequence[7] (y_n) in $(s, b]$ that converges to s and, just as before, we get

$$f(s) = \lim_{n\to\infty} f(y_n) \geq \lambda \ldots\ldots(2).$$

From (1) and (2) we conclude that $f(s) = \lambda$.

Notice that the variant of the IVT that uses *non-strict* inequalities is also perfectly valid (and doesn't count as a different theorem):

8.5.3 The intermediate value theorem (again) Let f be continuous on (at least) a closed bounded interval $[a, b]$ and suppose that either $f(a) \leq \lambda \leq f(b)$ or $f(a) \geq \lambda \geq f(b)$. Then there is a number $c \in [a, b]$ such that $f(c) = \lambda$.

Proof

If $f(a) < \lambda < f(b)$ or $f(a) > \lambda > f(b)$ then there is nothing new to prove, because the original IVT applies. Yet if λ equals $f(a)$ or $f(b)$, the result is trivial anyway: $c = a$ or b will satisfy.

8.5.4 Example To show that the equation $p(x) = x^2(x^2 - 4)(x - 3) - 1 = 0$ has at least one (real) solution.

Proof

Since p is continuous (being just a polynomial) and $p(0) = -1$ is negative, all we need to do is to find a value of x such that $p(x)$ is positive: for then zero will lie between two values of p, and the IVT will guarantee that zero actually is a value of p

---

[6] For the second case, apply the conclusion of the first case to the continuous function $(-f)$.
[7] For instance, $y_n = s + \frac{b-s}{n}$ will do.

– as required. A little experimentation will readily find, for example, that $p(1) = 5$ which is positive. Therefore by 8.5.2 there is some $c \in (0, 1)$ such that $p(c) = 0$.

8.5.5 EXERCISE

1. Fill in the details in the following alternative (but heavier-handed) solution of the last example: if n is a positive integer then

$$\frac{p(n)}{n^5} \to 1 \text{ as } n \to \infty$$

so $p(n)/n^5$ will be positive for all sufficiently large values of n. Therefore we can find a positive integer n_2 such that $p(n_2)$ is positive. Similarly we can find a negative integer n_1 such that $p(n_1)$ is negative. Now the IVT says that there is $c \in (n_1, n_2)$ such that $p(c) = 0$.

2. Think how you could modify the demonstration of part 1 to prove that every polynomial equation of odd degree (that is, one in which the highest power of x appearing is an odd power) has at least one real solution.

8.5.6 Example
To show that the equation $p(x) = x^2(x^2 - 4)(x - 3) - 1 = 0$ has five (real) solutions.

Roughwork

This needs rather more trial-and-error...more precisely, it requires enough number-crunching to find not just one but five intervals over which $p(x)$ changes sign. We eventually discovered this:

$$p(-2) = -1; \quad p(-1) = +11; \quad p(0) = -1;$$
$$p(1) = +5; \quad p(2) = -1; \quad p(4) = +191.$$

Now we can use the same argument as in the previous example:

Solution

Since p is a continuous polynomial and $p(-2) = -1 < 0 < 11 = p(-1)$, there exists c_1 in $(-2, -1)$ such that $p(c_1) = 0$.

Since p is a continuous polynomial and $p(-1) = 11 > 0 > -1 = p(0)$, there exists c_2 in $(-1, 0)$ such that $p(c_2) = 0$.

Since p is a continuous polynomial and $p(0) = -1 < 0 < 5 = p(1)$, there exists c_3 in $(0, 1)$ such that $p(c_3) = 0$.

Since p is a continuous polynomial and $p(1) = 5 > 0 > -1 = p(2)$, there exists c_4 in $(1, 2)$ such that $p(c_4) = 0$.

Since p is a continuous polynomial and $p(2) = -1 < 0 < 191 = p(4)$, there exists c_5 in $(2, 4)$ such that $p(c_5) = 0$.

Because of the intervals in which they lie, c_1, c_2, c_3, c_4 and c_5 must all be *different* solutions of the equation. (You probably also know that a polynomial equation of degree 5, such as this, can never have more than five solutions: indeed, that a polynomial equation of degree n can never have more than n distinct solutions.)

8.5.7 Example To show that a function that is continuous on an interval and all of whose values are rational must actually be constant on that interval.

Solution

Let $f : I \to C$ be continuous on the interval I. If it were not constant, we could find x_1, x_2 in I such that $f(x_1) \neq f(x_2)$. Suppose, to make the picture more definite, that $x_1 < x_2$ and that $f(x_1) > f(x_2)$ (the other cases will work out in a very similar manner). We know that $f(x_1), f(x_2)$ are rational, but we can choose an irrational number j that lies between them.[8] By the IVT, j must be a value of f at a point somewhere in the interval I: but this contradicts what we were told about its values all being rational.

8.5.8 Example Given a continuous function f on the interval $[0,3]$ such that $f(0) = -f(3)$, to show that the equation $f(x) + f(x+2) = f(x+1)$ has a solution in $[0,1]$.

Solution

We create a new function (suggested by the equation we are trying to solve) $g(x) = f(x) - f(x+1) + f(x+2)$. This is defined on $[0,1]$ and is continuous there (because it has been built from continuous components f, $x+1$ and $x+2$). Notice that
$$g(0) = f(0) - f(1) + f(2)$$
and
$$g(1) = f(1) - f(2) + f(3) = f(1) - f(2) - f(0) = -(f(0) - f(1) + f(2))$$
have opposite signs.[9] Thus 0 lies intermediately between two values $g(0), g(1)$ of continuous g and therefore *is* a value of it: $0 = g(t)$ for some $t \in [0,1]$, that is, $0 = f(t) - f(t+1) + f(t+2)$ or $f(t) + f(t+2) = f(t+1)$.

8.5.9 EXERCISE Given a continuous function $f : [0,1] \to [0,1]$, show that the equation
$$(f(x))^2 = x^5$$
must have at least one solution in $[0,1]$.

[8] For instance, $f(x_2) + \frac{f(x_1) - f(x_2)}{\sqrt{2}}$ would do.
[9] Unless, of course, both are equal to zero: but then the result is immediate.

Roughwork

Does $g(x) = (f(x))^2 - x^5$ define a continuous function? Also pay attention to the *codomain* of f this time.

8.5.10 Optional extra – another proof of IVT   This proof ought to remind you of how we proved Bolzano-Weierstrass . . . which is very timely: for very shortly we are going to be making *serious use* of Bolzano-Weierstrass at last.

8.5.11 Lemma   Suppose that $f : [a, b] \to \mathbb{R}$ is continuous on the interval $[a, b]$, and $f(a) < 0$, and $f(b) \geq 0$. Then there is $c \in (a, b]$ such that $f(c) = 0$.

Proof

As a piece of temporary jargon, let us call $[a, b]$ a *signchange interval* for f to mean that $f(a) < 0$ and $f(b) \geq 0$. We are going to look for smaller signchange intervals for this function.

Consider the midpoint $m = (a + b)/2$ of $[a, b]$. If f has a negative value here, then $[m, b]$ is a signchange interval for f; if not, then $[a, m]$ is a signchange interval for f; in each case we have found one half of $[a, b]$ – label this half $[a_1, b_1]$ – that is a signchange interval for f.

Repeat the process: we shall find one half of $[a_1, b_1]$ – label this half $[a_2, b_2]$ – that is a signchange interval for f.

Repeat the process: we shall find one half of $[a_2, b_2]$ – label this half $[a_3, b_3]$ – that is a signchange interval for f.

Continue indefinitely.

We are producing two sequences (a_n) and (b_n) in $[a, b]$ and, because each interval contains the next one, they satisfy

$$a_1 \leq a_2 \leq a_3 \leq a_4 \leq \cdots < b; \quad b_1 \geq b_2 \geq b_3 \geq b_4 \geq \cdots > a.$$

So these two sequences are monotonic and bounded, and therefore converge:

$$a_n \to c \text{ (as } n \to \infty); \quad b_n \to c' \text{ (as } n \to \infty)$$

for some c, c' in $[a, b]$. But also, since each interval has just half the length of the previous one:

$$b_n - a_n = (b - a)/2^n.$$

Taking limits there, we see that $c' - c = 0$: in other words, $c = c'$.

Now f is negative at the left endpoint of each signchange interval so $f(a_n) < 0$ for all n, whence (using continuity at last) $f(c) = \lim_{n \to \infty} f(a_n) \leq 0$.

Equally, $f(b_n) \geq 0$ for all n, whence (via continuity) $f(c') = \lim_{n \to \infty} f(b_n) \geq 0$. Since $c = c'$, when we combine these we get $f(c) = 0$ as desired.

(Also, since $f(a) < 0$ and $f(c) = 0$, a and c cannot be equal; so $c \in (a, b]$.)

8.5.12 Theorem – the IVT yet again   Let f be continuous on a closed bounded interval $[a, b]$ and suppose that either $f(a) \leq \lambda \leq f(b)$ or $f(a) \geq \lambda \geq f(b)$. Then there is a number $c \in [a, b]$ such that $f(c) = \lambda$.

Proof

If $\lambda = f(a)$ or $f(b)$, the result is immediate. Otherwise, apply the last lemma to the function $f(x) - \lambda$ (or to the function $-f(x) + \lambda$ for the case $f(a) > \lambda > f(b)$).

8.5.13 Definition   If A is a subset of the domain of a function $f : D \to C$, the notation $f(A)$ means the set of all the values $f(x)$ as x varies over A. More formally,

$$f(A) = \{f(x) : x \in A\}$$

and is called the *image of A* or, sometimes, the *range of f over the set A*. Since it is a set of real numbers, it *may* be bounded above, or bounded below, or bounded, or have a maximum, or a minimum, or a supremum, or an infimum. By convention, these various properties are then also ascribed to f: we speak of *the function* being bounded above, or bounded below, or bounded, or having a maximum value, or a minimum value, or a supremum, or an infimum *on the set A*. Once again it is perfectly possible for a well-behaved function not to have a maximum or a minimum value, or even not to be bounded at all: easy examples can be set up to show this. However, for a *continuous* function on a *closed, bounded* interval, no such eccentric behaviour can occur – as we shall now show.

8.5.14 Theorem – continuous functions on closed bounded intervals are bounded   If $f : [a, b] \to \mathbb{R}$ is continuous, then it is bounded on $[a, b]$.

Proof

Suppose firstly that it were not bounded above. Then, for each positive integer n, n cannot be an upper bound for the range $f([a, b])$, so there is a point $x_n \in [a, b]$ such that $f(x_n) > n$.

According to Bolzano-Weierstrass, the bounded sequence (x_n) thus created has a convergent subsequence: $x_{n_k} \to p$ for some $p \in [a, b]$. Now continuity tells us that $f(x_{n_k}) \to f(p)$ so, in particular, $(f(x_{n_k}))$ is a convergent sequence, and therefore bounded. Yet it is not: for we arranged that, for each positive integer k, $f(x_{n_k}) > n_k \geq k$. The contradiction shows that f must have been bounded above.

Secondly, much the same argument will yield a contradiction from supposing that f were not bounded below. (Alternatively, since we now know that continuous functions on $[a, b]$ are always bounded above, apply that fact to the continuous function $(-f)$: for '$(-f)$ *bounded above*' and 'f *bounded below*' say exactly the same thing.)

This theorem entitles us to speak of the supremum and the infimum of any continuous function on a closed bounded interval, and yet there is better news than that – the sup and the inf are actually values of the function: so we can safely speak of the function's *biggest* and *smallest* values instead.

8.5.15 **Theorem – sup and inf are attained** If $f : [a, b] \to \mathbb{R}$ is continuous, then it possesses a maximum and a minimum value on $[a, b]$.

Proof

Knowing from the previous theorem that $f([a, b])$ is bounded and therefore has a supremum f_{sup} and an infimum f_{inf}, we need to find x_0, x_1 in $[a, b]$ such that $f(x_0) = f_{inf}$ and $f(x_1) = f_{sup}$.

For each $n \in \mathbb{N}$, the definition of supremum tells us that there is a point y_n in $[a, b]$ such that

$$f_{sup} - \frac{1}{n} < f(y_n) \le f_{sup}.$$

Once more we have a bounded sequence (y_n), and once more Bolzano-Weierstrass promises us a convergent subsequence: say

$$y_{n_k} \to p \text{ as } k \to \infty$$

where also $p \in [a, b]$. Appealing to continuity, we find that $f(p) = \lim_{k\to\infty} f(y_{n_k})$.

Yet, when we take limits (as $k \to \infty$) across the inequality

$$f_{sup} - \frac{1}{n_k} < f(y_{n_k}) \le f_{sup}$$

we get $f_{sup} - 0 \le f(p) \le f_{sup}$, in other words, $f(p) = f_{sup}$ as required.

A similar argument will show that f_{inf} is a value (and therefore the *least* value) of f over the interval. (Alternatively, we could apply what we just proved to show that $(-f)$ attains a greatest value and, as is easily seen, minus the greatest value of $(-f)$ is, precisely, the least value of f.)

8.5.16 **Optional extra – an alternative proof that suprema are attained**   Again let $f : [a, b] \to \mathbb{R}$ be continuous, and suppose that it does not attain a maximum value. In that case, f_{sup} is always strictly greater than the values of f, so the function

$$g(x) = \frac{1}{f_{sup} - f(x)}$$

is defined and continuous everywhere on $[a, b]$. By above, $g(x)$ is bounded: there is a positive constant K such that, for every $x \in [a, b]$:

$$g(x) \le K, \text{ that is, } \frac{1}{f_{sup} - f(x)} \le K, \text{ that is, } f_{sup} - f(x) \ge \frac{1}{K},$$

$$\text{that is, } f(x) \le f_{sup} - \frac{1}{K}.$$

This, however, contradicts f_{sup} being the supremum (the *least* possible upper bound) of the values of f.

8.5.17 **EXERCISE**   Show by means of examples (preferably simple ones) that

1. A continuous function on a bounded open interval can fail to be bounded,
2. A continuous function on a bounded open interval, even if it is bounded, can fail to have a maximum value and can fail to have a minimum value,
3. A continuous function on an unbounded closed interval can fail to be bounded,
4. A continuous function on an unbounded closed interval, even if it is bounded, can fail to have a maximum value and can fail to have a minimum value,
5. A discontinuous function on a bounded closed interval can fail to be bounded,
6. A discontinuous function on a bounded closed interval, even if it is bounded, can fail to have a maximum value and can fail to have a minimum value.

8.5.18 **Example**   Given a continuous function $f : [0, \infty) \to [0, \infty)$ such that:

for each $\varepsilon > 0$ there is $K > 0$ such that $f(x) < \varepsilon$ whenever $x > K$,

show that f has a maximum value.

Solution

(Note that we cannot immediately use the 'supremum is attained' theorem since the domain of f here is an *unbounded* interval. But it would be good if we could somehow force the action to take place on a closed *bounded* interval and, since $f(x)$ is nearly 0 for very big values of x, that perhaps could be arranged ...)

In the special case where f is constant at 0, the result is trivial.

If not, then we can find $a \in [0, \infty)$ such that $f(a) > 0$.

Then the given condition on f tells us that we can find a positive number K so that:

for every $x > K$, we get $0 \le f(x) < f(a)/2$.

It should be obvious that $K \ge a$. Now on the closed bounded interval $[0, K]$, f must have a biggest value ($f(b)$, say), and this is at least as big as $f(a)$. It is therefore bigger than every value that f can take on (K, ∞) since they are all smaller than $f(a)$. In other words, $f(b)$ is the maximum value that f takes anywhere in $[0, \infty)$.

8.5.19 **EXERCISE**   (Assuming standard information about the exponential function) given a continuous function $f : [0, 10000] \to (0, \infty)$, show that there is a positive constant b such that

for every $x \in [0, 10000]$, $f(x) > be^x$.

(One approach is to consider $f(x)e^{-x}$. Pay attention to the codomain of f.)

8.5.20 EXERCISE Let $f : \mathbb{R} \to \mathbb{R}$ be any continuous function. We define a new continuous function $g : \mathbb{R} \to \mathbb{R}$ like this:

$$g(x) = f(5 \sin x + 7 \cos x), \quad (x \in \mathbb{R}).$$

Show that there is a positive constant K such that $|g(x)| < K$ for every $x \in \mathbb{R}$. (You can assume standard facts about the trig functions.)

8.6 Continuity of the inverse

Not all continuous functions have inverses but, for those that do, the inverse is also continuous in most of the cases that we encounter. Let us first revise the idea of inverse mapping. If $f : A \to B$ is a mapping or map or function (of any kind, not only one of the *real* functions that we are otherwise exclusively focusing upon) then:

- We call it *one-to-one* or *injective* if $x \neq x'$ in A implies that $f(x) \neq f(x')$ in B, that is, if $f(x) = f(x')$ happens *only* in the obvious special case where $x = x'$;

- We call it *onto* or *surjective* if its range is the whole of its codomain (not just a subset of it), that is, if every element y of B is $f(x)$ for some suitably chosen x in A;

- We call it *bijective* or *a bijection*[10] if it is both one-to-one and onto.

- An *identity mapping* is a map $id_A : A \to A$ whose domain and codomain are identical *and* which leaves every element unchanged: that is, $id_A(x) = x$ for every x in A.

- If (given a map $f : A \to B$) there is another map $g : B \to A$ such that $g \circ f = id_A$ and $f \circ g = id_B$, in other words, such that $g(f(x)) = x$ for every x in A and $f(g(y)) = y$ for every y in B then[11] f is said to be *invertible*, g is called *the inverse of* f, and g is usually written as f^{-1}.

- It is important not to confuse the inverse map f^{-1} with the reciprocal $\frac{1}{f}$: they are very different ideas. For instance, if f is the function described by the formula $x^2 + 1$ then its reciprocal $\frac{1}{f}$ is, of course, defined by the formula $\frac{1}{x^2 + 1}$; yet f is not bijective, so the inverse map f^{-1} does not exist (see the next bullet point). Again, if g is defined by the formula $x^3 + 1$ then, as is readily checked, the inverse map is that defined by $\sqrt[3]{x - 1}$ with domain $\mathbb{R}$, which is entirely different from the reciprocal function $\dfrac{1}{x^3 + 1}$ with domain $(-\infty, -1) \cup (-1, \infty)$.

---

[10] or just plain *one-to-one onto*
[11] less formally, such that f and g completely cancel out one another's effect

- The key result from the basic theory of sets and mappings is that the invertible mappings are *precisely* the bijective mappings: *f has an inverse if and only if it is both one-to-one and onto.*

That last result is important even in very elementary algebra. For instance, what is the inverse of $f(x) = x^2$? The short answer is that it doesn't have one ... if, by that brief formula, we mean the function

$$f : \mathbb{R} \to \mathbb{R} \text{ given by } f(x) = x^2 :$$

because it is obvious that this function (now that we have described it fully) is neither one-to-one nor onto.[12] If we modify the definition to

$$f : \mathbb{R} \to [0, \infty) \text{ given by } f(x) = x^2$$

then at least it becomes onto, since every element of $[0, \infty)$ is the square of some real number, but it is still not one-to-one. However, if we modify the definition again to read

$$f : [0, \infty) \to [0, \infty) \text{ given by } f(x) = x^2$$

then what we are looking at now is both one-to-one and onto, so it does possess an inverse.

The inverse, naturally, is the square root function – the function $g : [0, \infty) \to [0, \infty)$ given by $g(x) = \sqrt{x}$ – because it is clear that $\sqrt{x^2} = x$ and $(\sqrt{y})^2 = y$ for all non-negative x and y. This pattern of seeking an inverse for some important function that initially did not have one, by restricting the domain and/or codomain of its defining formula until it becomes one-to-one and onto, is common and valuable – we shall meet it again in Chapter 18.

Incidentally, it is important to keep in mind that the last three display lines defined three different functions, even though we (rather incorrectly) used the same letter f to stand for all of them.

Just as, amongst sequences, the monotonic ones were often easier to work with than the rest, *functions* whose values steadily increase or steadily decrease have some desirable and useful properties, and it will pay us to give clear definitions to these classes of function now:

8.6.1 **Definition**   A (real) function $f : D \to C$ is said to be

- *increasing* (on D) if $x < y$ in $D \Rightarrow f(x) \leq f(y)$,
- *strictly increasing* (on D) if $x < y$ in $D \Rightarrow f(x) < f(y)$,
- *decreasing* (on D) if $x < y$ in $D \Rightarrow f(x) \geq f(y)$,

---

[12] For example, because $f(1) = f(-1)$, and because -3 is in the codomain but is not in the range of f.

- *strictly decreasing* (on D) if $x < y$ in $D \Rightarrow f(x) > f(y)$,
- *monotonic* if it is either increasing or decreasing,
- *strictly monotonic* if it is either strictly increasing or strictly decreasing.

(These properties are usually easy to visualise in a sketch graph, simply by looking to see whether the graph climbs steadily up the page or down as we scan across from left to right.)

It turns out that a continuous function on an interval is one-to-one if and only if it is either strictly increasing or strictly decreasing. For that reason, when we set out to establish results concerning inverses of important functions, we really lose nothing in the way of generality if we deal only with strictly monotonic functions.

8.6.2 Lemma   If $f : D \to f(D)$ is strictly monotonic then it is also one-to-one and possesses an inverse. The inverse is strictly increasing if f is strictly increasing, and is strictly decreasing if f is strictly decreasing.

Proof

We'll consider only the case where f is strictly increasing – the other is very similar.

If $x \neq y$ in D then either $x < y$ or $y < x$. Accordingly either $f(x) < f(y)$ or $f(y) < f(x)$: and in both cases, $f(x) \neq f(y)$, as required for injectivity. The choice of codomain has ensured that f is also onto, so it is invertible.

Given $p < q$ in $f(D)$ choose x, y in D such that $p = f(x)$ and $q = f(y)$, that is, $x = f^{-1}(p)$ and $y = f^{-1}(q)$. If it were true that $x \geq y$ then the increasing nature of f would yield the contradiction $p = f(x) \geq f(y) = q$. Hence we must have $x < y$, that is, $f^{-1}(p) < f^{-1}(q)$. Therefore f^{-1} is a strictly increasing function.

8.6.3 Lemma   Let $f : I \to \mathbb{R}$ be a continuous function on an interval I. Then its range $f(I)$ is also an interval.

Proof

Any number y that lies between two elements $f(x_1), f(x_2)$ of the range is, according to the IVT, a value of f, that is, another element of the range: but this is the defining characteristic of an interval.

8.6.4 Lemma   Let $f : I \to \mathbb{R}$ be a continuous strictly monotonic function on an open interval I. Then its range $f(I)$ is also an open interval.

Proof

For any element $f(x)$ of the range, $x \in I$ cannot be an endpoint of the (open) interval I, so we can find x', x'' in I such that $x' < x < x''$. Depending on whether f is (strictly) increasing or decreasing, it follows that either $f(x') < f(x) < f(x'')$ or $f(x') > f(x) > f(x'')$. In both cases, $f(x)$ fails to be an endpoint for the interval $f(I)$, which therefore cannot include any of its endpoints: hence the result.

8.6.5 EXERCISE Confirm that

- if f is an increasing (respectively, decreasing) function then $-f$ is a decreasing (respectively, increasing) function,
- if f is a strictly increasing (respectively, strictly decreasing) function then $-f$ is a strictly decreasing (respectively, strictly increasing) function,
- if f and g are both decreasing (on the same domain) then so is $f + g$ but
- there is an example of two strictly decreasing functions f and g (on the same domain) whose product fg is strictly increasing.

8.6.6 Remark – optional extra Once we think in more detail about what a continuous strictly monotonic function can do to intervals of various types, a picture emerges that is rather more complicated than the last lemma suggests. As a way of building intuition about this, you could check out the details summarised in the following table:

Table 8.1. Possible forms of $f(I)$ for interval I and continuous strictly monotonic f

I	$f(I)$ if f is contin. and str. increasing	$f(I)$ if f is contin. and str. decreasing
$[a,b]$	$[f(a),f(b)]$	$[f(b),f(a)]$
(a,b)	(c,d) or (c,∞) or $(-\infty,d)$ or $(-\infty,\infty)$	(c,d) or (c,∞) or $(-\infty,d)$ or $(-\infty,\infty)$
$[a,b)$	$[f(a),d)$ or $[f(a),\infty)$	$(c,f(a)]$ or $(-\infty,f(a)]$
$(a,b]$	$(c,f(b)]$ or $(-\infty,f(b)]$	$[f(b),d)$ or $[f(b),\infty)$
$[a,\infty)$	$[f(a),d)$ or $[f(a),\infty)$	$(c,f(a)]$ or $(-\infty,f(a)]$
(a,∞)	(c,d) or (c,∞) or $(-\infty,d)$ or $(-\infty,\infty)$	(c,d) or (c,∞) or $(-\infty,d)$ or $(-\infty,\infty)$
$(-\infty,b]$	$(c,f(b)]$ or $(-\infty,f(b)]$	$[f(b),d)$ or $[f(b),\infty)$
$(-\infty,b)$	(c,d) or (c,∞) or $(-\infty,d)$ or $(-\infty,\infty)$	(c,d) or (c,∞) or $(-\infty,d)$ or $(-\infty,\infty)$
$(-\infty,\infty)$	(c,d) or (c,∞) or $(-\infty,d)$ or $(-\infty,\infty)$	(c,d) or (c,∞) or $(-\infty,d)$ or $(-\infty,\infty)$

8.6.7 EXERCISE – optional extra Confirm that the table above is complete and correct. That is, for each row, verify that when the interval I has the indicated form and the function $f : I \to \mathbb{R}$ is continuous and strictly increasing, $f(I)$ must take one of the listed forms, and confirm by examples that each listed form actually can occur; then repeat the exercise for f continuous and strictly decreasing.

Partial solution

Let us consider just the fifth row, the one in which I is of the form $[a,\infty)$.

When f is strictly increasing, since a is the least element of $I = [a, \infty)$ the interval $f(I)$ must have a least element, namely, $f(a)$. Also, no element $f(x)$ of $f(I)$ can be a greatest element of $f(I)$ because $x \in I \Rightarrow \exists x' \in I$ with $x < x' \Rightarrow f(x) < f(x') \in f(I)$. Therefore $[f(a), d)$ and $[f(a), \infty)$ are the only possible forms for $f(I)$.

The continuous strictly increasing function $f(x) = x^2$ on $[0, \infty)$ maps $[0, \infty)$ onto $[0, \infty)$.

The continuous strictly increasing function $f(x) = -x^{-1}$ on $[1, \infty)$ maps $[1, \infty)$ onto $[-1, 0)$.

The third column entries can be confirmed by noting that if f is continuous and strictly decreasing then $-f$ is continuous and strictly increasing, and then using what we just established in the second column.

8.6.8 **Proposition**   Suppose that $f : I \to f(I)$ is strictly increasing on the interval I, and continuous at a point $p \in I$. Then the inverse $f^{-1} : f(I) \to I$ is continuous at the point $f(p)$.

Proof

The inverse exists and is strictly increasing by 8.6.2.

Firstly, we shall look in detail at the case where p is *not* an endpoint of I (and, consequently, $f(p)$ is not an endpoint of the interval $f(I)$). Let (y_n) be any sequence in $f(I)$ that converges to $f(p)$; we need to show that $(f^{-1}(y_n))$ converges to $f^{-1}(f(p)) = p$.

For each $n \in \mathbb{N}, y_n = f(x_n)$ for some (unique) $x_n \in I$, namely $x_n = f^{-1}(y_n)$. Let $\varepsilon > 0$ be given. There is no loss of generality in assuming that ε is small enough to ensure that the interval $[p - \varepsilon, p + \varepsilon]$ lies inside I: for were it not so, we would replace ε by a smaller number that does ensure this. Consequently, we can talk about $f(p - \varepsilon)$ and $f(p + \varepsilon)$, and know that the first is smaller than $f(p)$ and the second is greater than $f(p)$. Since $y_n \to f(p)$, we see that[13] there will be a positive integer n_0 such that

$$n \geq n_0 \Rightarrow y_n \in (f(p - \varepsilon), f(p + \varepsilon))$$

which, in turn, implies that

$$x_n = f^{-1}(y_n) \in (p - \varepsilon, p + \varepsilon)$$

merely because f^{-1} is strictly increasing. So $n \geq n_0$ forces $|x_n - p| < \varepsilon$, and we have $x_n \to p$ as required.

---

[13] If this step is not sufficiently clear to you, try putting δ equal to the smaller of the two numbers $f(p) - f(p - \varepsilon)$ and $f(p + \varepsilon) - f(p)$, and notice that (for sufficiently large values of n) we shall have $|y_n - f(p)| < \delta$.

Secondly, if I possesses a left-hand endpoint and p happens to be that endpoint, we need to make small changes to the argument of the last paragraph. This time we ensure that the interval $[p, p + \varepsilon)$ lies inside I. Then $f(p + \varepsilon)$ makes sense and is greater than $f(p)$, so y_n must belong to $[f(p), f(p + \varepsilon))$ for all $n \geq$ some n_0. It follows that (for these large values of n) $x_n = f^{-1}(y_n)$ belongs to $[p, p + \varepsilon)$, and we again conclude that $x_n \to p$.

The third possibility – that I possesses a right-hand endpoint and p happens to be that endpoint – works out in a manner very similar to the second.

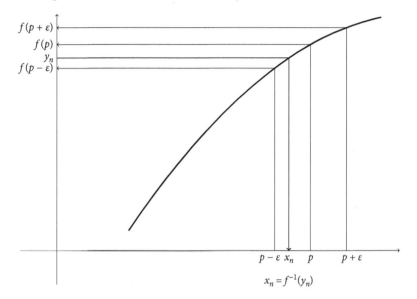

Continuity of an inverse mapping

Almost exactly the same argument will show that this also works for *decreasing* in place of *increasing*:

8.6.9 Proposition Suppose that $g : I \to g(I)$ is strictly decreasing on the interval I, and continuous at a point $p \in I$. Then the inverse $g^{-1} : g(I) \to I$ is continuous at the point $g(p)$.

(Alternatively, you may be able to see how to prove the second proposition for free, just by applying the first proposition to the increasing function $-g$.)

Combining the two propositions (invoked at each point of the domain) and Lemma 8.6.2, we obtain the continuous inverse theorem in the form that is usually most useful:

8.6.10 Theorem A real function, defined on an interval, that is continuous and strictly increasing on that interval, possesses an inverse (defined upon its range) that is also continuous and strictly increasing.

A real function, defined on an interval, that is continuous and strictly decreasing on that interval, possesses an inverse (defined upon its range) that is also continuous and strictly decreasing.

8.6.11 **EXERCISE**   Verify that

- for each odd positive integer n, the function $f(x) = \sqrt[n]{x}$ is continuous on $\mathbb{R}$,
- for each even positive integer n, the function $f(x) = \sqrt[n]{x}$ is continuous on $[0, \infty)$.

9 Limit of a function

9.1 Introduction

As in the previous chapter, it may be useful to begin by outlining informally and visually the next topic that we are going to define and investigate. Let's start by reviewing the sketch graphs we drew in our first attempt to explain continuity.

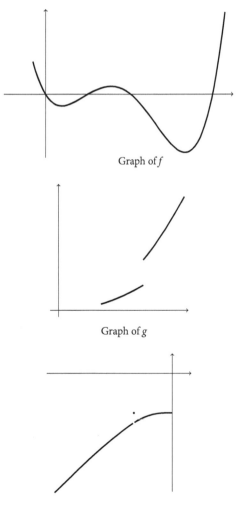

Graph of f

Graph of g

Graph of h

Undergraduate Analysis: A Working Textbook, Aisling McCluskey and Brian McMaster 2018.
© Aisling McCluskey and Brian McMaster 2018. Published 2018 by Oxford University Press

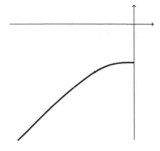

Graph of k

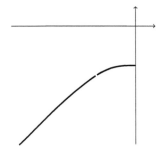

Graph of m

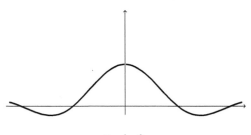

Graph of p

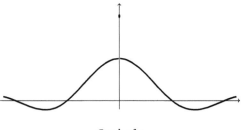

Graph of q

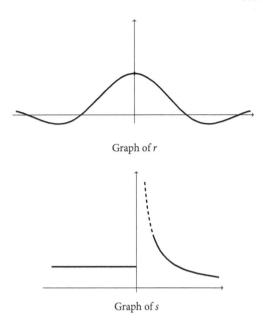

Graph of r

Graph of s

We initially presented continuity (at a particular point $x = a$) as a way to distinguish between functions such as f, k and r whose graphs did not seem to possess any sort of break at the point of interest, and the others whose graphs did – in some sense – break there. Amongst the discontinuous functions, however, some of those graphs are seen to be more severely broken than others: g, for instance, has a gap of 4 units between the left and right portions of its graph, and s has what appears to be an infinite gap. In contrast, h and m could be thought of as failing continuity on a mere technicality: either because the function has not been defined at all at $x = a$, or because it has, but in a way that is incompatible with its behaviour *close to* $x = a$. Similar remarks apply to p and q.

One way to make explicit the difference between g and s on the one hand, and h, m, p and q on the other, is to ask *how small a change* in the function's definition would alter it into a continuous function. In the case of m, simply writing in the extra phrase '$m(-1) = -5/4$' makes the modified function continuous (because it turns m into k); in the case of h, altering the value of $h(-1)$ from -1 to $-5/4$ would have the same effect. That is, altering (or creating) the value of the function *at one single point* is all it takes to turn m or h into a continuous function. Similar remarks – again – apply to p and q. In sharp contrast, there is no way to convert g or s to continuity by intervening at a single value – the gaps in their graphs simply cannot be bridged by moving (or filling in) just one point.

This insight also gives us a way to build a proper mathematical definition of the distinction between functions like g and s, and functions such as h, m, p and q. Since a function of the second type (let us now denote it by $f : D \to \mathbb{R}$) fails continuity at $x = a$ *only* because its value at a is either undefined or, in some sense, 'wrongly defined', we can use sequences to probe its behaviour near to a exactly as we did for ordinary continuity *but consistently avoiding x = a*. So, starting with Chapter 8's definition of continuity at the point a:

for every sequence (x_n) in D such that $x_n \to a$, we find that $f(x_n) \to f(a)$

we should *firstly* replace the (perhaps undefined) $f(a)$ by a symbol for the limiting number to which all the sequences $(f(x_n))$ need to converge, and *secondly* prevent the sequences (x_n) from including a as one or more of their terms. This suggests that the defining characteristic of functions of the second type is this: there is a real number ℓ such that

for every sequence (x_n) in $D\backslash\{a\}$ such that $x_n \to a$, we find that $f(x_n) \to \ell$.

When this is the case, the number ℓ which the values of f are approximating better and better as we approach a (but without actually reaching a, of course) is called the limit of $f(x)$ as x tends to a or the limit of f at a. In this language, looking back at our sketch graphs, we intend to say that both h and m have limits of $-5/4$ at -1, that both p and q have limits of 1 as we approach 0, but that g does not have a limit as we approach 2 and that s does not have a limit at 0. (Of course, we still need to show that these are *true* statements – but at least we now possess a logically sound definition against which to test their truthfulness.) We also gain from the discussion an alternative definition of continuity, namely: **a function f is continuous at a point a of its domain if the limit of f as we approach a is precisely $f(a)$.**

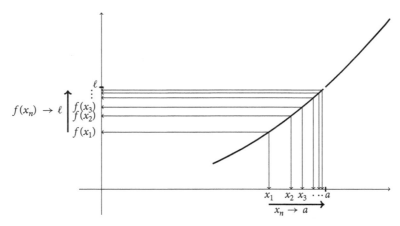

Suggested definition of function limit

Next, a technical warning: since the definition that we are setting up (of limit of $f(x)$ as x approaches a) is entirely dependent on what happens to sequences in $D \backslash \{a\}$ that converge to a, we must take care never to use it *if, in fact, there are no such sequences!* For instance, any attempt to find a limit of $\ln x$ as x approaches -1, or of $\sqrt{x}$ at $x = -0.3$, or of $\arcsin x$ as x tends to 2, is doomed to fail since these functions are undefined close to the point that we are claiming to approach. For a subtler example, think about the factorial function $f(x) = x!$. Now, we don't often consider 'factorial' as a *real* function at all since it only handles non-negative integers but, nevertheless, it does satisfy the requirements of our definition of a real function (with domain $\mathbb{N} \cup \{0\}$). Let us ask, then: what is the limit of this function $x!$ as x tends to, say, 2? The domain D of this function is $\{0, 1, 2, 3, 4, 5, \cdots\}$, so $D \setminus \{2\}$ is $\{0, 1, 3, 4, 5, \cdots\}$, and the definition needs us to look at a typical sequence in $\{0, 1, 3, 4, 5, \cdots\}$ that converges to 2 . . . *but no such sequences exist*: a sequence in that set never gets within the distance 1 of 2, so it cannot converge to 2. We conclude that the limit of $x!$ as we approach 2 (or, indeed, as we approach any other number) cannot be defined.

The final matter that we ought to stress before concluding this introductory section is that, although functions such as m and h appear somewhat artificial at first sight – *contrived* examples, designed to deliver a teaching point rather than practical, useful algebra or calculus – questions concerning the limiting values of non-continuous functions do turn up in a very large number of important application-oriented problems, and we should outline a couple of these before starting our serious study of function limits. The first is one that we have touched on already (see the functions p, q, r again), and the second introduces an idea that is fundamental to differential calculus, which we shall work on in Chapter 12.

9.1.1 Example When x is interpreted as an angle in radians, the ratio $\dfrac{\sin x}{x}$ calculates out very close to 1 when x is small. This is the basis of many arguments in Science as well as in Mathematics proper, in which $\sin x$ is replaced by (the much easier to handle) x as a high-quality approximation. Let us ask, then: when is $\dfrac{\sin x}{x}$ exactly 1?

Reply

Never. There is no value of x for which $\dfrac{\sin x}{x} = 1$. Of course, as is widely known, the ratio is very close to 1 provided that x is sufficiently close to 0...but a careful examination shows that the only solution of the equation $\sin x = x$ is $x = 0$, and we cannot replace x by 0 in the ratio $\dfrac{\sin x}{x}$ since $\dfrac{0}{0}$ is meaningless. What this example is informally expressing is that the limit of the function $\dfrac{\sin x}{x}$, as x approaches 0, is 1.

9.1.2 Example The point $P = (3, 9)$ lies on the graph of the quadratic function $f(x) = x^2$. In an attempt to evaluate the slope of the graph exactly at the point P, we take a nearby point on the graph – say, the point $Q = (3 + h, (3 + h)^2)$ – and work out the gradient of the straight line PQ. Provided that the number h is really small, that straight line should hug the curve closely enough to ensure that the gradient of PQ will be a good approximation to the gradient of the curve itself and – thinking imprecisely for a moment – when $h = 0$, the approximation ought to become perfect. What, then, actually does happen to the gradient of PQ when we allow h to become zero?

Reply

It ceases to have any meaning whatsoever. Since the gradient of the straight line
PQ (change in y-coordinate *divided by* change in x-coordinate) is actually

$$\frac{(3+h)^2 - 3^2}{(3+h) - 3} = \frac{6h + h^2}{h},$$

replacing h by 0 gives, once again, the meaningless symbol $\frac{0}{0}$. Notice that we cannot
simply cancel an h top and bottom in the previous display unless we write $h \neq 0$
into the contract, because cancelling just means dividing top and bottom lines by h,
and this is illegal precisely in the special case $h = 0$ that we really wanted to get
at. What the example is trying to work towards is not the value of $\dfrac{(3+h)^2 - 3^2}{(3+h) - 3}$
at $h = 0$, but its limit at that point: for this is what will give us the gradient of the
curve itself at the point P.

9.2 Limit of a function at a point

As the 'technical warning' pointed out, we must first think about the kind of point
at which calculation of a function limit makes sense:

9.2.1 Definition Let S be a set of real numbers and p a real number. We call p a
limit point[1] of S if there is at least one sequence of elements of $S \setminus \{p\}$ that converges
to p. (Note that p may or may not be an element of S.)

9.2.2 Notes

1. The only case that will occupy our attention is that in which S is the domain of
   some function. Then p being a limit point of S is exactly what is needed in
   order that we can sensibly try to find a limit of that function at p.

2. If S happens to be an interval then it is easy to see that the limit points of S are
   precisely the points of S and the endpoints of S (one or both of which, of
   course, might be elements of S already).

3. In nearly all the examples in this text, the domain of a function will be either a
   non-degenerate interval or the union of a finite list of non-degenerate
   intervals. For such a domain D it is again easy to see that the limit points are
   simply the points of D together with all the endpoints of those intervals. (The
   recent exception was $x!$, whose domain $\mathbb{N} \cup \{0\}$ had no limit points
   whatsoever.)

4. In the event that you might need to deal with some function whose domain is
   more complicated, the following lemma may be useful in identifying limit
   points:

[1] also called an accumulation point

9.2.3 Lemma　If $S \subseteq \mathbb{R}$ and $p \in \mathbb{R}$, then p is a limit point of S if and only if:

for each $\varepsilon > 0$, the interval $(p - \varepsilon, p + \varepsilon)$ includes
a point of S that is *different from p*.

Proof

If p is a limit point of S, choose a sequence (x_n) in $S \backslash \{p\}$ such that $x_n \to p$. Then for any choice of $\varepsilon > 0$ there are, in fact, infinitely many x_n in the interval $(p - \varepsilon, p + \varepsilon)$, and they are all different from p.

Conversely, suppose that the displayed condition holds. Then, choosing $\varepsilon = 1/n$ for each positive integer n in turn, we can find a point of S (call it x_n) in $(p - \varepsilon, p + \varepsilon)$ but distinct from p. The sequence (x_n) of elements of $S \backslash \{p\}$ thus created satisfies $p - 1/n < x_n < p + 1/n$ so, using the squeeze, it converges to p. Hence p is indeed a limit point of S.

9.2.4 Optional exercise　Verify that

- the limit points of the set of rational numbers include *all* real numbers,
- the set of integers has no limit points,
- every positive integer is a limit point of the set

$$\{n + (m + 1)^{-1} : n \in \mathbb{N}, m \in \mathbb{N}\}.$$

9.2.5 Definition　Suppose that $f : D \to C$, that p is a limit point of its domain D and that $\ell \in \mathbb{R}$. Then we say that $f(x)$ *converges* to the *limit* ℓ as $x \to p$ if:

for every sequence (x_n) in $D \backslash \{p\}$ such that $x_n \to p$, we find that $f(x_n) \to \ell$.

It is also common practice to call ℓ the *limit of f at p*, and to write all this as

$$\lim_{x \to p} f(x) = \ell$$

or to abbreviate it as $f(x) \to \ell$ (as $x \to p$).

9.2.6 Example　To show that the function f given by

$$f(x) = \frac{x^2 - 9}{x - 3}$$

converges to a limit of 6 as $x \to 3$.

Solution

The domain of f is $D = (-\infty, 3) \cup (3, \infty)$ so it has 3 as a limit point.

Let (x_n) be any sequence in $D \setminus \{3\}$ such that $x_n \to 3$. Then

$$f(x_n) = \frac{x_n^2 - 9}{x_n - 3} = \frac{(x_n - 3)(x_n + 3)}{x_n - 3} = x_n + 3 \to 6.$$

(There is an important point near the end of the line which it is all too easy to miss. When we cancel $x_n - 3$, what we are doing is *dividing* top and bottom by $x_n - 3$. Of course we dare not divide by zero ... but since x_n belongs to $D \setminus \{3\}$, we know that $x_n - 3$ is definitely non-zero; it is precisely this detail that allows us to cancel, and thus saves us from hitting a nonsensical conclusion, such as: that the limit is $\frac{0}{0}$.)

Hence $f(x) \to 6$ as $x \to 3$.

Alternative solution

To see more easily what is happening when x is close to 3, it often helps to put $x = 3 + h$ and then consider $h \to 0$ instead. Thus, given any sequence (x_n) in $D \setminus \{3\}$ such that $x_n \to 3$, if we set $x_n = 3 + h_n$ for each n, then we see that:

$$f(x_n) = f(3 + h_n) = \frac{(3 + h_n)^2 - 9}{(3 + h_n) - 3} = \frac{6h_n + h_n^2}{h_n} = 6 + h_n \to 6.$$

(Please note again that the cancellation is legal precisely because we are not cancelling or dividing zeros: h_n can never be exactly zero because x_n is never exactly 3.)

We conclude that $\lim_{x \to 3} f(x) = 6$.

For a fairly straightforward question such as the previous one, there was very little to choose between the two different solutions we demonstrated. However, if the algebra is more complicated and the possibility of cancelling less obvious, then the trick of putting $x = a + h$ to explore close to $x = a$ can save you quite a bit of time and effort.

9.2.7 Example   To determine the limit, as $x \to -1$, of the function

$$f(x) = \frac{x^5 + 1}{x^4 - 1}.$$

Solution

Since the only real numbers x for which $x^4 = 1$ are 1 and -1, the bottom line goes zero only at 1 and -1, the domain D of this function is $(-\infty, -1) \cup (-1, 1) \cup (1, \infty)$, and -1 is a limit point of D.

Given any sequence (x_n) in $D \setminus \{-1\}$ whose limit is -1, put $x_n = -1 + h_n$ for each n, and notice that

$$f(x_n) = f(-1 + h_n) = \frac{(-1 + h_n)^5 + 1}{(-1 + h_n)^4 - 1} = \frac{h_n^5 - 5h_n^4 + 10h_n^3 - 10h_n^2 + 5h_n}{h_n^4 - 4h_n^3 + 6h_n^2 - 4h_n}$$

$$= \frac{h_n^4 - 5h_n^3 + 10h_n^2 - 10h_n + 5}{h_n^3 - 4h_n^2 + 6h_n - 4}$$

(noting that $h_n \neq 0$ is what allows us that cancellation, and that to keep the new bottom line non-zero we also need to prevent $x_n = 1$, that is, avoid letting $h_n = 2$).

Now since $h_n \to 0$ we see that, by ignoring the first few terms if necessary, we can be sure that $h_n \neq 2$. Then

$$f(x_n) = \frac{h_n^4 - 5h_n^3 + 10h_n^2 - 10h_n + 5}{h_n^3 - 4h_n^2 + 6h_n - 4} \to \frac{5}{-4} = -\frac{5}{4}.$$

Hence $\lim_{x \to -1} f(x) = -5/4$.

9.2.8 EXERCISE The point $P = (2, 2)$ lies on the curve described by $y = f(x) = x^3 - 3x^2 + 6$. Determine the gradient of this curve at the point P. (*Hint:* set the problem up as in 9.1.2, and deal with the limit as in 9.2.6.)

9.2.9 Example The function f is defined as follows:

$$f(x) = 0 \text{ when } x = 2 \text{ or } x = 5/2; \quad \text{otherwise } f(x) = \frac{6x^2}{2x - 5}.$$

Determine its limiting behaviour as $x \to 2$.

Solution

The domain is $\mathbb{R}$ since, although the fraction formula fails to make sense at $x = 5/2$, a separate definition of $f(5/2)$ has been provided. (Why a separate definition of $f(2)$ has also been provided is not obvious, but the definition of f as a whole is unambiguous.)

We consider any sequence (x_n) in $\mathbb{R} \setminus \{2\}$ whose limit is 2, and we put $x_n = 2 + h_n$ for each n, where $h_n \to 0$. By ignoring (if necessary) the first few terms, we can arrange that $|x_n - 2| < 1/2$, that is, $3/2 < x_n < 5/2$: therefore the separate definition of f at $5/2$ is no longer relevant, and

$$f(x_n) = \frac{6x_n^2}{2x_n - 5} = \frac{6(2 + h_n)^2}{2(2 + h_n) - 5} = \frac{24 + 24h_n + 6h_n^2}{2h_n - 1}$$

– whose limit (using again the algebra of limits for sequences) is -24. Therefore

$$\lim_{x \to 2} f(x) = -24.$$

9.2.10 **Remark**   Two general points that emerge from the last example are worth stressing. Firstly, although the value ascribed to $f(2)$ was a distinctly odd choice, this had absolutely no effect upon the limit as x approached 2 because, when exploring limiting behaviour of $f(x)$ as x approaches a point p, *we don't care what (if anything) $f(x)$ does when x exactly equals p - only how it behaves when x is very close to but distinct from p.*

Secondly, the separate definition given to $f(5/2)$ was also essentially irrelevant to the problem since any investigative sequence (x_n) converging to 2 will eventually be closer to 2 than 5/2 is, so that the limiting behaviour of $(f(x_n))$ cannot be influenced by how f behaves at 5/2 or, indeed, at any significant distance from 2. This insight can even be rephrased into an occasionally useful theorem:

9.2.11 **Theorem: limits are locally determined**   Let $f : D \rightarrow C$ be a function, p be a limit point of its domain D and η a positive real number. Now let $g : D \cap (p - \eta, p + \eta) \rightarrow C$ be the same function[2] as f except that it is only defined on $D \cap (p - \eta, p + \eta)$. Then $f(x) \rightarrow \ell$ as $x \rightarrow p$ if and only if $g(x) \rightarrow \ell$ as $x \rightarrow p$.

Proof

Suppose that $g(x) \rightarrow \ell$ as $x \rightarrow p$. For any sequence (x_n) in $D \setminus \{p\}$ that converges to p, since $\eta > 0$ we can find n_0 such that $n \geq n_0$ implies that $p - \eta < x_n < p + \eta$, which in turn shows that

$$f(x_n) = g(x_n) \rightarrow \ell \text{ as } n \rightarrow \infty$$

and hence $f(x) \rightarrow \ell$ as $x \rightarrow p$.

The converse is simpler because any sequence in $(D \setminus \{p\}) \cap (p - \eta, p + \eta)$ (converging to p) already is a sequence in $D \setminus \{p\}$ (converging to p).

Comment

What that result says is that, if you wish to determine the limit of f as x approaches p, it is good enough to narrow your attention to what f does in *any* open interval centred on p. By way of illustration, if you were asked to find the limit at $x = 3.3$ of the following function

$$h(x) = \begin{cases} x^{0.37}\sqrt{5x^4 + 29}\,\sin(3x^2 + \pi/17) & \text{if } x < 3, \\ 2x & \text{if } x \geq 3 \end{cases}$$

---

[2] Such a function g is more properly called a *restriction* of f: in this case, the restriction of f to $D \cap (p - \eta, p + \eta)$.

then you could choose to work with $h(x)$ *as if* it were only defined on, say, $(3.1, 3.5)$. Yet on that interval, $h(x) = 2x$ is just twice an identity function, so its limit is almost immediately seen to be 6.6.

The usefulness of the insight, that the limit as $x \to 3.3$ is only influenced by what happens *locally* at 3.3, for example, on the interval $(3.1, 3.5)$, is that it allows us to ignore completely the more complicated behaviour of the function outside that locality.

9.2.12 **EXERCISE** Show that $\lfloor x \rfloor$, the floor of x,

1. possesses a limit as x tends to any real number p that is *not an integer*,

2. does not possess a limit as x tends to any integer. (*Hint*: if, for an integer m, the function f given by $f(x) = \lfloor x \rfloor$ does have a limit ℓ as $x \to m$, then for every sequence (x_n) in $\mathbb{R} \setminus \{m\}$ such that $x_n \to m$ we must have $f(x_n) \to \ell$. Try this with, for instance, $x_n = m + (n + 1)^{-1}$ and then again with $x_n = m - (n + 1)^{-1}$ to seek a contradiction.)

9.2.13 **EXERCISE** The function g is defined as follows:

$$g(x) = 0 \text{ when } x \text{ is any integer; otherwise } g(x) = \frac{x^3}{x + 2}.$$

Determine its limiting behaviour as $x \to \frac{1}{3}$.

The next example is something of a trick question but, nevertheless, it makes an important point.

9.2.14 **Example** To investigate the limit, as $x \to 0$, of the function $f(x) = (\sqrt{x^4 - x^2})^2$.

Solution

(*Aside*: yes, *of course* we want to 'cancel' the squaring and the square root, but this will be legal only when the square root exists as a real number.)

Now $x^4 - x^2$ factorises easily into $x^2(x + 1)(x - 1)$, from which we see that it is non-negative when $x \le -1$ and when $x = 0$ and when $x \ge 1$, but negative when x lies in the open interval $(-1, 0)$, and negative again when x lies in $(0, 1)$. Under our convention about the domain of a formula-defined function comprising all the real numbers for which the formula delivers a real answer, $f(x)$ exists and is $x^4 - x^2$ on $(-\infty, -1] \cup \{0\} \cup [1, \infty)$ but is undefined on $(-1, 0) \cup (0, 1)$.

Since 0 is the only member of the domain of f that lies in, for instance, the interval $(-0.5, 0.5)$, 0 is not a limit point of the domain, so the limit of f (as we approach 0) is not defined.

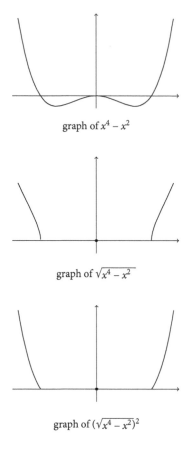

graph of $x^4 - x^2$

graph of $\sqrt{x^4 - x^2}$

graph of $(\sqrt{x^4 - x^2})^2$

9.2.15 EXERCISE Investigate the limit, as $x \to 1$, of the function $g(x) = e^{\ln(x-1)}$.
(*Hint:* begin by identifying the domain of g.)

At this stage we can start developing basic theorems about function limits that will allow us to handle them more efficiently than by the definition alone. Some of these will strike you as very predictable, given what we have already seen about sequences and continuous functions. We start with an observation that justifies our use of the word *the* whenever we talk about *the* limit of a function at a point:

9.2.16 Theorem   A function cannot converge to two or more different limits at a limit point of its domain.

Proof

With a view to a contradiction, suppose that $f : D \to C$, that p is a limit point of D, that $f(x) \to \ell_1$ and $f(x) \to \ell_2$ as $x \to p$, and that $\ell_1 \neq \ell_2$. Pick any sequence (x_n) in $D \setminus \{p\}$ that converges to p as limit. Then $f(x_n)$ has to converge both to ℓ_1 and to ℓ_2: which is impossible by 2.7.10.

9.2.17 The algebra of limits for functions   Suppose that $f : D \to C$ and $g : B \to A$ are two functions, that p is a limit point of the intersection $D \cap B$ of their domains, and that (as $x \to p$) $f(x) \to \ell$ and $g(x) \to m$. Then, as $x \to p$:

1. $f(x) + g(x) \to \ell + m$,

2. $f(x) - g(x) \to \ell - m$,

3. $f(x)g(x) \to \ell m$,

4. for constant k, $kf(x) \to k\ell$,

5. provided that $m \neq 0$,

$$\frac{f(x)}{g(x)} \to \frac{\ell}{m},$$

6. $|f(x)| \to |\ell|$.

Proof

These can all be proved in the same (and hopefully obvious) way. For example, let's do numbers (3) and (5):

Limit of a product In the above notation, let $(x_n)_{n \in \mathbb{N}}$ be an arbitrary sequence in $D \cap B \setminus \{p\}$ that converges to p. Then (via the algebra of sequence limits)

$$(fg)(x_n) = f(x_n)g(x_n) \to \ell m$$

and therefore $(fg)(x) \to \ell m$ as $x \to p$.

Limit of a quotient (This is the only proof among the six that needs a little extra caution: because we need to avoid any risk of dividing by zero.)

In the above notation, let $(x_n)_{n \in \mathbb{N}}$ be an arbitrary sequence in $D \cap B \setminus \{p\}$ that converges to p.

Because $g(x) \to m \neq 0$ as $x \to p$, we get $g(x_n) \to m \neq 0$ as $n \to \infty$, so there is $n_0 \in \mathbb{N}$ such that $n \geq n_0 \Rightarrow g(x_n) \neq 0$. Now (ignoring the first n_0 terms, which has no effect on sequence limits)

$$\left(\frac{f}{g}\right)(x_n) = \frac{f(x_n)}{g(x_n)} \to \frac{\ell}{m} \ (\text{as } n \to \infty)$$

and therefore

$$\left(\frac{f}{g}\right)(x) = \frac{f(x)}{g(x)} \to \frac{\ell}{m} \ (\text{as } x \to p)$$

as required.

9.2.18 EXERCISE   Choose two other parts of this theorem and write out proofs for them.

9.2.19 Theorem: a squeeze or sandwich rule for function limits   Suppose that $f:D \to \mathbb{R}$, $g:D' \to \mathbb{R}$, $h:D'' \to \mathbb{R}$ are three functions, that $D' \subseteq D \cap D''$ (that is, wherever g is defined, so are f and h), that p is a limit point of D', that $f(x) \le g(x) \le h(x)$ for each $x \in D'$ and that

$$\lim_{x \to p} f(x) = \lim_{x \to p} h(x) = \ell, \text{ say.}$$

Then also

$$\lim_{x \to p} g(x) = \ell.$$

9.2.20 EXERCISE   Construct a proof of this theorem.

We said in this chapter's introduction that a limit of a *continuous* function would always turn out to be simply the value of the function at the appropriate point. This, together with its converse, is a highly useful characterisation of continuity that we shall use many times. The only tricky detail in confirming its correctness is that, when we test for continuity at a point and for limit at that point, we need to use *two slightly different families of sequences*.

9.2.21 Theorem   Consider a real function $f : D \to C$ and a point p of D that is also a limit point of D. Then the following are equivalent:

1. f is continuous at p;
2. $f(x) \to f(p)$ as $x \to p$.

Solution

First, suppose that (1) holds. With a view to establishing (2), let (x_n) be any sequence in $D \setminus \{p\}$ that converges to p. Then merely because (x_n) is a sequence in D that converges to p, continuity gives $f(x) \to f(p)$ as $x \to p$ as we wanted.

The converse is a little less straightforward. Suppose this time that (2) holds, and let (x_n) be any sequence in D that converges to p.

- If there are only finitely many values of n for which $x_n = p$ then we can ignore them without implication for limiting processes, and thus regard (x_n) as a sequence in $D \setminus \{p\}$ that converges to p. By supposition, $f(x_n) \to f(p)$ as $n \to \infty$.

- At the other extreme, if there are only finitely many values of n for which $x_n \ne p$, then we can equally well ignore them, regard (x_n) as a constant sequence (p), and immediately have $f(x_n) = f(p) \to f(p)$ as $n \to \infty$.

- In the remaining case, (x_n) divides up into two (infinite) subsequences, one of which (call it (y_n)) lies in $D \setminus \{p\}$ and converges to p, while the other (call it (z_n)) is constant at p. Since both $(f(y_n))$ and $(f(z_n))$ converge to $f(p)$ – the first via condition (2) and the second because it is constant, an exercise in Chapter 5 (see 5.2.5, 5.2.6) tells us that $f(x_n) \to f(p)$ once again.

In each of the three possible scenarios, we have what we needed in order to conclude that f is continuous at p.

Once more we find that a benefit of possessing a battery of basic theorems is that we can tackle examples without having to fall back on the definitions:

9.2.22 Example A function $f : (0, \infty) \to \mathbb{R}$ is defined[3] thus: for each irrational x, put $f(x) = 0$; for each rational $x = \frac{p}{q}$ where the fraction has been expressed in its lowest terms, put $f(x) = \frac{5}{q}$ if q is prime but $f(x) = -\frac{7}{q}$ if q is not prime. To show that $f(x) \to 0$ as $x \to 0$.

Solution

Notice that $|f(x)| \leq 7x$ in all cases. That is:

$$-7x \leq f(x) \leq 7x, \quad \text{all } x \in (0, \infty).$$

Now the functions $7x$ and $-7x$ are both continuous (on $\mathbb{R}$) so, using 9.2.21, $\lim_{x \to 0} 7x = 0$ and $\lim_{x \to 0}(-7x) = 0$. Using the 'new squeeze' 9.2.19 on the previous display gives $\lim_{x \to 0} f(x) = 0$ as required.

9.2.23 Example To revisit the determination of the limit, as $x \to -1$, of the function

$$f(x) = \frac{x^5 + 1}{x^4 - 1}.$$

Alternative solution

Since what happens at -1 is irrelevant to this limit, we can assume that $x \neq -1$. Then (with a little effort of factorisation):

$$
\begin{aligned}
\lim f(x) = \lim \frac{x^5 + 1}{x^4 - 1} &= \lim \frac{(x+1)(x^4 - x^3 + x^2 - x + 1)}{(x+1)(x-1)(x^2 + 1)} \\
&= \lim \frac{x^4 - x^3 + x^2 - x + 1}{(x - 1)(x^2 + 1)} \\
&= \frac{\lim(x^4 - x^3 + x^2 - x + 1)}{\lim((x - 1)(x^2 + 1))} = \frac{5}{(-2)(2)} = -\frac{5}{4}
\end{aligned}
$$

---

[3] This is an example in which any attempt at sketching the graph is likely to waste quite a lot of time!

(noting that all the limits are as $x \to -1$, the cancellation was legitimate since $x + 1$ was non-zero, the algebra of limits allowed us to operate on the top and bottom lines separately, and the fact that top and bottom lines were continuous gave us the numerical values of those two limits immediately.)

9.2.24 Example To determine the limit, as $x \to 4$, of the function f described by

$$f(x) = \frac{\sqrt{x} - 2}{x - 4} \quad \text{when } x \neq 4; \quad f(4) = 1.$$

Solution

We can safely assume $x \neq 4$ since behaviour at 4 has no consequences for a limit while approaching 4. With that proviso:

$$f(x) = \frac{\sqrt{x} - 2}{x - 4} = \frac{\sqrt{x} - 2}{(\sqrt{x} - 2)(\sqrt{x} + 2)} = \frac{1}{\sqrt{x} + 2} \to \frac{1}{2 + 2} = \frac{1}{4}$$

because the function $\sqrt{x}$ is continuous (see 8.6.10) and so its limit as we approach 4 is merely its value $\sqrt{4}$ at 4 (and because we can use the algebra of limits theorem).

9.2.25 EXERCISE Evaluate the limits

$$\lim_{x \to 3} \frac{x^4 - 81}{x^4 - 8x^2 - 9} \quad \text{and} \quad \lim_{x \to 3} \frac{x^3 - 3x^2 - 9x + 27}{x^6 - 243}.$$

9.2.26 HARDER EXERCISE See if you can determine the limit of the function discussed in 9.2.22 as $x \to \frac{2}{3}$, and its limit as $x \to \pi$. Don't worry if this turns out to be difficult – we shall encounter, in the next chapter, an alternative method that works more easily for certain questions, including ones like this.

10 Epsilontics and functions

10.1 The epsilontic view of function limits

Up to this point, we have exclusively employed sequences to define continuity and function limits and to develop their theory. There is, however, an alternative approach that is very widely used and that works more smoothly for certain types of question. We need to become familiar with this before going any further.

The basic idea underlying limiting processes is that of approximation: to say that *some (variable) quantity X approaches a limit ℓ* asserts that values of X can be generated that are extremely good approximations to ℓ in the sense that the error $|X - \ell|$ is not merely small, but can be made smaller than any required tolerance just by continuing far enough with the approximation-generating procedure. We have dealt pretty thoroughly with the case in which the approximation process is a sequence – an unending list of one number after another – but how much has to change when X is instead a function, say, a function $f(x)$ where x is tending towards a limit point p of the domain of f? It still makes perfectly good sense to call $|f(x) - \ell|$ the error of the approximation and to insist that this can be made smaller than any given tolerance $\varepsilon > 0$; the only item that needs to be re-thought is what controls the approximation-generating procedure this time and how far we need to go with it: we now need the 'best' approximations to be the ones that correspond to values of x that are very close to p – say, those in which x lies within a sufficiently small distance δ from p. This discussion suggests a possible definition of $f(x)$ converging to the limit ℓ as $x \to p$, as follows:

for each $\varepsilon > 0$ there exists $\delta > 0$ such that $|x - p| < \delta$ implies that $|f(x) - \ell| < \varepsilon$

which captures the intuition of high-quality approximation without needing to call in sequences as a probing mechanism. Two details of this proposed definition still need refinement, however. Firstly, we need to write in the requirement that x shall belong to the domain of the function f before we even mention $f(x)$. Secondly, to acknowledge that what f does or does not do at the exact point $x = p$ has absolutely no effect on its limit, we should prevent $|x - p|$ from equalling zero. The epsilontic definition of limit of a function at a point therefore takes the following final form:

Undergraduate Analysis: A Working Textbook, Aisling McCluskey and Brian McMaster 2018.
© Aisling McCluskey and Brian McMaster 2018. Published 2018 by Oxford University Press

10.1.1 **Alternative definition of function limit**   If $f : D \to C$ is a real function and p is a limit point of its domain D, we say that $f(x)$ *converges* to the *limit* ℓ (as $x \to p$) if

for each $\varepsilon > 0$ there exists $\delta > 0$ such that

$x \in D$ and $0 < |x - p| < \delta$ together imply that $|f(x) - \ell| < \varepsilon$.

As before, we can write this symbolically either as $\lim_{x \to p} f(x) = \ell$ or as $f(x) \to \ell$ as $x \to p$, and we can leave out the phrase '$x \to p$' whenever the context makes it clear enough that this is what is intended.

Now it is perfectly possible to develop the entire theory of function limits and, in turn, the entire theory of continuous functions, starting from this definition. We shall do a little of this just to illustrate that it can be done, but it is a common student experience that 10.1.1 is a harder definition to use than 9.2.5 (and also a common lecturer experience that it is a harder definition to teach) which is why we opted to make 9.2.5 our primary definition here.

A task that we cannot shirk is to show that 9.2.5 and 10.1.1 really are logically equivalent. This is quite a sophisticated proof, and you might want to omit it on a first reading, but it is important and urgent to take on board what the result is saying: that despite their apparent differences, 10.1.1 and 9.2.5 are completely interchangeable: if either one of them is satisfied, then so must the other be. You are therefore free to use whichever of the two definitions you please (other things being equal) in any given argument.

10.1.2 **Theorem**   Let $f : D \to C$ be a function and p a limit point of its domain D. Then the following are equivalent:

1. for every $\varepsilon > 0$ there is $\delta > 0$ such that $x \in D, 0 < |x - p| < \delta$ together imply that $|f(x) - \ell| < \varepsilon$.
2. for every sequence $(x_n)_{n \in \mathbb{N}}$ of elements of $D \setminus \{p\}$ that converges to p, we have $f(x_n) \to \ell$.

Proof

(I): (1) implies (2).
- Suppose that condition (1) is satisfied.
- Let $(x_n)_{n \in \mathbb{N}}$ be an arbitrary sequence of elements of $D \setminus \{p\}$ that converges to p.
- For a given positive value of ε, use condition (1) to obtain a number $\delta > 0$ such that whenever $x \in D$ and $0 < |x - p| < \delta$, we have $|f(x) - \ell| < \varepsilon$.
- Because $x_n \to p$, there is a positive integer n_0 such that $n \geq n_0$ will guarantee that $|x_n - p| < \delta$.
- Also $0 < |x_n - p|$ because x_n is not equal to p.

- Therefore

$$n \geq n_0 \Rightarrow 0 < |x_n - p| < \delta \Rightarrow |f(x_n) - \ell| < \varepsilon.$$

- That is, $f(x_n) \rightarrow \ell$ as required.
- Since (x_n) was *any* sequence in $D \setminus \{p\}$ that happened to converge to p, condition (2) is now proved.

TIMEOUT

Although the mathematics in the upcoming proof of the converse is reasonably straightforward, the logical content is easily the most sophisticated that we have dealt with so far, so let us call 'timeout' and pick our way through it one small step at a time. We'll also continue, for the moment, to bullet-point out those steps, to try for a little added clarity.

- Instead of trying to show directly that (2) implies (1), we shall call in the logical device of contraposition and show instead (but equivalently) that NOT-(1) implies NOT-(2).
- Condition (1) says that, for every ε that we are challenged with, we can find a positive δ that 'works'.
- If this is not true then there is *some* special and awkward value of ε for which *no* value of δ that we choose to try will 'work'.
- In particular, if we pick a positive integer n and try $\delta = 1/n$, it will not 'work'.
- In other words, the implication

$$0 < |x - p| < \delta = 1/n \Rightarrow |f(x) - \ell| < \varepsilon$$

(for $x \in D$) is not true.

- So there must exist an x in D that satisfies $0 < |x - p| < \delta = 1/n$ and yet the conclusion $|f(x) - \ell| < \varepsilon$ is false.
- Thus $x \in D$ and $0 < |x - p| < \delta = 1/n$, but $|f(x) - \ell| \geq \varepsilon$.
- A refinement of notation: the x that we just discovered will almost certainly depend on the value of n that we picked several steps ago, so we had better label it in a way that renders that visible – say, instead of plain x we should write $x(n)$ or, more simply perhaps, x_n.
- Now we suddenly find ourselves in possession of a sequence $(x_n)_{n \in \mathbb{N}}$, and we know the following details about it:

    each $x_n \in D$,
    each x_n is distinct from p because $0 < |x_n - p|$,
    (x_n) converges to p because $|x_n - p| < 1/n$ and $1/n \rightarrow 0$ (look back at 2.7.14 here if necessary),
    the sequence $(f(x_n))$ does not converge to ℓ because $|f(x_n) - \ell| \geq \varepsilon$ for all n.

- That is to say, condition (2) – which claims validity for *every* appropriate sequence that converges to p, is NOT TRUE.
- The proof, that NOT-(1) implies NOT-(2), is therefore now complete.

Once you have managed to follow the details of that expanded argument, you should be able to grasp the sort of condensed version that typically appears in textbooks:

(II): (2) implies (1).

- Suppose that condition (1) is not satisfied.
- That is, there exists a value of $\varepsilon > 0$ for which *no* suitable $\delta > 0$ can be found.
- In particular, for each $n \in \mathbb{N}$, $\delta = 1/n$ is not suitable...
- ...and so there is $x_n \in D$ such that $0 < |x_n - p| < 1/n$ and yet $|f(x_n) - \ell| \geq \varepsilon$.
- Therefore $(x_n)_{n \in \mathbb{N}}$ converges to p (with each $x_n \in D \setminus \{p\}$), and yet $(f(x_n))_{n \in \mathbb{N}}$ does not converge to ℓ.
- That is, condition (2) is not satisfied.
- By contraposition, (2) implies (1).

10.1.3 Example   To show that

$$f(x) = \frac{x^3 - 1000}{x^2 - 100} \to 15 \text{ as } x \to 10$$

directly from the epsilon-delta definition 10.1.1.

Solution

We can assume that $x \neq 10$ since behaviour *at* 10 is irrelevant to the limit, and so

$$\frac{x^3 - 1000}{x^2 - 100} - 15 = \frac{(x-10)(x^2 + 10x + 100)}{(x-10)(x+10)} - 15 = \frac{x^2 + 10x + 100}{x + 10} - 15$$

$$= \frac{x^2 - 5x - 50}{x + 10} = \frac{(x+5)(x-10)}{x + 10}.$$

Let $\varepsilon > 0$ be given. We need to decide: how close to 10 must we take x in order that this error term shall be less than ε in modulus?

- If we make $|x - 10| < 1$, that is, $9 < x < 11$, then $|x + 5| < 16$ and $|x + 10| > 19$, so $|f(x) - 15| < (16/19)|x - 10|$.
- If we also make $|x - 10| < 19\varepsilon/16$ then $(16/19)|x - 10| < \varepsilon$.

Therefore choose $\delta = \min\{1, 19\varepsilon/16\}$ and we find that

$$0 < |x - 10| < \delta \Rightarrow |f(x) - 15| < (16/19)|x - 10| < \varepsilon, \text{ as required.}$$

10.1.4 Example   To use the epsilon-delta definition of limit to show that a function cannot converge to two or more different limits at a limit point of its domain.

Proof

With a view to a contradiction, suppose that $f : D \to C$, that p is a limit point of D, that $f(x) \to \ell_1$ and $f(x) \to \ell_2$ as $x \to p$, and that $\ell_1 \neq \ell_2$. Arrange the labelling so that $\ell_1 < \ell_2$ for convenience, and put $\varepsilon = \dfrac{\ell_2 - \ell_1}{2} > 0$. Using 10.1.1, there exist two positive numbers δ_1 and δ_2 such that

$$x \in D, 0 < |x - p| < \delta_1 \text{ together imply } |f(x) - \ell_1| < \varepsilon, \text{ and}$$
$$x \in D, 0 < |x - p| < \delta_2 \text{ together imply } |f(x) - \ell_2| < \varepsilon.$$

Now since δ_1 and δ_2 are each positive, so is the lesser of the two (whichever one it is). Call that lesser δ_3. Because p is a limit point of D there must actually be[1] an element of D – let us call it x', for instance – that satisfies $0 < |x' - p| < \delta_3$. So both of the displayed lines apply to x', and we therefore know that

$$|f(x') - \ell_1| < \varepsilon \text{ and } |f(x') - \ell_2| < \varepsilon.$$

Now invoke the triangle inequality:

$$|\ell_1 - \ell_2| = |\ell_1 - f(x') + f(x') - \ell_2| \leq |\ell_1 - f(x')| + |f(x') - \ell_2| < \varepsilon + \varepsilon,$$

that is, $2\varepsilon < 2\varepsilon$, which is as crisp a contradiction as one can ask for.

10.1.5 EXERCISE   Directly use the epsilon-delta definition of limit (10.1.1) to show that

$$\lim_{x \to -6} \left(\frac{x^2 - 36}{x^3 + 216} \right) = -\frac{1}{9}.$$

10.1.6 EXERCISE   Directly use the epsilon-delta definition of limit to prove that if p is a limit point of D, and the functions f, g both have D as their domain, and $f(x) \to \ell_1$ and $g(x) \to \ell_2$ as $x \to p$, then $f(x) + g(x) \to \ell_1 + \ell_2$ as $x \to p$.

In all fairness to the epsilon-delta definition of limit, here is an example in which it works more easily and more naturally than our primary, sequence-based definition. The function involved is one we began to study in paragraph 9.2.22.

---

[1] See 9.2.3

10.1.7 **Example**   The function $f : (0, \infty) \to \mathbb{R}$ is defined thus: for each irrational x, put $f(x) = 0$; for each rational $x = \frac{p}{q}$ where the fraction has been expressed in its lowest terms, put $f(x) = \frac{5}{q}$ if q is prime but $f(x) = -\frac{7}{q}$ if q is not prime. To show that $f(x) \to 0$ as $x \to \pi$.

Solution

Let $\varepsilon > 0$ be given. Consider a positive integer N. The interval $(\pi - 1, \pi + 1)$ includes π of course, and it includes only a *finite*[2] number of rationals whose (lowest-terms) denominators lie in the range 2 to N so, putting $\delta = $ the shortest distance from π to one of these, we get $\delta > 0$ and every rational r in $(\pi - \delta, \pi + \delta)$ has a denominator greater than N. It follows that $|f(r)| < 7/N$ and, since f is exactly zero at each irrational, we get

$$0 < |x - \pi| < \delta \Rightarrow x \in (\pi - \delta, \pi + \delta) \Rightarrow |f(x) - 0| < \frac{7}{N}.$$

Therefore if we choose N so large that $7/N < \varepsilon$, we have what is required in order to show that the limit of f at π is 0.

10.1.8 **EXERCISE**   Use 10.1.1 to investigate the limit of this function at $\frac{2}{3}$.

10.2 The epsilontic view of continuity

Intuitively, it is not difficult to guess what the epsilon-delta alternative definition of continuity at a point should be: for f to be continuous at p, we need $f(x)$ to have a limit equal to $f(p)$, and there is no longer any need to prevent x from equalling p since, if that happens, it certainly doesn't prevent $f(x)$ from being a good approximation to $f(p)$. We should therefore expect the formal alternative definition to take this shape:

10.2.1 **Alternative definition of continuity**   If $f : D \to C$ is a real function and $p \in D$, we say that f is *continuous at the point p* when

for each $\varepsilon > 0$ there exists $\delta > 0$ such that

$x \in D$ and $|x - p| < \delta$ together imply that $|f(x) - f(p)| < \varepsilon$.

If you carefully compare this with 10.1.1, it will strike you that we have left out any reference to p being a limit point of D. This is not a casual oversight, for it

---

[2] Since the length of this open interval is 2, it cannot include more than 4 rationals of denominator 2, nor more than 6 rationals of denominator 3, nor more than 8 rationals of denominator 4, and so on.

turns out that 10.2.1 is entirely equivalent to our original definition of continuity whether or not p is a limit point of D. The following proof of this assertion is so like that of 10.1.2 that, were it a bit shorter or a bit less complicated, we would have asked you to check it out as an exercise (and you might still decide to try that if you have already properly grasped the argument of 10.1.2).

10.2.2 Theorem Let $f : D \to C$ be a function and $p \in D$. Then the following are equivalent:

1. for every $\varepsilon > 0$ there is $\delta > 0$ such that $x \in D$, $|x - p| < \delta$ together imply that $|f(x) - f(p)| < \varepsilon$.

2. for every sequence $(x_n)_{n \in \mathbb{N}}$ of elements of D that converges to p, we have $f(x_n) \to f(p)$.

Proof

(I): (1) implies (2).

- Suppose that condition (1) is satisfied.
- Let $(x_n)_{n \in \mathbb{N}}$ be an arbitrary sequence of elements of D that converges to p.
- For a given positive value of ε, use condition (1) to obtain a number $\delta > 0$ such that whenever $x \in D$ and $|x - p| < \delta$, we have $|f(x) - f(p)| < \varepsilon$.
- Because $x_n \to p$, there is a positive integer n_0 such that $n \geq n_0$ will guarantee that $|x_n - p| < \delta$.
- Therefore

$$n \geq n_0 \Rightarrow |x_n - p| < \delta \Rightarrow |f(x_n) - f(p)| < \varepsilon.$$

- That is, $f(x_n) \to f(p)$ as required.
- Since (x_n) was *any* sequence in D that happened to converge to p, condition (2) is now proved.

(II): (2) implies (1).

- Suppose that condition (1) is not satisfied.
- That is, there exists a value of $\varepsilon > 0$ for which *no* suitable $\delta > 0$ can be found.
- In particular, for each $n \in \mathbb{N}$, $\delta = 1/n$ is not suitable . . .
- . . . and so there is $x_n \in D$ such that $|x_n - p| < 1/n$ and yet $|f(x_n) - f(p)| \geq \varepsilon$.
- Therefore $(x_n)_{n \in \mathbb{N}}$ converges to p (with each $x_n \in D$), and yet $(f(x_n))_{n \in \mathbb{N}}$ does not converge to $f(p)$.
- That is, condition (2) is not satisfied.
- By contraposition, (2) implies (1).

We shall again present a couple of samples of how to use the alternative definition in arguments:

10.2.3 Example   To show, using the epsilontic definition, that a composite of continuous functions is continuous.

Solution

Suppose that $f : D \to C$ and $g : C \to B$ are both continuous. We need to verify that their composite $g \circ f : D \to B$ is continuous. So let $p \in D$ and $\varepsilon > 0$ be given.

Since g is continuous at the point $f(p)$ of its domain, there is $\delta_1 > 0$ such that $y \in C, |y - f(p)| < \delta_1$ together imply $|g(y) - g(f(p))| < \varepsilon$.

Since f is continuous at p (and $\delta_1 > 0$), there is $\delta > 0$ such that $x \in D, |x - p| < \delta$ together imply $|f(x) - f(p)| < \delta_1$.

Thus, for $x \in D$ and $|x - p| < \delta$, we get $|f(x) - f(p)| < \delta_1$ and, consequently, $|g(f(x)) - g(f(p))| < \varepsilon$, that is, $|(g \circ f)(x) - (g \circ f)(p)| < \varepsilon$. So $g \circ f$ is (by the epsilon-delta definition) continuous at each element of its domain.

10.2.4 Example   We define a function $f : [0, \infty) \to \mathbb{R}$ by:

$$f(0) = 1, \quad f(x) = x \left\lfloor \frac{1}{x} \right\rfloor \quad \text{for } x > 0.$$

To show (using 10.2.1) that f is continuous at 0.

Solution

We start from the fact that (for all real t) $t \geq \lfloor t \rfloor > t - 1$ simply by the definition of floor or 'integer part'. So, for $x > 0$,

$$1 = x \left(\frac{1}{x} \right) \geq x \left\lfloor \frac{1}{x} \right\rfloor = f(x) > x \left(\frac{1}{x} - 1 \right) = 1 - x.$$

In particular, $|f(x) - 1| < x$. Therefore, given $\varepsilon > 0$, if we choose $\delta = \varepsilon$, we get

$$|x - 0| < \delta \Rightarrow |f(x) - f(0)| = |f(x) - 1| < x < \delta = \varepsilon$$

which is our requirement for continuity at 0.

10.2.5 EXERCISE   Using the epsilon-delta definition of continuity, show that

$$f(x) = \begin{cases} 7x - 5 & \text{if } x \text{ is rational}, \\ 4x + 4 & \text{if } x \text{ is irrational} \end{cases}$$

defines a function that is continuous at 3.

10.2.6 **EXERCISE** Verify, using the epsilon-delta definition of continuity, that the function discussed in paragraph 10.1.7 is continuous at every positive irrational number, but discontinuous at every positive rational number.

10.2.7 **EXERCISE** Show, without using sequences, that the following function

$$g(x) = \begin{cases} x^2 + 4x - 1 & \text{if } x < 3, \\ x^3 - x^2 + 3 & \text{if } x \geq 3 \end{cases}$$

does not have a limit as $x \to 3$.

10.3 One-sided limits

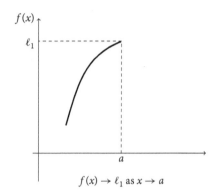

$f(x) \to \ell_1$ as $x \to a$

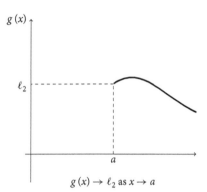

$g(x) \to \ell_2$ as $x \to a$

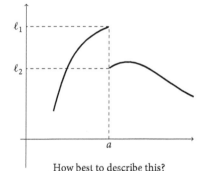

How best to describe this?

There were a few examples in the preceding section ($\lfloor x \rfloor$ close to an integer value for x and 10.2.7 are cases in point) where the reader can be forgiven for thinking informally along the following lines: 'this function $f(x)$ does not have a limit as $x \to p$ ... and yet, if we were allowed to look only at values of x close to *but just less than* p, it does appear to be settling to a limiting value ... and likewise if we look only at values of x close to *but just greater than* p.' This is a perfectly legitimate idea, and to explore and develop it we need first to formulate a clear definition of such *one-sided limits*. By this point in the text you are probably able to guess even the definitions with high accuracy.

10.3.1 **Definition**   Suppose that $f : D \to C$ and that p is a limit point of $D \cap (p, \infty)$. Then we call a number ℓ the *right-hand limit* of $f(x)$ at p, or the *limit on the right* of $f(x)$ at p, and say that $f(x)$ converges to ℓ as $x \to p$ *from the right* (or *from above*) if, for each $\varepsilon > 0$, there is $\delta > 0$ such that

$$x \in D, p < x < p + \delta \ \text{ together imply } \ |f(x) - \ell| < \varepsilon.$$

This is also written as

$$\lim_{x \to p^+} f(x) = \ell \ \text{ or as } \ f(x) \to \ell \text{ as } x \to p^+.$$

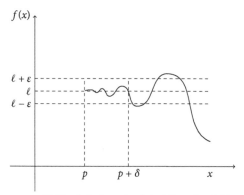

$f(x)$ tends to ℓ as x tends to p from the right

10.3.2 **Definition**   Suppose that $f : D \to C$ and that p is a limit point of $D \cap (-\infty, p)$. Then we call a number ℓ the *left-hand limit* of $f(x)$ at p, or the *limit on the left* of $f(x)$ at p, and say that $f(x)$ converges to ℓ as $x \to p$ *from the left* (or *from below*) if, for each $\varepsilon > 0$, there is $\delta > 0$ such that

$$x \in D, p - \delta < x < p \ \text{ together imply } \ |f(x) - \ell| < \varepsilon.$$

This is also written as

$$\lim_{x \to p^-} f(x) = \ell \ \text{ or as } \ f(x) \to \ell \text{ as } x \to p^-.$$

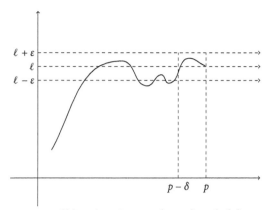

$f(x)$ tends to ℓ as x tends to p from the left

10.3.3 Remark You should keep in mind that one-sided limits, just like other types of limit, can fail to exist. For instance, the function

$$f(x) = \begin{cases} x^{-1} & \text{if } x < 0, \\ \sin(x^{-1}) & \text{if } x > 0 \end{cases}$$

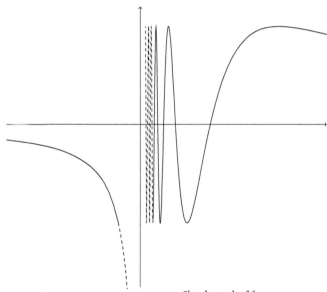

Sketch graph of f

does not have a left-hand limit as $x \to 0$ because the function is unbounded close to 0 on the left, and also does not have a right-hand limit here (informally speaking, because $\sin(1/x)$ oscillates wildly as $x \to 0^+$ rather than settling towards a stable value). To prove that second point properly, you can use an argument involving two sequences of positive numbers both homing in on $x = 0$ but at which the sine-one-over-x function gives two streams of values with different limits. The overall cautionary comment is that we must **never assume that a function has limits of any kind unless there is enough information given** to guarantee that it has. In this connection, also note Exercise 10.3.5.

10.3.4 Example   In reviewing the work we did earlier on the floor or integer part $\lfloor x \rfloor$ of x we notice that, in the new notation, our observations amounted to:

$$\lim_{x \to p^+} \lfloor x \rfloor = p, \quad \lim_{x \to p^-} \lfloor x \rfloor = p - 1$$

for every integer p.

10.3.5 EXERCISE   Recall that a function f on an interval (a, b) is said to be *increasing* if (throughout the interval)

$$x < y \Rightarrow f(x) \le f(y).$$

Show that if f is both increasing and bounded on (a, b) then $\lim_{x \to b^-} f(x)$ and $\lim_{x \to a^+} f(x)$ must both exist.

Partial solution

Think about the supremum and the infimum of the values of $f(x)$ on the interval, and argue as in the proof that a bounded monotonic *sequence* has to converge.

10.3.6 Theorem – a sequence description of right-hand limits   Let $f : D \to C$ be a function and p a limit point of $D \cap (p, \infty)$. Then the following are equivalent:

1. $$\lim_{x \to p^+} f(x) = \ell,$$

2. for every sequence $(x_n)_{n \in \mathbb{N}}$ of elements of $D \cap (p, \infty)$ that tends to p, we have $f(x_n) \to \ell$.

Proof

The proof amounts to little more than a re-run of that of 10.1.2 but ensuring that every x_n shall be greater than p.

10.3.7 The algebra of limits for functions as $x \to p^+$ Suppose that $f : D \to C$ and $g : B \to A$ are two functions, that p is a limit point of the intersection $D \cap B \cap (p, \infty)$, and that (as $x \to p^+$) $f(x) \to \ell$ and $g(x) \to m$. Then (as $x \to p^+$):

1. $f(x) + g(x) \to \ell + m$,
2. $f(x) - g(x) \to \ell - m$,
3. $f(x)g(x) \to \ell m$,
4. for constant k, $kf(x) \to k\ell$,
5. provided that $m \neq 0$,

$$\frac{f(x)}{g(x)} \to \frac{\ell}{m},$$

6. $|f(x)| \to |\ell|$.

Partial proof

As a sample, let us set up a proof of part (3). For any sequence (x_n) in $D \cap B \cap (p, \infty)$ such that $x_n \to p$ we know (from 10.3.6) that $f(x_n) \to \ell$ and that $g(x_n) \to m$. The algebra of limits for sequences tells us that $f(x_n)g(x_n) \to \ell m$. Thus we see (from 10.3.6 again) that $f(x)g(x) \to \ell m$ as $x \to p^+$.

10.3.8 Theorem: a squeeze or sandwich rule for function limits as $x \to p^+$ Suppose that $f : D \to \mathbb{R}$, $g : D' \to \mathbb{R}$, $h : D'' \to \mathbb{R}$ are three functions, that $D' \cap (p, \infty) \subseteq D \cap D''$ (that is, wherever g is defined at a number greater than p, so are f and h), that p is a limit point of $D' \cap (p, \infty)$, that $f(x) \leq g(x) \leq h(x)$ for each $x \in D' \cap (p, \infty)$ and that

$$\lim_{x \to p^+} f(x) = \lim_{x \to p^+} h(x) = \ell, \text{ say.}$$

Then also

$$\lim_{x \to p^+} g(x) = \ell.$$

EXERCISE

Write out a proof of 10.3.8.

All of this material can be tweaked routinely to apply also to left-hand limits, and the proofs are routine variations of what we have already done.

The last result we set out in this chapter concerns how the two one-sided limits (of a function, at a point), if they both exist, can either agree to create a limit in the full sense, or disagree to prevent a limit (in the full sense) existing.

10.3.9 Theorem   Let $f : D \to C$ be a function and p be a limit point both of $D \cap (-\infty, p)$ and of $D \cap (p, \infty)$. Then $f(x) \to \ell$ as $x \to p$ if, and only if, both of the one-sided limits

$$\lim_{x \to p^-} f(x), \quad \lim_{x \to p^+} f(x)$$

exist and are equal to ℓ.

Proof

Suppose that $f(x) \to \ell$ as $x \to p$. Given positive ε,[3] it is therefore possible to find $\delta > 0$ such that

$$x \in D, 0 < |x - p| < \delta \text{ together imply } |f(x) - \ell| < \varepsilon.$$

In particular,

$$x \in D, p < x < p + \delta \text{ together imply } |f(x) - \ell| < \varepsilon,$$

and since ε was arbitrary, we deduce that $f(x) \to \ell$ as $x \to p^+$. Yet it is equally a consequence that

$$x \in D, p - \delta < x < p \text{ together imply } |f(x) - \ell| < \varepsilon,$$

so, ε being arbitrary, we also get $f(x) \to \ell$ as $x \to p^-$.

Conversely, suppose that both $\lim_{x \to p^-} f(x)$ and $\lim_{x \to p^+} f(x)$ exist and equal ℓ. Given $\varepsilon > 0$, we use these facts to identify $\delta > 0$ and $\delta' > 0$ for which

$$x \in D, p - \delta < x < p \Rightarrow |f(x) - \ell| < \varepsilon \text{ and } x \in D, p < x < p + \delta' \Rightarrow |f(x) - \ell| < \varepsilon.$$

Put $\delta'' = \min\{\delta, \delta'\}$ and we see that

$$x \in D, 0 < |x - p| < \delta'' \Rightarrow |f(x) - \ell| < \varepsilon$$

whether x is less than p or greater. Therefore $f(x) \to \ell$ as $x \to p$.

10.3.10 Example We show that the following function

$$g(x) = \begin{cases} 3x^2 + 5x - 1 & \text{if } x < -2, \\ 5x + 11 & \text{if } x \geq -2 \end{cases}$$

possesses a limit at -2.

Solution

For any sequence (x_n) of numbers less than -2 that converges to -2, we have

$$g(x_n) = 3x_n^2 + 5x_n - 1 \to 3(-2)^2 + 5(-2) - 1 = 1$$

---

[3] Somewhat unusually, the ε-style definition is more convenient in this demonstration than the sequence-style alternative.

by the algebra of limits for sequences, and therefore $\lim_{x \to -2^-} g(x)$ exists and equals 1 via the natural sequence description of left-hand limits. Likewise, for any sequence (x_n) of numbers greater than -2 that converges to -2, we have

$$g(x_n) = 5x_n + 11 \to 5(-2) + 11 = 1.$$

By the preceding Theorem 10.3.9, the 'two-sided limit' $\lim_{x \to -2} g(x)$ exists (and equals 1).

Alternative solution

For $x < -2$, the function g and the (continuous) quadratic $3x^2 + 5x - 1$ are identical, so

$$\lim_{x \to -2^-} g(x) = \lim_{x \to -2^-} (3x^2 + 5x - 1) = \lim_{x \to -2} (3x^2 + 5x - 1) = 1$$

(where we tacitly used 10.3.9 to switch from $\lim_{x \to -2^-}$ to $\lim_{x \to -2}$ for the quadratic).
    Similarly

$$\lim_{x \to -2^+} g(x) = \lim_{x \to -2^+} (5x + 11) = \lim_{x \to -2} (5x + 11) = 1.$$

Now 10.3.9 shows that $g(x) \to 1$ as $x \to -2$.

10.3.11 **EXERCISE** Given that the function

$$f(x) = \begin{cases} \frac{1}{x^2+1} & \text{if } x < 2, \\ ax + b & \text{if } 2 \le x \le 5, \\ \frac{x^2}{x^2-21x+110} & \text{if } 5 < x < 10 \end{cases}$$

has limits at $x = 2$ and at $x = 5$, determine the values of the constants a and b.

11 Infinity and function limits

To speak of *infinity* in connection with limits can seem almost contrary to the basic meaning of the word, since its usual import is the complete absence of any limit. Nevertheless there are many natural and simple functions f in which the value $f(x)$ settles towards some kind of stable or equilibrium state *not* as x approaches a particular (finite) number p, but as x becomes enormously big and positive (or enormously big and negative). In many ways this is actually closer to the limiting behaviour of sequences that we first studied, where the focus of attention was on how the typical term x_n behaved as n tended to infinity and, indeed, it is scarcely possible to draw sketch graphs of functions such as x^{-1} or x^{-2} or $\arctan x$ or $\dfrac{x^2 - 1}{x^2 + 1}$ without some phrase about their behaviour *as x tends to infinity* coming to mind.

Our first objective in this chapter is to formulate clear definitions of these ideas and to develop some basic theory concerning them. This will offer us very little difficulty provided we resist the temptation to regard ∞ and $-\infty$ as numbers, or to use senseless symbols such as $|x - \infty|$ to assess how 'close' x is to infinity.

11.1 Limit of a function as *x* tends to infinity or minus infinity

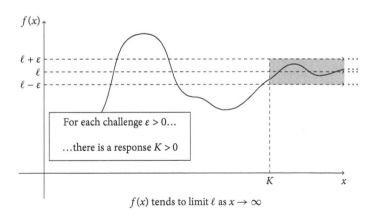

$f(x)$ tends to limit ℓ as $x \to \infty$

Undergraduate Analysis: A Working Textbook, Aisling McCluskey and Brian McMaster 2018.
© Aisling McCluskey and Brian McMaster 2018. Published 2018 by Oxford University Press

11.1.1 Definition   Suppose that $f : D \rightarrow C$ and that its domain D is not bounded above.[1] Then we say that $f(x)$ *converges* to the limit ℓ (or *tends* to ℓ) as $x \rightarrow \infty$ if, for each $\varepsilon > 0$, there is $K \in \mathbb{R}$ such that

$$x \in D, x > K \text{ together imply } |f(x) - \ell| < \varepsilon.$$

This is also written as

$$\lim_{x \to \infty} f(x) = \ell \text{ or as } f(x) \rightarrow \ell \text{ (as } x \rightarrow \infty).$$

It is always safe to assume that K is *positive* in the above definition: for if K were negative or zero, then $x > |K| + 1 \Rightarrow x > K \Rightarrow |f(x) - \ell| < \varepsilon$ while $x \in D$, and $|K| + 1$ certainly is positive.

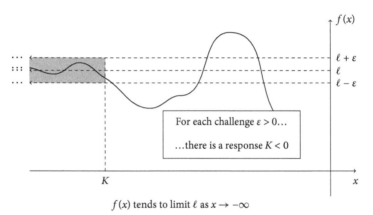

For each challenge $\varepsilon > 0$...

...there is a response $K < 0$

$f(x)$ tends to limit ℓ as $x \rightarrow -\infty$

11.1.2 Definition   Suppose that $f : D \rightarrow C$ and that its domain D is not bounded below.[2] Then we say that $f(x)$ *converges* to the limit ℓ (or *tends* to ℓ) as $x \rightarrow -\infty$ if, for each $\varepsilon > 0$, there is $K \in \mathbb{R}$ such that

$$x \in D, x < K \text{ together imply } |f(x) - \ell| < \varepsilon.$$

This is also written as

$$\lim_{x \to -\infty} f(x) = \ell \text{ or as } f(x) \rightarrow \ell \text{ (as } x \rightarrow -\infty).$$

---

[1] This is just to ensure that there are arbitrarily big (positive) values of x for which $f(x)$ makes sense.
[2] This is just to ensure that there are arbitrarily big negative values of x for which $f(x)$ makes sense.

It is always safe to assume that K is *negative* in this second definition: for if K were positive or zero, then $x < -|K| - 1 \Rightarrow x < K \Rightarrow |f(x) - \ell| < \varepsilon$ while $x \in D$, and $-|K| - 1$ certainly is negative.

11.1.3 Example To show that the function f described by the formula

$$f(x) = \frac{x^3}{1 - x^3}$$

converges to -1 as $x \to \infty$, and also converges to -1 as $x \to -\infty$.

Solution

The domain of f is $(-\infty, 1) \cup (1, \infty)$ which is neither bounded above nor bounded below, so both questions are legitimate.

Provided that $x > 1$, we have

$$|f(x) - (-1)| = \left| \frac{1}{1 - x^3} \right| < \varepsilon \iff x^3 - 1 > \varepsilon^{-1} \iff x > \sqrt[3]{1 + \varepsilon^{-1}}$$

so, for a given $\varepsilon > 0$, if we choose $K = \max\{1, \sqrt[3]{1 + \varepsilon^{-1}}\}$,[3] we get $|f(x) - (-1)| < \varepsilon$ whenever $x > K$, as required to show $f(x) \to -1$ as $x \to \infty$.

Provided that $x < 1$, we next have

$$|f(x) - (-1)| = \left| \frac{1}{1 - x^3} \right| < \varepsilon \iff 1 - x^3 > \varepsilon^{-1} \iff x < \sqrt[3]{1 - \varepsilon^{-1}}$$

so, with $\varepsilon > 0$ given, if we choose $K = \min\{1, \sqrt[3]{1 - \varepsilon^{-1}}\}$,[4] we find $|f(x) - (-1)| < \varepsilon$ whenever $x < K$, as needed to demonstrate that $f(x) \to -1$ as $x \to -\infty$.

11.1.4 EXERCISE Confirm that the function $\tanh x$ defined by the formula

$$\tanh x = \frac{e^x - e^{-x}}{e^x + e^{-x}}$$

has a limit of 1 as $x \to \infty$, and has a limit of -1 as $x \to -\infty$. (You may assume the well-known properties of exponential and logarithmic functions.)

We could explore more technical examples of function convergence as x tends to infinity or minus infinity but, once again, it is largely unnecessary to do so because these ideas can be converted conveniently into sequence convergence instead. In the case of limits as $x \to \infty$, the key theorem is as follows:

---

[3] Actually, this piece of notation is heavier than it need be, since $\sqrt[3]{1 + \varepsilon^{-1}}$ is greater than 1.
[4] Again, this notation is unnecessarily heavy-handed, since $\sqrt[3]{1 - \varepsilon^{-1}}$ is clearly less than 1.

11.1.5 Theorem   Let $f : D \to C$ be a function whose domain D is not bounded above. Then the following are equivalent:

1.
$$\lim_{x \to \infty} f(x) = \ell,$$

2. for every sequence $(x_n)_{n \in \mathbb{N}}$ of elements of D that tends to ∞, we have $f(x_n) \to \ell$.

Proof

(I): (1) implies (2).

- Suppose that condition (1) is satisfied.
- Let $(x_n)_{n \in \mathbb{N}}$ be an arbitrary sequence of elements of D such that $x_n \to \infty$.
- For a given positive value of ε, use condition (1) to obtain a number K such that whenever $x \in D$ and $x > K$, we have $|f(x) - \ell| < \varepsilon$.
- Because $x_n \to \infty$, there is a positive integer n_0 such that $n \geq n_0$ will guarantee that $x_n > K$.
- Therefore
$$n \geq n_0 \Rightarrow x_n > K \Rightarrow |f(x_n) - \ell| < \varepsilon.$$
- That is, $f(x_n) \to \ell$ as required.
- Since (x_n) was *any* sequence in D that happened to tend to ∞, condition (2) is now proved.

(II): (2) implies (1).

- Suppose that condition (1) is not satisfied.
- That is, there exists a value of $\varepsilon > 0$ for which *no* suitable K can be found.
- In particular, for each $n \in \mathbb{N}$, $K = n$ is not suitable...
- ... and so there is $x_n \in D$ such that $x_n > n$ and yet $|f(x_n) - \ell| \geq \varepsilon$.
- Therefore $(x_n)_{n \in \mathbb{N}}$ tends to ∞ (with each $x_n \in D$), and yet $(f(x_n))_{n \in \mathbb{N}}$ does not converge to ℓ.
- That is, condition (2) is not satisfied.
- By contraposition, (2) implies (1).

Use of this theorem is often a convenient way to deal with a problem on function limits as $x \to \infty$, and also to prove the (predictable) theorems about those limits, such as:

11.1.6 The algebra of limits for functions as $x \to \infty$ Suppose that $f : D \to C$ and $g : B \to A$ are two functions, that the intersection $D \cap B$ of their domains is not bounded above, and that (as $x \to \infty$) $f(x) \to \ell$ and $g(x) \to m$. Then (as $x \to \infty$):

1. $f(x) + g(x) \to \ell + m$,
2. $f(x) - g(x) \to \ell - m$,

3. $f(x)g(x) \to \ell m$,

4. for constant k, $kf(x) \to k\ell$,

5. provided that $m \neq 0$,

$$\frac{f(x)}{g(x)} \to \frac{\ell}{m},$$

6. $|f(x)| \to |\ell|$.

11.1.7 **EXERCISE** Select any part of this theorem and give a proof of it using sequences.

11.1.8 **Remark** It is, of course, perfectly possible for the limit of a function $f(x)$ as $x \to \infty$ or as $x \to -\infty$ not to exist at all. For instance (and assuming the basic behaviour of trigonometric functions) both of the sequences

$$(n\pi)_{n \in \mathbb{N}} \quad \text{and} \quad ((2n + 0.5)\pi)_{n \in \mathbb{N}}$$

tend to ∞, while $(\sin(n\pi))_{n \in \mathbb{N}}$ converges to 0 whereas $(\sin((2n + 0.5)\pi))_{n \in \mathbb{N}}$ converges to 1. In the light of Theorem 11.1.5, this shows that $\lim_{x \to \infty} \sin x$ does not exist.

11.1.9 **Theorem: a squeeze or sandwich rule for function limits as $x \to \infty$** Suppose that $f : D \to \mathbb{R}$, $g : D' \to \mathbb{R}$, $h : D'' \to \mathbb{R}$ are three functions, that $D' \subseteq D \cap D''$ (that is, wherever g is defined, so are f and h), that D' is not bounded above, that $f(x) \leq g(x) \leq h(x)$ for each $x \in D'$ and that

$$\lim_{x \to \infty} f(x) = \lim_{x \to \infty} h(x) = \ell, \text{ say.}$$

Then also

$$\lim_{x \to \infty} g(x) = \ell.$$

11.1.10 **EXERCISE** Construct a proof of this theorem.

The analogous results for function limits as $x \to -\infty$ are almost too obvious even to state but, for the sake of completeness, here is the basic set. If you wish, feel free to prove any of them as an additional exercise. (The first theorem is the one that would be most worthwhile to try proving, since the others are very routine.)

11.1.11 **Theorem** Let $f : D \to C$ be a function whose domain D is not bounded below. Then the following are equivalent:

1.
$$\lim_{x \to -\infty} f(x) = \ell,$$

2. for every sequence $(x_n)_{n \in \mathbb{N}}$ of elements of D that tends to $-\infty$, we have $f(x_n) \to \ell$.

11.1.12 The algebra of limits for functions as $x \to -\infty$ Suppose that $f : D \to C$ and $g : B \to A$ are two functions, that the intersection $D \cap B$ of their domains is not bounded below, and that (as $x \to -\infty$) $f(x) \to \ell$ and $g(x) \to m$. Then (as $x \to -\infty$):

1. $f(x) + g(x) \to \ell + m$,
2. $f(x) - g(x) \to \ell - m$,
3. $f(x)g(x) \to \ell m$,
4. for constant k, $kf(x) \to k\ell$,
5. provided that $m \neq 0$,

$$\frac{f(x)}{g(x)} \to \frac{\ell}{m},$$

6. $|f(x)| \to |\ell|$.

11.1.13 Theorem: a squeeze or sandwich rule for function limits as $x \to -\infty$
Suppose that $f : D \to \mathbb{R}, g : D' \to \mathbb{R}, h : D'' \to \mathbb{R}$ are three functions, that $D' \subseteq D \cap D''$ (that is, wherever g is defined, so are f and h), that D' is not bounded below, that $f(x) \leq g(x) \leq h(x)$ for each $x \in D'$ and that

$$\lim_{x \to -\infty} f(x) = \lim_{x \to -\infty} h(x) = \ell, \text{ say.}$$

Then also

$$\lim_{x \to -\infty} g(x) = \ell.$$

It is sometimes useful to convert a problem concerning a limit 'at infinity' or 'at minus infinity' into a one-sided limit at an ordinary real number. For instance, to examine the limiting behaviour of $\sin\left(\frac{1}{x}\right)$ as $x \to \infty$, it is tempting to argue informally that when x is very big, $\frac{1}{x}$ is very small (but positive) and so we are actually dealing with $\sin(t)$ as $t \to 0^+$, and it is now easy to determine this limit. (See 11.1.15 for another illustration of why this can be convenient.) Here is a suggested exercise – there are several such possibilities – that legitimises this kind of translation between function limits 'at infinity' and 'at zero'.

11.1.14 EXERCISE (I) Given a function $f : D \to C$ whose domain D is unbounded above, we define an associated function g as follows:

$$g(x) = f\left(\frac{1}{x}\right).$$

Show that the following are equivalent:

1. $f(x) \to \ell$ as $x \to \infty$,
2. $g(x) \to \ell$ as $x \to 0^+$.

(II) Given a function $f : D \to C$ whose domain D is unbounded both above and below, we define an associated function g as follows:

$$g(x) = f\left(\frac{1}{x}\right).$$

Show that the following are equivalent:

1. $f(x) \to \ell$ as $x \to \infty$ and $f(x) \to \ell$ as $x \to -\infty$,
2. $g(x) \to \ell$ as $x \to 0$.

11.1.15 Example   To show that $\dfrac{1}{x^2} \to 0$ as $x \to \infty$, and to deduce the limit (as $x \to \infty$) of the function

$$q(x) = \frac{x^2 + \sin x}{x^2 + \cos x}, \quad x \in (1, \infty).$$

Solution

By part (I) of 11.1.14, the statement $x^{-2} \to 0$ as $x \to \infty$ is equivalent to the statement $(1/x)^{-2} \to 0$ as $x \to 0^+$, that is, to $x^2 \to 0$ as $x \to 0^+$. Yet the truth of the latter is immediate from the continuity of x^2.

Next, for the usual trigonometric reasons,[5] we have (for $x > 1$)

$$\frac{x^2 - 1}{x^2 + 1} \le q(x) \le \frac{x^2 + 1}{x^2 - 1}.$$

Also, using parts of the algebra of limits theorem 11.1.6:

$$\frac{x^2 - 1}{x^2 + 1} = \frac{1 - x^{-2}}{1 + x^{-2}} \to \frac{1 - 0}{1 + 0} = 1 \quad \text{as } x \to \infty,$$

$$\frac{x^2 + 1}{x^2 - 1} = \frac{1 + x^{-2}}{1 - x^{-2}} \to \frac{1 + 0}{1 - 0} = 1 \quad \text{as } x \to \infty.$$

Lastly, the 11.1.9 version of the squeeze gives us the desired conclusion $q(x) \to 1$ as $x \to \infty$.

11.1.16 EXERCISE   Determine the limit as $x \to \infty$, and the limit as $x \to -\infty$, of

$$\frac{x^3 + 5 + (3x^2 - 7x - 2) \sin x}{x^3 + 5 + (3x^2 - 7x - 2) \cos x}.$$

---

[5] and using multiplication of inequalities, as flagged up in the checklist of 1.2

11.2 Functions tending to infinity or minus infinity

Just as, at the end of Chapter 2, we needed to give proper definitions to the ideas of sequences tending to infinity or to minus infinity, it is now useful to do the same for functions; indeed, it is even more obviously necessary for functions than it was for sequences – it is barely possible to draw a sketch graph of, say, $\tan x$ just to the left of $x = \pi/2$, or x^{-1} just to the right or just to the left of zero, or e^x for large values of x, without at least some informal notion of the expression 'heading off to infinity or minus infinity' arising.

A complicating factor this time is that, when we want to discuss limiting behaviour of a function $f(x)$, there are (as we have seen) at least five different kinds of thing that the 'control' variable x might be doing: tending to p, one-sidedly tending to p from the left, or from the right, tending to infinity, or tending to minus infinity. Combine those with the two outcome scenarios in which $f(x)$ tends to infinity and in which it tends to minus infinity, and we face the prospect of an array of ten more very similar definitions! We shall, of course, list all ten of them for the sake of completeness, but from the reader's point of view it is much more important to understand how they are put together, rather than to spend time attempting to memorise them.

It may be useful to imagine that *restricting the permitted values of x is a 'cause'* and *observing the resulting range of possible values of $f(x)$ is an 'effect'*. The interplay between cause and effect is then a mental image for any limiting scenario, including the ones that we first discussed in Chapter 2 and Chapter 9. When we defined the idea '$f(x) \to \ell$ as $x \to p$', what we effectively said was that the effect '$f(x)$ shall be as close to ℓ as is demanded' can be brought about by the cause 'x is sufficiently close to, but distinct from, p'. More precisely, for any challenge $\varepsilon > 0$ we can somehow determine a response $\delta > 0$ so that the desired effect

$$|f(x) - \ell| < \varepsilon$$

can be brought about by the (carefully designed) cause

$$0 < |x - p| < \delta.$$

Now that we have divided out the two aspects of any limiting process, we can look separately at the conditions that can be imposed upon them to express various convergence/divergence behaviours of functions, including those that we have already studied.

For the effects:

- $f(x)$ converges to ℓ means $f(x)$ gets very close to ℓ: $|f(x) - \ell| < \varepsilon$ for any given positive ε;
- $f(x)$ diverges to ∞ means $f(x)$ gets extremely big and positive: $f(x) > K$ for any given real K;

- $f(x)$ diverges to $-\infty$ means $f(x)$ gets extremely big but negative: $f(x) < K$ for any given real K.

For the causes:

- x tends to p means x gets very close to but distinct from p: $0 < |x - p| < \delta$ for a suitably chosen positive δ;
- x tends to p^- means x gets very close to p from the left but remains distinct from p: $p - \delta < x < p$ for a suitably chosen positive δ;
- x tends to p^+ means x gets very close to p from the right but remains distinct from p: $p < x < p + \delta$ for a suitably chosen positive δ;
- x tends to infinity means x becomes very big and positive: $x > K'$ for a suitably chosen real number K' (which we can assume to be positive);
- x tends to minus infinity means x becomes very big but negative: $x < K'$ for a suitably chosen real number K' (which we can assume to be negative).

(Be careful not to use the same symbol K in cause and effect if you are combining an infinity-type cause with an infinity-type effect. This is why we used K' instead of K on the last few lines.)

Now let's assemble one of the new definitions: say, that of $f(x)$ diverging to minus infinity as x approaches p one-sidedly from the right. The desired effect is $f(x) < K$ for any given K. The appropriate cause is $p < x < p + \delta$ for some suitable δ chosen in response to the challenge K. Combining:

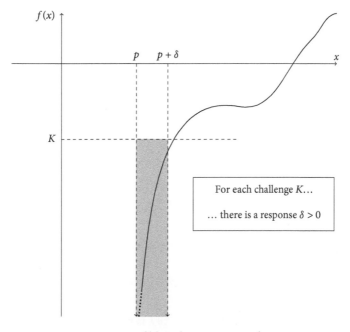

$f(x)$ tends to $-\infty$ as $x \to p^+$

11.2.1 Definition We say that $f(x) \to -\infty$ as $x \to p^+$ if, for each $K \in \mathbb{R}$, there is some $\delta > 0$ such that

$$p < x < p + \delta \Rightarrow f(x) < K.$$

We also then write $\lim_{x \to p^+} f(x) = -\infty$.

(The definition assumes that the domain of f includes points in every open interval that has p as its left-hand endpoint, in order that $f(x)$ shall make sense for some x such that $p < x < p + \delta$ no matter how small we choose δ. In other words, p is assumed to be a limit point of $D \cap (p, \infty)$ where D is the domain of f.)

Another: what should be the official definition of $f(x)$ diverging to infinity as x tends to minus infinity? The desired effect is $f(x) > K$ for any given K. The appropriate cause is $x < K'$ for some suitable K' chosen in response to the challenge K. Combining:

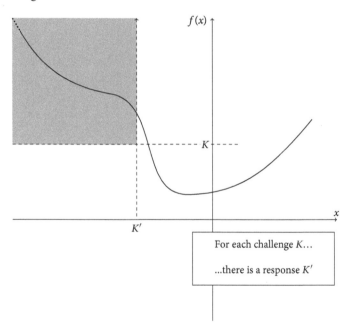

$f(x)$ tends to ∞ as $x \to -\infty$

11.2.2 Definition We say that $f(x) \to \infty$ as $x \to -\infty$ if, for each $K \in \mathbb{R}$, there is some $K' \in \mathbb{R}$ such that
$$x < K' \Rightarrow f(x) > K.$$
We also then write $\lim_{x \to -\infty} f(x) = \infty$.

(The definition assumes that the domain of f is not bounded below, in order that $f(x)$ shall make sense for some values of x that are arbitrarily big and negative.)

11.2.3 Example Verify that $\dfrac{x}{1-x} \to -\infty$ as $x \to 1^+$.

Roughwork

Since $\dfrac{x}{1-x} = -1 + \dfrac{1}{1-x}$, we ask, given K and thinking of x as being slightly greater than 1, how do we contrive that $-1 + \dfrac{1}{1-x} < K$? That is, $\dfrac{1}{1-x} < K+1$? Make sure for a start that K is strictly less than -1 to keep the signs unambiguous, and this is then the same as $1-x > (1+K)^{-1}$, that is, $x < 1 - \dfrac{1}{1+K} = 1 + \dfrac{1}{|K|-1}$.

Solution

Given K, we first choose $K^* = \min\{-2, K\}$ (just to guarantee that the 'new' K shall be strictly less than -1). Then, as the roughwork shows, the choice $\delta = \dfrac{1}{|K^*|-1}$ will ensure that

$$1 < x < 1 + \delta \Rightarrow \frac{x}{1-x} < K^* \leq K$$

in line with the requirements.

11.2.4 **Example** Verify that $x^2 \to \infty$ as $x \to -\infty$.

Solution

Given $K \in \mathbb{R}$ (and arranging if necessary that $K > 0$ so that the next step makes sense), we choose $K' = -\sqrt{K}$ (< 0). Then

$$x < K' \ (< 0) \Rightarrow x^2 > (K')^2 = K$$

in line with the requirements.

Provided that you understand how we engineered the last two definitions, you probably don't even need to read the other eight. For the sake of completeness, however, we shall set out the full collection of ten. In each case, the definition tacitly assumes that the function f is defined on a subset of $\mathbb{R}$ that is extensive enough for the conclusion about $f(x)$ to make sense.

11.2.5 **Definition** We say that $f(x) \to \infty$ as $x \to p$ if, for each $K \in \mathbb{R}$, there is some $\delta > 0$ such that

$$0 < |x - p| < \delta \Rightarrow f(x) > K.$$

We also then write $\lim_{x \to p} f(x) = \infty$.

11.2.6 **Definition** We say that $f(x) \to \infty$ as $x \to p^+$ if, for each $K \in \mathbb{R}$, there is some $\delta > 0$ such that
$$p < x < p + \delta \Rightarrow f(x) > K.$$
We also then write $\lim_{x \to p^+} f(x) = \infty$.

11.2.7 Definition   We say that $f(x) \to \infty$ as $x \to p^-$ if, for each $K \in \mathbb{R}$, there is some $\delta > 0$ such that

$$p - \delta < x < p \Rightarrow f(x) > K.$$

We also then write $\lim_{x \to p^-} f(x) = \infty$.

11.2.8 Definition   We say that $f(x) \to \infty$ as $x \to \infty$ if, for each $K \in \mathbb{R}$, there is some $K' \in \mathbb{R}$ such that

$$x > K' \Rightarrow f(x) > K.$$

We also then write $\lim_{x \to \infty} f(x) = \infty$.

11.2.9 Definition   We say that $f(x) \to \infty$ as $x \to -\infty$ if, for each $K \in \mathbb{R}$, there is some $K' \in \mathbb{R}$ such that

$$x < K' \Rightarrow f(x) > K.$$

We also then write $\lim_{x \to -\infty} f(x) = \infty$.

11.2.10 Definition   We say that $f(x) \to -\infty$ as $x \to p$ if, for each $K \in \mathbb{R}$, there is some $\delta > 0$ such that

$$0 < |x - p| < \delta \Rightarrow f(x) < K.$$

We also then write $\lim_{x \to p} f(x) = -\infty$.

11.2.11 Definition   We say that $f(x) \to -\infty$ as $x \to p^+$ if, for each $K \in \mathbb{R}$, there is some $\delta > 0$ such that

$$p < x < p + \delta \Rightarrow f(x) < K.$$

We also then write $\lim_{x \to p^+} f(x) = -\infty$.

11.2.12 Definition   We say that $f(x) \to -\infty$ as $x \to p^-$ if, for each $K \in \mathbb{R}$, there is some $\delta > 0$ such that

$$p - \delta < x < p \Rightarrow f(x) < K.$$

We also then write $\lim_{x \to p^-} f(x) = -\infty$.

11.2.13 Definition   We say that $f(x) \to -\infty$ as $x \to \infty$ if, for each $K \in \mathbb{R}$, there is some $K' \in \mathbb{R}$ such that

$$x > K' \Rightarrow f(x) < K.$$

We also then write $\lim_{x \to \infty} f(x) = -\infty$.

11.2.14 Definition   We say that $f(x) \to -\infty$ as $x \to -\infty$ if, for each $K \in \mathbb{R}$, there is some $K' \in \mathbb{R}$ such that

$$x < K' \Rightarrow f(x) < K.$$

We also then write $\lim_{x \to -\infty} f(x) = -\infty$.

11.2.15 EXERCISE Show that

$$\lim_{x \to 2} \frac{x+1}{(x-2)^2} = \infty.$$

11.2.16 EXERCISE (Assuming basic facts about the sine function, including that it is continuous) show that the function

$$\operatorname{cosec} x = \frac{1}{\sin x}$$

diverges to $-\infty$ as $x \to 0^-$, and diverges to ∞ as $x \to 0^+$.

By analogy with our key theorems about characterising convergence of functions in terms of limits of sequences, every one of these ten divergence definitions can be translated into sequential limit terms. Rather than plough through ten more theorems of high similarity, we shall set out details of a couple, and invite the interested reader to explore one or two further exemplars.

11.2.17 Theorem Let $f : D \to C$ be a function whose domain D is not bounded above. Then the following are equivalent:

1. $$\lim_{x \to \infty} f(x) = \infty,$$

2. for every sequence $(x_n)_{n \in \mathbb{N}}$ of elements of D that tends to ∞, we have $f(x_n) \to \infty$.

Proof

(I): (1) implies (2).
- Suppose that condition (1) is satisfied.
- Let $(x_n)_{n \in \mathbb{N}}$ be an arbitrary sequence of elements of D such that $x_n \to \infty$.
- For a given K, use condition (1) to obtain a number K' such that whenever $x \in D$ and $x > K'$, we have $f(x) > K$.
- Because $x_n \to \infty$, there is a positive integer n_0 such that $n \geq n_0$ will guarantee that $x_n > K'$.
- Therefore
$$n \geq n_0 \Rightarrow x_n > K' \Rightarrow f(x_n) > K.$$
- That is, $f(x_n) \to \infty$ as required.
- Since (x_n) was *any* sequence in D that happened to tend to ∞, condition (2) is now proved.

(II): (2) implies (1).

- Suppose that condition (1) is not satisfied.
- That is, there exists a value of $K \in \mathbb{R}$ for which *no* suitable $K' \in \mathbb{R}$ can be found.
- In particular, for each $n \in \mathbb{N}$, $K' = n$ is not suitable...
- ...and so there is $x_n \in D$ such that $x_n > n$ and yet $f(x_n) \leq K$.
- Therefore $(x_n)_{n \in \mathbb{N}}$ tends to ∞ (with each $x_n \in D$), and yet $(f(x_n))_{n \in \mathbb{N}}$ does not tend to ∞.
- That is, condition (2) is not satisfied.
- By contraposition, (2) implies (1).

11.2.18 Theorem   Let $f : D \to C$ be a function and p a limit point of $D \cap (p, \infty)$. Then the following are equivalent:

1.
$$\lim_{x \to p^+} f(x) = -\infty,$$

2. for every sequence $(x_n)_{n \in \mathbb{N}}$ of elements of $D \cap (p, \infty)$ that tends to p, we have $f(x_n) \to -\infty$.

Proof

(I): (1) implies (2).

- Suppose that condition (1) is satisfied.
- Let $(x_n)_{n \in \mathbb{N}}$ be an arbitrary sequence of elements of $D \cap (p, \infty)$ such that $x_n \to p$.
- For a given value of K, use condition (1) to obtain a number $\delta > 0$ such that whenever $x \in D$ and $p < x < p + \delta$, we have $f(x) < K$.
- Because $x_n \to p$, there is a positive integer n_0 such that $n \geq n_0$ will guarantee that $x_n < p + \delta$.
- Therefore
$$n \geq n_0 \Rightarrow p < x_n < p + \delta \Rightarrow f(x_n) < K.$$
- That is, $f(x_n) \to -\infty$ as required.
- Since (x_n) was *any* appropriate sequence, condition (2) is now proved.

(II): (2) implies (1).

- Suppose that condition (1) is not satisfied.
- That is, there exists a value of K for which *no* suitable $\delta > 0$ can be found.
- In particular, for each $n \in \mathbb{N}$, $\delta = n^{-1}$ is not suitable...
- ...and so there is $x_n \in D$ such that $p < x_n < p + 1/n$ and yet $f(x_n) \geq K$.
- Therefore $(x_n)_{n \in \mathbb{N}}$ tends to p (with each $x_n \in D$), and yet $(f(x_n))_{n \in \mathbb{N}}$ does not diverge to $-\infty$.

- That is, condition (2) is not satisfied.
- By contraposition, (2) implies (1).

11.2.19 EXERCISE Select two more of the various definitions for $f(x)$ diverging to infinity or minus infinity under constraints upon x, and formulate for each a theorem (like the last two) characterising this divergence in terms of the behaviour of sequences. Give a detailed proof of one of them. Do not expect to enjoy it disproportionately.

11.2.20 EXERCISE Give sequence-based proofs of the last two worked examples 11.2.3 and 11.2.4.

11.2.21 Example To establish the following variant of 11.1.14: if the domain of f is not bounded above, and we define $g(x) = f(1/x)$, to show that the following are equivalent:

- $f(x) \to \infty$ as $x \to \infty$,
- $g(x) \to \infty$ as $x \to 0^+$.

Solution

Suppose firstly that $f(x) \to \infty$ as $x \to \infty$. Given $K \in \mathbb{R}$, we can therefore find $K' \in \mathbb{R}$ such that $x > K' \Rightarrow f(x) > K$. Without loss of generality, $K' > 0$. Then put $\delta = 1/K' > 0$. We have

$$0 < x < \delta \Rightarrow x^{-1} > K' \Rightarrow g(x) = f(x^{-1}) > K,$$

therefore $g(x) \to \infty$ as $x \to 0^+$.

Suppose secondly that $g(x) \to \infty$ as $x \to 0^+$. Given $K \in \mathbb{R}$, we can find $\delta > 0$ such that $0 < x < \delta \Rightarrow g(x) > K$. Put $K' = \delta^{-1}$. Then

$$x > K' \Rightarrow 0 < x^{-1} < (K')^{-1} = \delta \Rightarrow g(x^{-1}) > K$$

$$\Rightarrow f(x) = f\left((x^{-1})^{-1}\right) = g(x^{-1}) > K.$$

That is, $f(x) \to \infty$ as $x \to \infty$.

11.2.22 Example To establish this variant of 11.1.13: given that $f(x) \geq Cg(x)$ for all $x \in (a, b)$ where C is a positive constant, and that $g(x) \to \infty$ as $x \to a^+$, to show that $f(x) \to \infty$ as $x \to a^+$.

Solution

Given $K \in \mathbb{R}$, we use the fact that $g(x) \to \infty$ to find $\delta > 0$ such that $a < x < a + \delta \Rightarrow g(x) > K/C$. It follows that $a < x < a + \delta \Rightarrow f(x) \geq Cg(x) > C(K/C) = K$. Thus $f(x) \to \infty$ as $x \to a^+$ as required.

11.2.23 **EXERCISE** Use the preceding material 11.2.21 and 11.2.22 to show that

$$\lim_{x \to 0^+} \left(\frac{1}{x^2} - \frac{1}{x} \right) = \infty.$$

(*Hint*: on the interval $(0, \frac{1}{2})$, $x^{-2} > 2x^{-1}$.)

12 Differentiation – the slope of the graph

12.1 Introduction

In everyday English, we call a line *straight* if its direction of travel is the same wherever we choose to inspect it, and *curved* if the apparent direction of travel varies as we shift the focus of our attention from one part of it to another. As usual, those ideas need to be sharpened up before we can do significant work with them but, at least in the case of straight lines on a plane surface, this is very elementary: impose a grid of (cartesian) coordinates on the surface, identify two distinct points (x_1, y_1) and (x_2, y_2) on the line, define the gradient or slope between them to be

$$\frac{y_2 - y_1}{x_2 - x_1},$$

and the informal idea of straightness corresponds tidily to the fact that the numerical value of this ratio is the same no matter which two points you have chosen.

In the case of a curved line in the coordinate plane – in particular, of the graph $y = f(x)$ of some function – we can still define the gradient *between* two points on the graph in exactly the same way, but the notion of gradient *at* a typical point is harder to make precise, partly because it is expected to vary with the point. One approach is to draw, or to imagine drawing, a 'tangent' straight line that just skims the curve at a chosen point, and whose gradient then gives a reasonable interpretation of 'gradient of the curve' at that point . . . but such a procedure will always be subject to error, to our limited drawing skill, even to the precision of our instruments and the sharpness of our pencil. Besides, it is very time-consuming to implement, even at a dozen or so points.

What would be really useful here is some routine procedure that could be applied to the formula $f(x)$ – if the curve indeed has such a formula – and that could derive from it another formula for the gradient at any point. Now it is highly probable that you have done enough calculus to be aware of such a procedure, called differentiation, that works excellently upon a wide range of formulas and has several different routines for dealing with the internal structure of formulas that are modestly complicated, and also of important applications such as the identification of regions where a function is increasing or decreasing, and of stationary points,

Undergraduate Analysis: A Working Textbook, Aisling McCluskey and Brian McMaster 2018.
© Aisling McCluskey and Brian McMaster 2018. Published 2018 by Oxford University Press

and maximizing or minimizing variable quantities. Our task in the present chapter is to revisit this idea and look under the bonnet: Why does it work? What functions does it work on? How do we proceed if we cannot access a suitable formula? Are there functions where not only the familiar procedures fail, but the very idea of gradient loses its meaning? What further applications can be anticipated? We make no pretence to give complete answers to such questions, for calculus is a very large field, but we shall make a start (and later chapters will continue aspects of the study).

To determine the slope of the graph at a particular point (say, at the point $P = (a, f(a))$) without having to depend on the uncontrolled approximation of our limited draftsmanship, we consider a nearby point on the same graph (say, $Q = (a + h, f(a + h))$) and the exact gradient of the straight line PQ. This is

$$\frac{f(a+h) - f(a)}{(a+h) - a} = \frac{f(a+h) - f(a)}{h}$$

and, if the curve is reasonably smooth (an idea which we also need to clarify) we should expect this ratio to be an approximation to the gradient of the curve (that is, of the ideal tangent line to the curve) at P, and that it should become a better and better approximation as the horizontal width h of the segment PQ becomes smaller and smaller. This leads us to *define* the gradient of the curve at P to be the limit of this ratio as $h \to 0$, and to replace the intuitive term 'smooth' by the simple mathematical demand that this limit shall exist. (You may also find it useful to re-read an earlier example – see 9.1.2 and 9.2.6 – in which we actually did all this with the curve $y = f(x) = x^2$ at the point $(3, 9)$.) This is precisely what we do with a general function f:

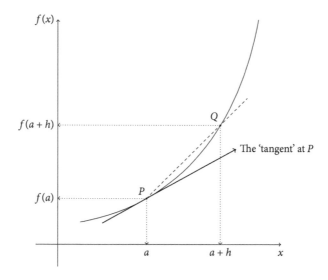

Line PQ approximates 'tangent' at P

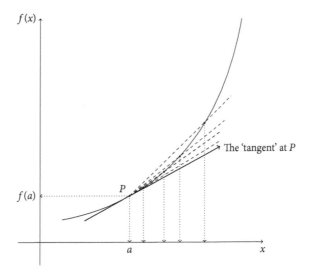

The approximation improves as $h \to 0$

12.2 The derivative

12.2.1 Definition   Given a function f whose domain includes a bounded open interval of the form $(p-d, p+d)$ centred on a point p, we say that f is *differentiable at p* if the limit[1]

$$\lim_{h \to 0} \frac{f(p+h) - f(p)}{h}$$

exists. We then call this limit the *derivative of f at p*, and it is denoted by $f'(p)$ or $\frac{df}{dx}(p)$ (or, indeed, several other notations depending on context or on the preferences of the writer). Any reference to the gradient or slope or direction of the curve or of its tangent at P means this derivative; indeed, we could then *define* the tangent to be the straight line through P whose gradient is this derivative. The process of obtaining a derivative is called *differentiation*.

Many functions turn out to be differentiable not merely at one or several points, but at every point of an interval or even of the entire domain of the function. When this is the case, we say that f is differentiable *on* the interval or domain in question, and the derivative $f'(x)$ at a typical point x where it makes sense is then itself a function (sometimes also called the *derived function* of f) defined on that interval or on that domain. When no interval or domain is mentioned, the phrase 'f is

---

[1] or, equivalently, the limit

$$\lim_{x \to p} \frac{f(x) - f(p)}{x - p}$$

differentiable' usually means that f is differentiable at every point of its domain of definition.

We establish differentiability of a handful of simple functions:

12.2.2 **Proposition**   A constant function is differentiable, and its derivative is (constantly) zero.

Proof

When $f(x) = k$ for every x in some open interval centred on p, k being a constant, the limit

$$\lim_{h\to 0} \frac{f(p+h) - f(p)}{h} = \lim_{h\to 0} \frac{k - k}{h} = \lim_{h\to 0} \frac{0}{h} = 0$$

as predicted. (Notice that we had to assume that $|h|$ was less than the half-width of the open interval, but that this is not a problem since h is tending to zero.)

12.2.3 **Proposition**   An identity function $f(x) = x$ is differentiable, and its derivative is (constantly) 1.

Proof

When $f(x) = x$ for every x in some open interval centred on p, we get

$$\lim_{h\to 0} \frac{f(p+h) - f(p)}{h} = \lim_{h\to 0} \frac{(p+h) - p}{h} = \lim_{h\to 0} \frac{h}{h} = 1$$

as predicted. (Again, there is the tacit assumption that $|h|$ is small enough to put $p + h$ inside the interval where we know f is identity.)

12.2.4 **Proposition**   The function $f(x) = x^2$ is differentiable, and its derived function is $f'(x) = 2x$.

Proof

Calculation of the relevant limit at a typical point p gives

$$\lim_{h\to 0} \frac{f(p+h) - f(p)}{h} = \lim_{h\to 0} \frac{(p+h)^2 - p^2}{h} = \lim_{h\to 0} \frac{2ph + h^2}{h} = \lim_{h\to 0}(2p+h) = 2p.$$

Re-writing that to express the discovery that the derivative at any point is twice the x-value there, we normally write $f'(x) = 2x$ (although $f'(p) = 2p$ for each $p \in \mathbb{R}$ is equally correct).

12.2.5 **Proposition**   For any positive integer n, the function $f(x) = x^n$ is differentiable, and its derived function is $f'(x) = nx^{n-1}$.

Proof

It's really the same argument as the others, except that this time we need the binomial theorem to unscramble the algebra:

$$\frac{(p+h)^n - p^n}{h} = \frac{p^n + np^{n-1}h + \frac{1}{2}n(n-1)p^{n-2}h^2 + \cdots + h^n - p^n}{h}$$

$$= \frac{np^{n-1}h + \frac{1}{2}n(n-1)p^{n-2}h^2 + \cdots + h^n}{h}$$

$$= np^{n-1} + \frac{1}{2}n(n-1)p^{n-2}h + \cdots + h^{n-1}$$

whose limit, as $h \to 0$, is

$$np^{n-1} + 0 + 0 + \cdots + 0 = np^{n-1}.$$

So the derived function, the derivative, is $f'(p) = np^{n-1}$ for each $p \in \mathbb{R}$ or, in slightly more familiar function notation, $f'(x) = nx^{n-1}$.

We could continue this catalogue of special cases, but once again it will be more profitable and efficient to develop rules for processing 'built-up' functions based on the derivatives of the simple components out of which they have been assembled. Most of these rules should already be familiar to you, but possibly not the reasons why they are valid, so we shall discuss them in fair detail.

12.2.6 Differentiation of a sum, of a difference, and of a scaled function If f and g are both differentiable at p, and k is a constant, then

1. $f + g$ is differentiable at p, and its derivative is $f'(p) + g'(p)$;
2. kf is differentiable at p, and its derivative is $kf'(p)$;
3. $f - g$ is differentiable at p, and its derivative is $f'(p) - g'(p)$.

Proof

For 1, notice that (for any sufficiently small $h \neq 0$)

$$\frac{(f+g)(p+h) - (f+g)(p)}{h} = \frac{f(p+h) + g(p+h) - f(p) - g(p)}{h}$$

$$= \frac{f(p+h) - f(p)}{h} + \frac{g(p+h) - g(p)}{h}$$

whose limit, as $h \to 0$, is $f'(p) + g'(p)$, as expected.

EXERCISE: check part 2 (on very similar lines).

Part 3 follows from parts 1 and 2, by expressing $f - g$ as $f + (-1)g$.

To prepare for the next two rules, we need to connect with material from Chapter 8 (and to keep in mind that a function f is continuous at a suitable point p precisely when the limit of $f(x)$ as $x \to p$ coincides with the value $f(p)$ of f at p – see 9.2.21 for this important insight):

12.2.7 **Theorem – differentiable implies continuous**   If a function f is differentiable at a point p, then f is continuous at p.

Proof

Given that the limit

$$\lim_{h \to 0} \frac{f(p+h) - f(p)}{h} = f'(p)$$

exists, all we need do is to notice that

$$\left(\frac{f(p+h) - f(p)}{h} \right) \times h + f(p) = f(p+h),$$

and now take limits (as $h \to 0$) across this equality. We discover that

$$f(p+h) \to f'(p) \times 0 + f(p) = f(p) \text{ as } h \to 0,$$

which, letting x stand for $p+h$, is the same as saying that $f(x) \to f(p)$ as $x \to p$.

12.2.8 **EXERCISE**   Show that the converse of this result is not true, by checking that the function $m(x) = |x|$ is continuous at $x = 0$ but *not* differentiable at $x = 0$. You will almost certainly use one-sided limits (both of $m(x)$ itself, and of $(m(0+h) - m(0))/h$) since the behaviour of m (that is, $m(x) = x$ if $x \geq 0$, but $m(x) = -x$ if $x < 0$) is different on the two sides of $x = 0$.

12.2.9 **Differentiation – the product rule**   If f and g are both differentiable at p, then so is their product fg, and its derivative there is

$$(fg)'(p) = f(p)g'(p) + f'(p)g(p).$$

Proof

We need to evaluate the limit of

$$\frac{f(p+h)g(p+h) - f(p)g(p)}{h}$$

and our initial problem is that the two halves of the top line have nothing in common. Remembering what happened in a similar *impasse* many pages ago, we bring in an extra term that does have a common factor with each half, thus:

$$= \frac{f(p+h)g(p+h) - f(p+h)g(p) + f(p+h)g(p) - f(p)g(p)}{h}$$

$$= f(p+h)\left(\frac{g(p+h)-g(p)}{h}\right) + g(p)\left(\frac{f(p+h)-f(p)}{h}\right).$$

Here is the moment when we need the 'differentiable implies continuous' theorem. As $h \to 0, f(p+h) \to f(p)$, and the other components of the last display have their more obvious limits; so we get

$$\frac{f(p+h)g(p+h) - f(p)g(p)}{h} \to f(p)g'(p) + g(p)f'(p)$$

as expected.

12.2.10 **Differentiation – the quotient rule**   If f and g are both differentiable at p and $g(p) \neq 0$, then their quotient f/g is differentiable at p, and its derivative there is

$$\left(\frac{f}{g}\right)'(p) = \frac{g(p)f'(p) - f(p)g'(p)}{(g(p))^2}.$$

Partial proof

We need to evaluate the limit of

$$\frac{\frac{f(p+h)}{g(p+h)} - \frac{f(p)}{g(p)}}{h} = \frac{f(p+h)g(p) - f(p)g(p+h)}{hg(p+h)g(p)}$$

and, firstly, we must make sure that no division by zero can happen. It is built into the definition of differentiability that there are already two open intervals centred on p in which f and g are defined. Also, using again the 'differentiable implies continuous' theorem, g is continuous at p and $g(p) \neq 0$, so there is[2] a third open interval centred on p throughout which g does not take the value zero. Take now the smallest of the three intervals: here, f and g are defined and g is non-zero so, once h is small enough to put $p + h$ into that interval, no risk remains of our piece of algebra failing to make sense.

Next, we again introduce an extra term that has something in common with each half of the top line:

[2] If you do not find this to be sufficiently convincing, here is a fuller argument. Since $|g(p)| > 0$ and $g(p+h) \to g(p)$ as $h \to 0$, there is $\delta > 0$ such that whenever $|h| < \delta$ we get $|g(p+h)-g(p)| < |g(p)|$. That last inequality forces $g(p+h) \neq 0$, and it holds whenever $p + h$ lies in the interval $(p - \delta, p + \delta)$.

$$\cdots = \frac{f(p+h)g(p) - f(p)g(p) + f(p)g(p) - f(p)g(p+h)}{hg(p+h)g(p)}$$
$$= \frac{\{f(p+h)g(p) - f(p)g(p)\} + \{f(p)g(p) - f(p)g(p+h)\}}{hg(p+h)g(p)}.$$

The rest of the proof proceeds on the same lines as did that of the product rule.

12.2.11 **EXERCISE** Complete this proof. (You can expect to use the 'differentiable implies continuous' theorem yet again.)

Next, corresponding to the fact that the composite of two continuous functions is continuous, we have that the composite of two differentiable functions is differentiable.

12.2.12 **The chain rule – differentiating a composite function**   If f is differentiable at p, and $q = f(p)$ and g is differentiable at q, then $g \circ f$ is differentiable at p and its derivative there is $g'(f(p))f'(p)$.

Partial proof

As $h \to 0$, the quantity $k = f(p+h) - f(p)$ also converges to 0 since the (differentiable) function f is continuous. Consider the equation

$$\frac{(g \circ f)(p+h) - (g \circ f)(p)}{h} = \frac{(g \circ f)(p+h) - (g \circ f)(p)}{f(p+h) - f(p)} \times \frac{f(p+h) - f(p)}{h}$$
$$= \frac{g(f(p+h)) - g(f(p))}{f(p+h) - f(p)} \times \frac{f(p+h) - f(p)}{h}$$
$$= \frac{g(f(p) + k) - g(f(p))}{k} \times \frac{f(p+h) - f(p)}{h}$$
$$= \frac{g(q+k)) - g(q)}{k} \times \frac{f(p+h) - f(p)}{h}.$$

Now as $h \to 0$ and, in consequence, $k \to 0$ also, this converges to $g'f'$ where the derivative g' is taken at the point $q = f(p)$, and the derivative f' is taken at p. In other words, the derivative of the composite has been established as $g'(f(p))f'(p)$ as expected.

Unfortunately, this is not yet a general proof – because at our first step of multiplying by

$$\frac{f(p+h) - f(p)}{f(p+h) - f(p)}$$

we tacitly assumed that $f(p+h) - f(p)$ was not zero. This is a safe assumption *provided* that $f'(p) \neq 0$ because, in that case, the ratio

$$\frac{f(p+h) - f(p)}{h}$$

will be converging to a non-zero limit, and must therefore itself be non-zero for sufficiently small h. In the general case where $f'(p)$ might be zero, the result is still true, but we shall need to find a different way of proving it.[3] (See the upcoming 12.5.2 for a 'tidy' proof that works in all cases.)

12.2.13 **Note** The chain rule, as we have presented it so far, applies only to the composite of *two* differentiable functions, but it extends readily to three or more. This is a place where the older style of notation df/dx, rather than $f'(p)$, makes it easier to explain what is going on. Suppose we are given three functions f, g, h with domains such that the composite $h \circ g \circ f$ makes sense, and each of the three is differentiable at appropriate points (starting with p for f). Temporarily write $y = f(x)$, $z = g(y), u = h(z)$. Then

$$\frac{dy}{dx} \text{ means } f', \quad \frac{dz}{dy} \text{ means } g', \quad \frac{du}{dz} \text{ means } h'.$$

---

[3] Meanwhile, here is an *ad hoc* proof for the awkward case $f'(p) = 0$. The proof is really intricate, and it will probably be wise to skip it on a first (and, indeed, on a second and a third) reading.

Suppose $f'(p) = 0, q = f(p), g'(q) = M$. For small values of h, put $k = f(p+h) - f(p)$, so that $f(p+h) = f(p) + k$, that is, $f(p+h) = q + k$. Let $\varepsilon > 0$ be given. Then:

$$\text{there is } \delta_1 > 0 \text{ such that } 0 < |k| < \delta_1 \Rightarrow \frac{g(q+k) - g(q)}{k} \in [M-1, M+1]$$

$$\Rightarrow |g(q+k) - g(q)| \le (1 + |M|) \times |k|.$$

This *also* holds when $k = 0$. Continuing:

$$\text{there is } \delta_2 > 0 \text{ such that } 0 < |h| < \delta_2 \Rightarrow \left| \frac{f(p+h) - f(p)}{h} \right| < \frac{\varepsilon}{1 + |M|}$$

$$\Rightarrow |f(p+h) - f(p)| < \frac{\varepsilon |h|}{1 + |M|}$$

$$\Rightarrow |k| < \frac{\varepsilon |h|}{1 + |M|}.$$

Now f is continuous at p, so

$$\text{there is } \delta_3 > 0 \text{ such that } |h| < \delta_3 \Rightarrow |f(p+h) - f(p)| < \delta_1 \text{ that is, } |k| < \delta_1.$$

Finally, if $0 < |h| < \min\{\delta_2, \delta_3\}$, we get:

$$|k| < \delta_1, \text{ therefore } |g(q+h) - g(q)| < (1 + |M|) \times \frac{\varepsilon |h|}{1 + |M|}$$

which implies

$$\left| \frac{g(f(p+h)) - g(f(p))}{h} \right| < \varepsilon.$$

Hence, since ε was arbitrary,

$$\lim_{h \to 0} \frac{g(f(p+h)) - g(f(p))}{h} = 0 = g'(f(p))f'(p).$$

(The weakness of this notation is that it does not explicitly name the individual points at which these derivatives are evaluated – these have to be judged by context.)

Then, using the chain rule,

$$z = g(f(x)) \text{ so } \frac{dz}{dx} \text{ is } (g \circ f)'(p) = g'(f(p))f'(p)$$

and, continuing,

$$u = h((g \circ f)(x)) \text{ so } \frac{du}{dx} \text{ is } h'(g(f(p)))(g \circ f)'(p) = h'(g(f(p)))g'(f(p))f'(p).$$

Thus the derivative of the three-way composite exists, and the two alternative notations that we have for it are

$$(h \circ g \circ f)'(p) = h'(g(f(p)))g'(f(p))f'(p) \text{ and } \frac{du}{dx} = \frac{du}{dz}\frac{dz}{dy}\frac{dy}{dx}.$$

Despite its reluctance to name points, the second strikes many people as much more readable. It is also useful in helping us to remember what the chain rule says, because it *looks as if* the two dz and the two dy cancel out – of course, that is emphatically not what is really happening, since du/dx and its cousins are not fractions. As an *aide memoire*, however, the fact that *they multiply as if they were fractions* makes it easy to keep track of what the chain rule is telling us.

12.2.14 Example (Assuming for the moment that the derivatives of $\sin x$ and e^x are $\cos x$ and e^x respectively) we differentiate the function $\sin((x^3 + e^x)^7)$.

Solution – modern notation

Unpick the given formula into its components so that we can perceive it as a composite: if $f(x) = x^3 + e^x$ and $g(x) = x^7$ and $h(x) = \sin x$ then the function described by the given formula is $j(x) = h(g(f(x)))$, so (for any x)

$$j'(x) = h'(g(f(x)))g'(f(x))f'(x) = \cos(g(f(x))) \times 7(f(x))^6 \times (3x^2 + e^x)$$
$$= \cos((x^3 + e^x)^7) \times 7(x^3 + e^x)^6 \times (3x^2 + e^x).$$

Solution – 'heritage' notation

Unpick $y = \sin((x^3 + e^x)^7)$ into its components

$$y = \sin u, u = v^7, v = x^3 + e^x.$$

Then

$$\frac{dy}{du} = \cos u, \quad \frac{du}{dv} = 7v^6, \quad \frac{dv}{dx} = 3x^2 + e^x, \quad so$$

$$\frac{dy}{dx} = \frac{dy}{du}\frac{du}{dv}\frac{dv}{dx} = \cos u \times 7v^6 \times (3x^2 + e^x)$$

$$= \cos((x^3 + e^x)^7) \times 7(x^3 + e^x)^6 \times (3x^2 + e^x).$$

12.2.15 Note   For the sake of (relative) completeness we should also discuss here the rule for differentiating an inverse function (see Chapter 8 for more detailed comments on inverse functions, including their continuity) and we shall first do so rather informally.

If $f : (a, b) \to \mathbb{R}$ is strictly increasing or strictly decreasing and $f'(p) \neq 0$ for a point $p \in (a, b)$ then, merely because f is one-to-one from (a, b) onto its range $f((a, b))$, the inverse mapping $f^{-1} : f((a, b)) \to (a, b)$ exists. Better than that, though, f^{-1} is differentiable at $f(p)$ and its derivative is

$$(f^{-1})'(f(p)) = \frac{1}{f'(p)}.$$

In what we called 'heritage' notation in the last paragraph, this can be expressed by saying that if $y = f(x)$ is strictly monotonic on an open interval and dy/dx is non-zero at a point, then the inverse $x = g(y)$ is also differentiable at the corresponding point and

$$\frac{dx}{dy} = \frac{1}{\left(\frac{dy}{dx}\right)}$$

(provided we carefully keep track of the points at which the derivatives are calculated). Once again, although these symbols are not fractions, we see that they can be manipulated as if they were, and the observation helps us to hold the result in mind.

As a small illustration, we determine the derivative of $\sqrt[3]{x}$ at any point. Starting with the function $f(x) = x^3$, which is strictly increasing on $\mathbb{R}$ and has derivative $3p^2$ at each point p, we see that the inverse is given by $f^{-1}(x) = \sqrt[3]{x}$, and its derivative at $f(p) = p^3$ is

$$\frac{1}{f'(p)} = \frac{1}{3p^2} = \frac{p}{3p^3}$$

provided we avoid division by zero, of course. Putting $x = p^3$, that is, $p = \sqrt[3]{x}$, this says (more readably) that the derivative of $\sqrt[3]{x}$ at any x except 0 is

$$\frac{p}{3p^3} = \frac{\sqrt[3]{x}}{3x} = \frac{1}{3}x^{-2/3}.$$

Switching to the alternative view using heritage notation: if $y = \sqrt[3]{x}$ then $x = y^3$ and so $dx/dy = 3y^2$. Provided that this is non-zero, we therefore get

$$\frac{dy}{dx} = \frac{1}{\left(\frac{dx}{dy}\right)} = \frac{1}{3y^2} = \frac{1}{3}y^{-2} = \frac{1}{3}(\sqrt[3]{x})^{-2} = \frac{1}{3}x^{-\frac{2}{3}} \quad (x \neq 0)$$

(if we take care to track the points at which the derivatives are calculated).

Here is a proof of the result that we have outlined over the last page.

12.2.16 **Theorem**   If $f : I \to f(I)$ is a continuous strictly increasing function on an interval I, and f is differentiable at a non-endpoint point p of I, and $f'(p) \neq 0$, then the inverse function f^{-1} is differentiable at $f(p)$, and

$$\left(f^{-1}\right)'(f(p)) = \frac{1}{f'(p)}.$$

Proof

Differentiability at p tells us that f is defined (and continuous) on a small open interval centred on p, so there is a small open interval centred on $f(p)$ contained in the range (see Lemma 8.6.4 if this is not clear). Thus, for all sufficiently small non-zero k, $f(p) + k$ is in that range and we can find $h \neq 0$ such that

$$f(p) + k = f(p + h).$$

(To be fussy, h depends on k so we should really write it as $h(k)$, but that would make the algebra harder to read.)

Now, again using g to stand for the inverse of f, to investigate g' at $f(p)$ we must look for a limit (as $k \to 0$) of:

$$\frac{g(f(p) + k) - g(f(p))}{(f(p) + k) - f(p)}$$
$$= \frac{g(f(p + h)) - g(f(p))}{k} = \frac{p + h - p}{f(p + h) - f(p)} = \frac{h}{f(p + h) - f(p)}$$
$$= \left(\frac{f(p + h) - f(p)}{h}\right)^{-1}.$$

As $k \to 0$, so does h because[4] g is continuous at $f(p)$. Then the content of the final large pair of brackets tends to (non-zero) $f'(p)$, and the overall answer is $(f'(p))^{-1}$, as predicted.

---

[4] $h = g(f(p) + k) - g(f(p))$ which $\to 0$ as $k \to 0$

12.3 Up and down, maximum and minimum: for differentiable functions

Since the derivative $f'(p)$ gives the slope of the curve $y = f(x)$ at a typical point $(p, f(p))$ on its graph, it is hardly surprising that positive gradients are associated with upward-sloping, rising graphs and with functions that increase as you increase the input value x, and negative gradients with downward-sloping, falling graphs and functions that decrease. Analysing the precise nature of these associations requires some care and, in the course of exercising that care, we shall encounter a couple of theorems that have much wider application and significance than our short-term aims suggest.

You may find it useful at this point to review the ideas of increasing and decreasing functions as discussed in Chapter 8: especially paragraphs 8.6.1 and 8.6.3.

12.3.1 Lemma

1. If f is increasing on an open interval I and differentiable at a point p of I, then $f'(p) \geq 0$.
2. If f is decreasing on an open interval I and differentiable at a point p of I, then $f'(p) \leq 0$.

Proof

By assumption,

$$\frac{f(x) - f(p)}{x - p}$$

converges to the limit $f'(p)$ as $x \to p$. Choose a sequence (x_n) of numbers *greater than* p that converges to p and we get both $f(x_n) - f(p) \geq 0$ and $x_n - p > 0$, and therefore also

$$\frac{f(x_n) - f(p)}{x_n - p} \geq 0$$

for all n. Therefore $f'(p)$ is the limit of a sequence of non-negative numbers, and so is non-negative itself (*via* 'taking limits across an inequality', Theorem 4.1.17).

The proof of part 2 is very similar.

12.3.2 Notes

1. Although it is tempting to believe that if f is strictly increasing on an open interval I and differentiable at a point p of I, then $f'(p)$ should be strictly greater than 0, it is not true. For example, the function $f(x) = x^3$ is strictly increasing on $(-1, 1)$ and differentiable at 0, and yet $f'(0) = 0$ exactly.

2. Although it is reasonable and, indeed, correct, to expect some sort of converse to the lemma, the sign of the derivative at a *single* point is not enough to guarantee that the function shall increase or decrease over a small enough interval. For a relatively difficult exercise, you might like to investigate the following assertion: that the function f defined by

$$f(x) = x^2 \sin(x^{-1}) + 0.5x \text{ if } x \neq 0, \quad f(0) = 0$$

is differentiable at every point of the real line, and $f'(0) = +0.5$, and yet in every interval of the form $(-\delta, \delta)$ there are points at which f' is strictly less than 0, and sub-intervals on which f is strictly decreasing.

3. We shall find converses along the lines of: if $f'(x) \geq 0$ at *every* point of an interval, then f is increasing on that interval (and so on).

12.3.3 **Definitions**   Suppose that f is defined on an interval (open or closed) I.

- We say that f *has a local maximum* at $p \in I$ if, for some $\delta > 0$,

$$f(p) \geq f(x) \text{ for every } x \in I \cap (p - \delta, p + \delta);$$

we also call the point $(p, f(p))$ *a local maximum point on the graph* of f.
- We say that f *has a local minimum* at $p \in I$ if, for some $\delta > 0$,

$$f(p) \leq f(x) \text{ for every } x \in I \cap (p - \delta, p + \delta);$$

we also call the point $(p, f(p))$ *a local minimum point on the graph* of f.

Bear in mind that a *local* maximum may very well not show us an *overall* maximum value that the function might reach in the interval: for one thing, there may be several local maxima; for another, it is possible (depending partly on the type of interval) that the function is unbounded above, or never attains the supremum of its values. Consider, for instance, the functions whose graphs are sketched here:

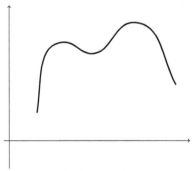

Two local maxima

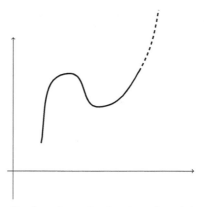

Local maximum, but function unbounded

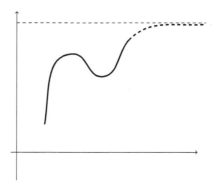

Local maximum, but overall maximum not attained

In the same way, the value at a local minimum might not be an overall, *global* minimum value on the interval.

Also be aware that, at an endpoint of the interval I (in the case where it has an endpoint that belongs to I), the local-maximum and local-minimum criteria only pay attention to what happens on one side of the point (since the function is not defined on the other side). For instance, $f(x) = x^2$ on the interval $[-2, 3]$ has a local maximum at $x = -2$ because, for instance, $f(-2) \geq f(x)$ for every x within $[-2, 3] \cap (-2 - 0.5, -2 + 0.5)$ even though that intersection, namely $[-2, -1.5)$, only contains points at and on the right of -2.

12.3.4 Lemma   If $f : I \to \mathbb{R}$ has a local maximum (or a local minimum) at p, and is differentiable at p, then $f'(p) = 0$.

Proof

Remember that differentiability at p includes the fact that f is defined throughout some open interval $(p - \delta, p + \delta)$ centred on p and, if f also has a local maximum at p, we can make that number δ small enough to ensure that $f(p) \geq f(x)$ for every

x in $(p - \delta, p + \delta)$. Pick a sequence (x_n) of numbers greater than p in that interval that converges to p, and note that (for each n)

$$f(x_n) - f(p) \leq 0, x_n - p > 0 \text{ and so } \frac{f(x_n) - f(p)}{x_n - p} \leq 0.$$

Since $f'(p)$ is the limit of that last fraction, we now get $f'(p) \leq 0$ also.

Repeat the argument with a sequence (y_n) of numbers *less* than p that converges to p, and we find that $f'(p) \geq 0$. To reconcile the two findings, we must have $f'(p) = 0$.

To establish the result for a local minimum, apply what we have just discovered to the function $(-f)$.

Recall that if a function f is continuous on a bounded closed interval $[a, b]$, then it must reach a greatest value (and a smallest value) somewhere in that interval. From the previous lemma there are three possibilities about a point where it does this: this point could be a point where f' takes the value zero, or a point where f is not differentiable, or an endpoint of the interval. Each of the three possibilities can actually occur, as even a few rough sketch graphs will readily indicate:

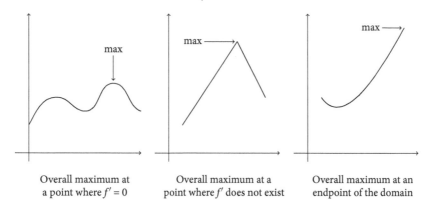

Overall maximum at   Overall maximum at a   Overall maximum at an
a point where $f' = 0$ point where f' does not exist   endpoint of the domain

12.3.5 **Definition**   A *stationary point* (also called *critical point*) of a function f is a number p for which $f'(p) = 0$. (The term is also sometimes applied to the point $(p, f(p))$ on the graph of f.) We can re-cast the preceding discussion as follows:

12.3.6 **Theorem**   Let f be continuous on $[a, b]$ and differentiable on (a, b). Then f takes its greatest value (and its smallest value) either at a stationary point of f, or at an endpoint of the interval.

12.3.7 **Note**   Stationary points do not have to be local (or global) maxima nor minima: for instance, $x = 0$ is a stationary point for $f(x) = x^3$ on the interval $[-1, 1]$, but it is not a maximum/minimum of any kind. For a more complicated example,

$$f(x) = x^2 \sin(x^{-1}) \text{ when } x \neq 0; \quad f(0) = 0$$

gives a function differentiable on the whole of $\mathbb{R}$ that has $f'(0) = 0$, and yet there are points arbitrarily near to $x = 0$ at which f is greater than $0 = f(0)$, *and* points arbitrarily near to $x = 0$ at which f is less than 0.

12.3.8 Example (Assuming for the moment that the derivative of e^x is e^x) find the greatest and least values of $f(x) = xe^{-x}$ on the interval $[0.5, 2]$.

Solution

Using both the product rule and the chain rule, we find that $f'(x) = 1e^{-x} + xe^{-x}(-1) = (1 - x)e^{-x}$ which is zero only at $x = 1$. The greatest and least values of f can therefore only occur at 1 or at an endpoint of the interval.

Since the values of $f(x)$ at $x = 0.5, 1, 2$ (respectively) are $0.5e^{-0.5}, e^{-1}, 2e^{-2}$ and evaluate approximately to $0.303, 0.368, 0.271$, it is clear that the maximum value on the interval is e^{-1} and that the minimum value is $2e^{-2}$.

12.3.9 EXERCISE Let $f(x) = \dfrac{x}{x^2 + 1}$ for each real number x. Find the largest and smallest values that $f(x)$ can attain while x ranges over:

1. the interval $[-10, 0]$,
2. the interval $[-\frac{1}{2}, 2]$,
3. the interval $[2, 6]$,
4. the whole real line (if indeed such largest and smallest values exist: note that 12.3.6 does not directly apply on this *unbounded* interval).

Now we meet the two theorems that will have significant roles to play later in the text, as well as being useful in getting the converse implications that we mentioned earlier. The first is actually a special case of the second, but is the version that we are able to prove almost immediately from our earlier work.

12.3.10 Rolle's theorem ('RT') Let f be continuous on $[a, b]$ and differentiable on (a, b). If also $f(a) = f(b)$, then there is at least one point c in (a, b) such that $f'(c) = 0$ (a stationary point in the open interval).

Proof

In the special case where f is constant on the whole of $[a, b]$, this result is trivial and immediate: any point $c \in (a, b)$ will do equally well.

If not, then *either* f takes somewhere in (a, b) values that are strictly greater than $f(a)$, *or* f takes somewhere in (a, b) values that are strictly smaller than $f(a)$ (or both, of course). If the former, then the greatest value of f on $[a, b]$ is not attained at a nor b, but at a point c in (a, b): and an earlier result (12.3.4) tells us that $f'(c) = 0$.

The latter case is similar.

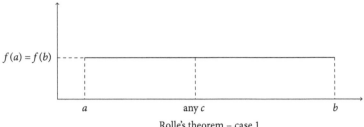

Rolle's theorem – case 1

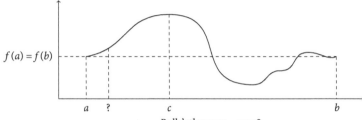

Rolle's theorem – case 2

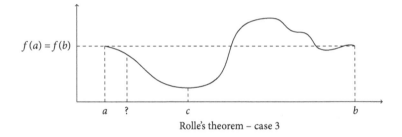

Rolle's theorem – case 3

12.3.11 Example   To show (assuming basic results about trig functions) that the equation

$$(1 + 3x^2) \sin x + (x + x^3) \cos x = 0$$

has at least one solution in the interval $(0, \pi)$.

Solution

The function $(x + x^3) \sin x$ is continuous on $[0, \pi]$ and differentiable on $(0, \pi)$, and takes equal values (zero) at 0 and at π, so RT applies and tells us that its derivative, namely $(1 + 3x^2) \sin x + (x + x^3) \cos x$, is zero at some point between 0 and π, as the question required.

12.3.12 Example   To show using Rolle's theorem (and assuming basic results about trig functions) that the equation $\tan x = \dfrac{2}{x}$ has at least one solution in the interval $(0, \pi/2)$.

Solution

(The difficulty here is to decide what function to apply the theorem to. Re-writing the equation as $x \tan x = 2$, and then as $x^2 \tan x = 2x$, and then as $x^2 \sin x / \cos x = 2x$, and finally as $x^2 \sin x - 2x \cos x = 0$ reveals $x^2 \cos x$ as the key formula.)

The function $x^2 \cos x$ is continuous on $[0, \pi/2]$ and differentiable on $(0, \pi/2)$, and takes equal values (zero) at 0 and at $\pi/2$, so RT applies and tells us that its derivative, namely $-x^2 \sin x + 2x \cos x$, is zero at some point strictly between 0 and $\pi/2$, whence the result follows (noting that, on the open interval $(0, \pi/2)$, neither x nor $\cos x$ is zero, so dividing by them as we unpick the roughwork is legitimate).

12.3.13 EXERCISE Using Rolle's theorem (and proof by contradiction) show that, whatever constants a and b are selected, the graph of

$$y = f(x) = x^4 + 4x^3 + 12x^2 + ax + b$$

cannot have more than one stationary point.

12.3.14 EXERCISE Show that there is a sequence (c_n) in the interval $(0, \pi/2)$ such that, for each $n \in \mathbb{N}$:

$$\tan(c_n) = \frac{n}{c_n}.$$

(*Hint:* for each positive integer n, consider the function described by the formula $x^n \cos x$.)

12.3.15 The first mean value theorem ('FMVT') Let f be continuous on $[a, b]$ and differentiable on (a, b). Then there is at least one point c in (a, b) such that

$$f'(c) = \frac{f(b) - f(a)}{b - a}.$$

Proof

(The idea is to modify the given function, by subtracting a multiple of x, to make it satisfy all three conditions of Rolle's theorem instead of just the first two.)

We seek a constant λ for which the function $g(x) = f(x) - \lambda x$ (which will at least be continuous on $[a, b]$ and differentiable on (a, b) because f was) also obeys $g(a) = g(b)$. Easy algebra gives the answer that

$$\lambda = \frac{f(b) - f(a)}{b - a}$$

and now Rolle applied to g gives us $c \in (a, b)$ such that $0 = g'(c) = f'(c) - \lambda$, as was required.

12.3.16 Remark   Since $\dfrac{f(b)-f(a)}{b-a}$ is precisely the gradient of the straight line
that joins the first and last points $P = (a, f(a))$ and $Q = (b, f(b))$ on the graph of f
over $[a, b]$, this result has easy geometrical interpretations: there is a point on the
graph (not at its first or last points) where the tangent to the curve runs parallel to
the straight line PQ; since PQ gives a kind of 'overall slope' for this section of the
graph, that says there is a point where the 'instantaneous slope' (of the curve itself)
equals the average slope, the mean value of the gradient over the whole interval.

There may, of course, be more than one such point.

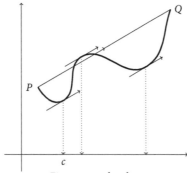

First mean value theorem

12.3.17 Example   To show (assuming standard facts about the trig functions) that
the equation

$$\pi(1+x^3)\cos(\pi x) + 3x^2\sin(\pi x) = \frac{9}{4}$$

has at least one solution in the interval $(0, \frac{1}{2})$.

Solution

We notice that the left-hand side of the equation is the derivative of
$f(x) = (1+x^3)\sin(\pi x)$ (using the product and chain rules). Now $f(x)$ is continuous
on $[0, \frac{1}{2}]$ and differentiable on $(0, \frac{1}{2})$, and $f(0) = 0$ and $f(\frac{1}{2}) = (\frac{9}{8})\sin(\frac{\pi}{2}) = \frac{9}{8}$, so
the FMVT guarantees at least one solution in $(0, \frac{1}{2})$ of the equation

$$f'(x) = \frac{\frac{9}{8} - 0}{\frac{1}{2} - 0},$$

which is equivalent to what we were asked.

12.3.18 EXERCISE   Let n be a positive integer. Use the *two different methods*
indicated to verify that the equation

$$nx^{n-1} = \sqrt{1 - x^n}$$

has a solution in $(0, 1)$:

1. by applying the FMVT to a suitable function,
2. by applying the intermediate value theorem to a suitable function.

12.3.19 Theorem: sign of derivative indicates monotonicity of function Let f be continuous on $[a, b]$ and differentiable on (a, b). Then

1. if $f'(x) > 0$ for each $x \in (a, b)$ then f is strictly increasing on $[a, b]$,
2. if $f'(x) \geq 0$ for each $x \in (a, b)$ then f is increasing on $[a, b]$,
3. if $f'(x) < 0$ for each $x \in (a, b)$ then f is strictly decreasing on $[a, b]$,
4. if $f'(x) \leq 0$ for each $x \in (a, b)$ then f is decreasing on $[a, b]$.

Proof

All four proofs are really the same argument.[5] For instance, if $f'(x) > 0$ for each $x \in (a, b)$ then, for any choice of p, q in $[a, b]$ we can apply FMVT to f on the interval $[p, q]$ to obtain

$$\frac{f(q) - f(p)}{q - p} = f'(c), \text{ some } c \in (p, q) \subseteq (a, b)$$

and so $f(q) - f(p) = (q - p)f'(c)$ is strictly positive; hence $f(q) > f(p)$ and we conclude that f is strictly increasing on $[a, b]$.

Here is a rather classy application of the FMVT that uses *limit* of a derivative to prove *existence* of a derivative at a single point at which differentiability was in doubt:

12.3.20 Theorem Suppose we are told that a function f is continuous on $(a - \delta, a + \delta)$ (for some positive δ), differentiable on $(a - \delta, a)$ and on $(a, a + \delta)$ (but not necessarily *at* the point a), and that $\lim_{x \to a} f'(x)$ exists (equal to ℓ, say). Then f is also differentiable at a, and $f'(a) = \ell$.

Proof

For h positive and smaller than δ, the function f satisfies the FMVT conditions on the interval $[a, a + h]$ so there is a point c_h in $(a, a + h)$ such that

$$\frac{f(a + h) - f(a)}{h} = f'(c_h).$$

---

[5] With a little care, the results can be extended to apply to *unbounded* intervals also: see Example 12.3.21.

Now ℓ is the limit of f' as we approach a so, given $\varepsilon > 0$, we can find positive δ_1 smaller than δ such that, in particular, $|f'(x) - \ell| < \varepsilon$ whenever $x \in (a, a + \delta_1)$. If now $0 < h < \delta_1$, we see that c_h lies in that interval, so $|f'(c_h) - \ell| < \varepsilon$. Thus $f'(c_h) \to \ell$ as $h \to 0^+$. In other words,

$$\lim_{h \to 0^+} \frac{f(a + h) - f(a)}{h} = \ell.$$

Essentially the same argument, applied to small negative values of h, yields

$$\lim_{h \to 0^-} \frac{f(a + h) - f(a)}{h} = \ell,$$

and the equality of the one-sided limits gives (see Theorem 10.3.9) what we wanted.

12.3.21 **Example**   To show that a function f that is continuous on $[a, \infty)$ and has positive derivative at every point of (a, ∞) must be strictly increasing on $[a, \infty)$.

Solution

For any p, q in $[a, \infty)$ such that $p < q$, we observe that f is continuous on $[p, q]$ and differentiable (with positive derivative) on (p, q). By 12.3.19, f is strictly increasing on $[p, q]$. In particular, $f(p) < f(q)$: hence the result.

12.3.22 **A specimen theorem on bounded monotonic functions**   If f is increasing on (a, b) and bounded above, then the one-sided limit $\lim_{x \to b^-} f(x)$ exists. Likewise if f is decreasing on (a, b) and bounded below.

Proof

(This result and the next exercise are simply extensions of paragraph 10.3.5.) Put $M = \sup\{f(x) : x \in (a, b)\}$ (which must exist since the set of values here is non-empty and bounded above). Given $\varepsilon > 0$, by the definition of supremum we can find $x' \in (a, b)$ for which $f(x') > M - \varepsilon$. Now since f is increasing:

$$x' < x < b \Rightarrow M - \varepsilon < f(x') \leq f(x) \leq M$$

and it follows that

$$b - (b - x') < x < b \Rightarrow |M - f(x)| < \varepsilon$$

so the one-sided limit exists, and equals M.
   The proof of the second claim is similar.

12.3.23 **EXERCISE**   Show that if f is increasing on (a, b) and bounded below, or if f is decreasing on (a, b) and bounded above, then the one-sided limit $\lim_{x \to a^+} f(x)$ exists.

12.3.24 Example   (Assuming for the moment that $\ln x$ is differentiable on $(0, \infty)$ and that its derivative is x^{-1}) find the smallest value of the function $f(x) = x \ln x$ on $(0, \infty)$, and verify that

$$\lim_{x \to 0^+} f(x)$$

exists.

Proof

Since x^{-1} is always positive on $(0, \infty)$, $\ln x$ is strictly increasing there. By the product rule, $f'(x) = x(x^{-1}) + (\ln x) \times 1 = 1 + \ln x$ which is positive on $(1/e, \infty)$ since $\ln x > -1 = \ln(e^{-1})$ there, therefore f is strictly increasing on $[1/e, \infty)$.

Likewise, $f'(x)$ is negative on $(0, 1/e)$ since $\ln x < -1 = \ln(e^{-1})$ there, therefore f is strictly decreasing on $(0, 1/e)$. That already tells us that the smallest value that f can take is $f(1/e) = -1/e$.

(The fact that $f'(1/e) = 0$ flags up $1/e$ as the only stationary point of f is also informative; on its own, though, it doesn't indicate what *sort* of a stationary point occurs here.)

Secondly, the observation that f is negative on $(0, 1/e)$, that is, bounded above by 0 (as well as decreasing) allows us to invoke the previous Exercise 12.3.23 to conclude that $\lim_{x \to 0^+} f(x)$ exists.

12.4 Higher derivatives

12.4.1 Definitions   When, as often happens, the derivative f' of a function f exists at every point of an open interval or of a union of open intervals, so that $f'(x)$ emerges as a properly defined (derived) function in its own right, it may well be worth asking whether f' can itself be differentiated. If it can (either at a point or across a union of one or more open intervals) then its derivative is called the *second derivative* of f and is written as f'' or, occasionally, as $\dfrac{d^2 f}{dx^2}$. In turn, if f'' turns out to be differentiable, then its derivative f''' is known as the *third derivative* of f, and so on.

Beyond the third, the multiple dashes notation becomes clumsy and hard to read, and very few texts use the notation f'''', preferring either to borrow roman numerals (f^{iv}) or additional brackets ($f^{[4]}$ or $f^{(4)}$) or even just f^4. In all cases one must be careful not to confuse a symbol for a higher derivative with a symbol for a power of the function, either by repeated multiplication or by repeated composition. Indeed, this is an area in which mathematical notation is by no means set in concrete, and will vary from writer to writer and from text to text depending on precisely what mathematical construction is being written about. If, for example, $f(x) = \sin x$, then the notation $f^5(x)$ could conceivably refer to any of $f'''''(x)$, $(\sin x)^5$ or $\sin(\sin(\sin(\sin(\sin x))))$, and so could $f^{[5]}(x)$ or $f^v(x)$, so it will pay you to read carefully what a particular text actually means by it.

In the present work, we shall explicitly refer to higher derivatives before using any such cluster of symbols. The higher derivatives are especially important in Taylor's theorem (see Chapter 16), but they also have a neat and well-known application to the classification of stationary points which we shall deal with now.

12.4.2 Theorem: the second derivative test for local extrema   Suppose that a function f is twice differentiable on an open interval including the number p, that $f'(p) = 0$ and that $f''(x)$ is continuous at p. Then

- if $f''(p) > 0$ then f has a local minimum at p,
- if $f''(p) < 0$ then f has a local maximum at p.

(This is the result that has embedded itself in the minds of generations of mnemonic-laden school pupils through the slogan 'POS MIN, NEG MAX'.)

Proof

We shall deal only with the first scenario since the second is so similar.

Since f'' is continuous at p and $f''(p) > 0$, we can choose $\delta > 0$ such that f'' is positive at every point of the interval $(p - \delta, p + \delta)$.

Now we work backwards from knowledge of f'' to knowledge of f': for each $x \in (p - \delta, p)$ apply FMVT to f' on $[x, p]$ and we find that

$$\frac{f'(p) - f'(x)}{p - x} = f''(y) \quad \text{for some} \quad y \in (x, p)$$

that is, $0 - f'(x) = f''(y)(p - x)$ is positive, and so $f'(x)$ itself is negative at every point in (x, p). This implies (see 12.3.19) that f is (strictly) decreasing on $[x, p]$ so, in particular, $f(x) > f(p)$.

In the same fashion we show that for every $x \in (p, p + \delta)$ we again get $f(x) > f(p)$. Hence the claimed result.

The reader will almost certainly be familiar with standard exercises such as the following:

12.4.3 Example   Find and classify the stationary points on the graph of the function $f(x) = x^4 - 8x^3 - 8x^2 + 96x + 144$.

Solution

It is easy to differentiate f two (or more) times:

$$f'(x) = 4x^3 - 24x^2 - 16x + 96 = 4(x^3 - 6x^2 - 4x + 24),$$
$$f''(x) = 12x^2 - 48x - 16 = 4(3x^2 - 12x - 4).$$

To find the stationary points we solve $f'(x) = 0$, that is, $x^3 - 6x^2 - 4x + 24 = 0$. A little back-of-the-envelope work yields $x = 2$ as one solution, so $x - 2$ is a factor of the left-hand side, which now factorises into

$$x^3 - 6x^2 - 4x + 24 = (x - 2)(x^2 - 4x - 12) = (x - 2)(x + 2)(x - 6)$$

and at this stage, we know that $2, -2$ and 6 are the x-coordinates of the three stationary points. We find their y-coordinates by substituting these numbers into the $f(x)$ formula, and identify

$$A = (-2, 0), \quad B = (2, 256), \quad C = (6, 0)$$

as the points in question. Now f'' is certainly continuous, and

$$f''(-2) = +128, f''(2) = -64, f''(6) = +128.$$

The second derivative test informs us that $A = (-2, 0)$ and $C = (6, 0)$ are local minima, while $B = (2, 256)$ is a local maximum.

12.4.4 **EXERCISE**   Devise and solve a similar question based on a fourth-degree polynomial of your own choice. (If you are wise, you will reverse-engineer the problem a little by making sure that the cubic equation that you will have to solve for the three x-coordinates actually does have three real and arithmetically simple solutions.)

12.5 Alternative proof of the chain rule

To try to reduce the amount of algebra in this proof, let us introduce some temporary jargon. We shall call a function $j(x)$ a *junk function* if

- it is defined on (at least) some open interval centred on 0,
- $j(0) = 0$, and
- as $x \to 0, j(x) \to 0$.

(This is less unprofessional than it probably looks: it is quite common in analytic arguments to refer to some complicated piece of algebra as 'junk' if we cannot be bothered to work out its value exactly *and* we are sure that it will be too small to make any difference to the answer that we seek.)

For instance, if f is defined at and on both sides of a number p and is continuous at p, then $f(p + h) - f(p)$ is a junk function (of h). Again, if f is differentiable at p, then

$$\epsilon(h) = \frac{f(p+h) - f(p)}{h} - f'(p)$$

would be a junk function of h, *except* that it fails to be defined at $h = 0$. We can easily fix this: the improved description

$$\epsilon(h) = \frac{f(p+h) - f(p)}{h} - f'(p) \ \text{ if } h \neq 0; \ \ \epsilon(0) = 0$$

remedies this flaw, and makes ϵ into a junk function of h. This, in turn, gives rise to an alternative description of differentiability:

12.5.1 Lemma A function f is differentiable at a point p, and its derivative there is ℓ, if and only if (for all sufficiently small values of h)

$$f(p+h) = f(p) + h(\ell + \epsilon(h))$$

for some junk function ϵ.

Proof

If f is differentiable, then the above discussion shows how to define a suitable junk function. Conversely, if such an ϵ does exist, then (for sufficiently small non-zero h)

$$\frac{f(p+h) - f(p)}{h} = \ell + \epsilon(h),$$

which does converge to ℓ as $h \to 0$, that is, f is differentiable at p and ℓ is its derivative.

For most people, the condition in this lemma is less intuitive as a description of differentiability than the one we first gave, but it has one important and occasionally redeeming feature: it does not involve dividing by h: and for that reason, on the few occasions when we opt to use it, we do not have to treat $h = 0$ as a special case.

A few moments' thought should be enough to persuade you that

- if ϵ and η are junk functions, then so is $\epsilon + \eta$,
- if ϵ is a junk function, then so is $K\epsilon$ for any constant K,
- if ϵ and η are junk functions, then so is their product $\epsilon\eta$,
- if ϵ and η are junk functions, then so is their composite $\epsilon \circ \eta$ (that is, the function $\epsilon(\eta(x))$).

12.5.2 The chain rule (again) If f is differentiable at p, and $q = f(p)$ and g is differentiable at q, then $g \circ f$ is differentiable at p and its derivative there is $g'(f(p))f'(p)$.

Alternative proof

Let ℓ and m stand for the numbers $f'(p)$ and $g'(q)$. By the lemma, there are two junk functions ϵ and η such that (for sufficiently small h and k)

$$f(p + h) = f(p) + h(\ell + \epsilon(h))$$

and

$$g(q + k) = g(q) + k(m + \eta(k)).$$

We put $k = f(p + h) - f(p)$, noting that k is itself a junk function of h because the differentiable function f is necessarily continuous. Now

$$
\begin{aligned}
(g \circ f)(p + h) &= g(f(p + h)) \\
&= g(f(p) + k) \\
&= g(q + k) \\
&= g(q) + k(m + \eta(k)) \\
&= g(f(p)) + (f(p + h) - f(p))(m + \eta(k)) \\
&= (g \circ f)(p) + h(\ell + \epsilon(h))(m + \eta(k))
\end{aligned}
$$

that is,

$$(g \circ f)(p + h) = (g \circ f)(p) + h(\ell m + \ junk)$$

where $junk = m\epsilon(h) + \ell\eta(k) + \epsilon(h)\eta(k)$ is a junk function by the comments made above. Using the lemma again, this completes the demonstration that $g \circ f$ is differentiable *and* that its derivative is $\ell m = f'(p)g'(q) = g'(f(p))f'(p)$.

13 The Cauchy condition – sequences whose terms pack tightly together

13.1 Cauchy equals convergent

As we saw in Chapter 2, establishing convergence of a sequence by the definition alone is viable only when we know in advance – or can make a well-informed guess at – what number its limit is. In many cases this is difficult or practically impossible. The insight that a bounded *monotonic* sequence must always converge allowed us to bypass this difficulty in many important examples, but the fact remains that a wide variety of sequences are not monotonic and, nevertheless, do converge. How can we handle such sequences? What is needed is a way to recognise convergence or divergence that works for all sequences (not just special cases like the monotonic ones) but which makes no explicit mention of the limit. The most effective recognition criterion of this kind is the so-called Cauchy condition, which we shall now investigate.

13.1.1 Definition A sequence $(x_n)_{n\in\mathbb{N}}$ is called a *Cauchy* sequence if

for each $\varepsilon > 0$ there is $n_\varepsilon \in \mathbb{N}$ such that

$|x_m - x_n| < \varepsilon$ for all $m \geq n_\varepsilon$ and all $n \geq n_\varepsilon$.

13.1.2 Notes

- By convention, the last line here is usually written as

$$|x_m - x_n| < \varepsilon \text{ for all } m, n \geq n_\varepsilon$$

which is safe provided you understand that *both* m and n are restricted to be $\geq n_\varepsilon$.

- Carefully compare this definition with the definition of convergence, and you will see that there is only one significant difference. To be specific: if you replace

Undergraduate Analysis: A Working Textbook, Aisling McCluskey and Brian McMaster 2018.
© Aisling McCluskey and Brian McMaster 2018. Published 2018 by Oxford University Press

x_m by ℓ (and, in consequence, remove the reference to m in the final line), you recover the original definition of $x_n \to \ell$. In other words:

- – for a sequence to converge, we must be able to force its terms to lie within any specified distance *from the limit* (just by going far enough along the sequence);

- – for a sequence to be Cauchy, we must be able to force its terms to lie within any specified distance *from one another* (just by going far enough along the sequence).

- Now here is the reason why this is important: Cauchy sequences and convergent sequences turn out to be exactly the same objects (as we shall show). It is in this sense that Cauchyness allows us to recognise exactly which sequences have limits, without at any stage mentioning the limit itself.

13.1.3 Lemma   Every convergent sequence is Cauchy.

Proof

If $x_n \to \ell$ then, given $\varepsilon > 0$, we can find n_0 such that $|x_n - \ell| < \varepsilon/2$ whenever $n \geq n_0$. Then, if both m and n are $\geq n_0$, we see that

$$|x_m - x_n| = |x_m - \ell + \ell - x_n|$$
$$\leq |x_m - \ell| + |\ell - x_n| \quad (why?^1)$$
$$< \varepsilon/2 + \varepsilon/2 = \varepsilon$$

and so (x_n) is also Cauchy.

The first step in establishing the converse is to show that all Cauchy sequences are bounded. (The following proof is very like the one we gave to show that all convergent sequences are bounded.)

13.1.4 Lemma   Every Cauchy sequence is bounded.

Proof

Suppose that $(x_n)_{n \in \mathbb{N}}$ is Cauchy. By the definition (and choosing $\varepsilon = 1$ for convenience) there is a positive integer n_0 such that all the terms of the sequence from the n_0^{th} one onwards are separated by less than 1 unit so, in particular, they are less than 1 unit distant from x_{n_0}. The earlier terms $x_1, x_2, x_3, \cdots, x_{n_0-1}$ may well be further away from x_{n_0}, but there are only a finite number of them: so we can find the *biggest* distance from one of them to x_{n_0} ... call it M. If we now put $M' = \max\{M, 1\}$ then *every* x_n lies within the distance M' from x_{n_0}, so $(x_n)_{n \in \mathbb{N}}$ is bounded.

---

[1] via the triangle inequality

13.1.5 Lemma A Cauchy sequence cannot have two subsequences with different limits.

Proof

Suppose it did: that is, suppose $(x_n)_{n \in \mathbb{N}}$ is Cauchy, that subsequences $(x_{n_k})_{k \in \mathbb{N}}$ and $(x_{m_j})_{j \in \mathbb{N}}$ converge (respectively) to ℓ and m, and that $\ell < m$. Take $\varepsilon = \dfrac{m - \ell}{3} > 0$.

Subsequences of a Cauchy sequence

Cauchyness tells us that we can find n_ε such that $|x_m - x_n| < \varepsilon$ whenever $m, n \geq n_\varepsilon$. Yet all but finitely many of the x_{n_k} belong to the interval $(\ell - \varepsilon, \ell + \varepsilon)$ and all but finitely many of the x_{m_j} belong to the interval $(m - \varepsilon, m + \varepsilon)$. So we can find such values of n_k and m_j both bigger than n_ε, and then the gap between x_{n_k} and x_{m_j} must exceed ε: contradiction.

13.1.6 Theorem A sequence converges if and only if it is Cauchy.

Proof

From a result in Chapter 5 (Proposition 5.3.6) that we have had little need to use until now, in order to be divergent a sequence must either be unbounded or must possess two subsequences that have different limits. By the second and third of the above lemmata, a Cauchy sequence cannot do either of these, so it must converge. The converse (convergent implies Cauchy) was the first lemma above.

We offer an alternative method of proof for this theorem, partly because it is such a central result, and partly because it uses the following lemma which is useful in its own right.

13.1.7 Lemma If a Cauchy sequence has even one subsequence that converges, then the entire sequence converges also (and to the same limit).

Proof

Suppose that $(x_n)_{n \in \mathbb{N}}$ is a Cauchy sequence and that one of its subsequences $(x_{n_k})_{k \in \mathbb{N}}$ converges to ℓ. Given $\varepsilon > 0$ we can use these two facts to find positive integers n_0 and k_ε such that

$$m, n \geq n_0 \Rightarrow |x_m - x_n| < \varepsilon/2, \text{ and}$$
$$k \geq k_\varepsilon \Rightarrow |x_{n_k} - \ell| < \varepsilon/2.$$

Now for any $n \geq n_0$, choose a value k' of k that is greater than both n_0 and k_ε. Then $n_{k'} \geq k' \geq n_0$ and we find

$$|x_n - \ell| = |x_n - x_{n_{k'}} + x_{n_{k'}} - \ell| \leq |x_n - x_{n_{k'}}| + |x_{n_{k'}} - \ell| < \varepsilon/2 + \varepsilon/2 = \varepsilon.$$

Hence the result.

13.1.8 Theorem - revisited   A sequence converges if and only if it is Cauchy.

Proof - alternative

Suppose that $(x_n)_{n\in\mathbb{N}}$ is Cauchy (and therefore bounded, by Lemma 13.1.4). According to Bolzano-Weierstrass, it has a convergent subsequence. According to Lemma 13.1.7, it is itself convergent. The converse is handled as previously.

13.1.9 Example   Given a sequence $(x_n)_{n\in\mathbb{N}}$ with the property that, for each $n \in \mathbb{N}, |x_n - x_{n+1}| < 0.9^n$, we show that (x_n) must be convergent.

Solution

We clearly do not have enough information to determine or guess the limit, nor to use monotonicity, so the Cauchy criterion is the only trick available to us. Whenever $n < m$, we see that

$$|x_n - x_m| = |x_n - x_{n+1} + x_{n+1} - x_{n+2} + x_{n+2} - x_{n+3} + \cdots + x_{m-1} - x_m|$$
$$\leq |x_n - x_{n+1}| + |x_{n+1} - x_{n+2}| + |x_{n+2} - x_{n+3}| + \cdots + |x_{m-1} - x_m|$$
$$< 0.9^n + 0.9^{n+1} + 0.9^{n+2} + \cdots + 0.9^{m-1} < \sum_{k=n}^{\infty} 0.9^k = \frac{0.9^n}{1 - 0.9} = 10(0.9)^n.$$

Now $0.9^n \to 0$ so, given $\varepsilon > 0$, we can locate $n_0 \in \mathbb{N}$ such that, whenever $n \geq n_0$: $10(0.9)^n < \varepsilon$. It follows that, provided that $n \geq n_0$ (and therefore also $m \geq n_0$), we shall have $|x_m - x_n| < \varepsilon$. Therefore $(x_n)_{n\in\mathbb{N}}$, being Cauchy, must also be convergent.

13.1.10 Note   Fairly slight changes to the above argument will show that if the *step* $|x_n - x_{n+1}|$ *from one term of a sequence to the immediately next one* is less than a constant times a power of t for some constant $t \in (0, 1)$, then the sequence (x_n) is Cauchy. Of course, in that circumstance, $|x_n - x_{n+1}|$ tends to zero. It is important, however, to realise that the condition $|x_n - x_{n+1}| \to 0$ on its own is *not enough* to guarantee that (x_n) shall be Cauchy. One straightforward way to illustrate this is to let x_n be the nth partial sum of the harmonic series

$$x_n = 1 + \frac{1}{2} + \frac{1}{3} + \frac{1}{4} + \frac{1}{5} + \cdots + \frac{1}{n}.$$

Then we know that (x_n) is not convergent (since the harmonic series diverges) and is therefore not Cauchy, and yet $|x_n - x_{n+1}| = \dfrac{1}{n+1}$ which certainly converges to zero.

Informally, we sometimes say that the step in the latter case is tending to zero *but not rapidly enough*, whereas in instances like Example 13.1.9 the step is *decaying geometrically or exponentially* and that this *is fast enough* to force Cauchyness. You can explore this area a little further in some of the additional Exercises (numbers 184 to 190).

13.1.11 EXERCISE Given a sequence $(x_n)_{n \in \mathbb{N}}$ with the properties that, for each $n \in \mathbb{N}$:

$$|x_n - x_{n+2}| < 10(0.6)^n \text{ and } |x_n - x_{n+5}| < 20(0.7)^n,$$

show that (x_n) converges.
[*Suggestion*: $x_n - x_{n+5} + x_{n+5} - x_{n+3} + x_{n+3} - x_{n+1} = x_n - x_{n+1}$.]

13.1.12 EXERCISE Show that the series

$$\sum_{k=1}^{\infty} \frac{\sin(k^3 + 5k^2 - 3) - 4\cos(k^2 + 3k - 7)}{(1.1)^k}$$

is convergent, by establishing that the sequence of its partial sums is Cauchy (and assuming basic trigonometric facts).

'Showing that the sequence of partial sums is Cauchy' will turn out, in Chapter 14, to be pretty much *the* fundamental tool for establishing convergence of a series.

13.1.13 Example We use the Cauchy condition to revisit the proof that the harmonic series $\sum n^{-1}$ diverges.

Solution

For any given positive integer n_0, first find a positive integer m such that 2^m is bigger than n_0. Then, with the usual notation for partial sums

$$S_n = \sum_{k=1}^{n} \frac{1}{k}$$

we see that

$$S_{2^{m+1}} - S_{2^m} = \sum_{k=2^m+1}^{2^{m+1}} \frac{1}{k}$$

is the total of 2^m fractions of which the smallest one is $1/2^{m+1}$. This total therefore exceeds $1/2$. In other words, no matter how large we choose n_0, there will be partial sums *later* than the n_0^{th} that differ by more than 0.5; so the sequence of partial sums here is not Cauchy, and cannot converge.

13.1.14 **Example** Given that $(x_n)_{n\in\mathbb{N}}$ and $(y_n)_{n\in\mathbb{N}}$ are both Cauchy sequences, to show that their 'term-by-term product' $(x_n y_n)_{n\in\mathbb{N}}$ is also Cauchy.

Solution

(This can become quite messy if we try to argue from the definition of Cauchy, so we won't.)

Since (x_n) and (y_n) are both Cauchy, they must each be convergent to some limit: say, $x_n \to \ell$ and $y_n \to \ell'$ (as $n \to \infty$). Algebra of limits now tells us that $x_n y_n \to \ell\ell'$, so the product sequence converges, and is therefore Cauchy too.

13.1.15 **EXERCISE** Of the following two statements, just one is true in general. Give a proof for the one that is true, and find a counterexample that disproves the false one.

1. If $(x_n)_{n\in\mathbb{N}}$ and $(y_n)_{n\in\mathbb{N}}$ are both Cauchy sequences, and there is a (strictly) positive number $\delta > 0$ such that $|y_n| \geq \delta$ for all $n \geq 1$, then the 'term-by-term quotient' sequence

$$\left(\frac{x_n}{y_n}\right)_{n\in\mathbb{N}}$$

   must also be Cauchy.

2. If $(x_n)_{n\in\mathbb{N}}$ and $(y_n)_{n\in\mathbb{N}}$ are both Cauchy sequences, and $|y_n| > 0$ for all $n \geq 1$, then the 'term-by-term quotient' sequence

$$\left(\frac{x_n}{y_n}\right)_{n\in\mathbb{N}}$$

   must also be Cauchy.

13.1.16 **Example** If $f : [a, b] \to \mathbb{R}$ is a continuous function on a closed bounded interval $[a, b]$, and $(x_n)_{n\in\mathbb{N}}$ is a Cauchy sequence of elements of $[a, b]$, we show that $(f(x_n))_{n\in\mathbb{N}}$ is also Cauchy (less formally, that the function f *preserves Cauchyness*.)

Solution

Because $(x_n)_{n\in\mathbb{N}}$ is Cauchy, it must converge: $x_n \to \ell$ for some ℓ. Also, since $a \leq x_n \leq b$ for all n, we have $a \leq \lim x_n = \ell \leq b$ also, that is, $\ell \in [a, b]$. Because continuous functions preserve convergence, it follows that $f(x_n) \to f(\ell)$. Thus $(f(x_n))$ is a convergent sequence, and consequently Cauchy.

13.1.17 EXERCISE   Of the following three statements, at least one is true in general and at least one is false. Give a proof for each that is true, and find a counterexample to disprove each false one.

1. If $f : (a, \infty) \to \mathbb{R}$ is a continuous function on an unbounded open interval (a, ∞), and $(x_n)_{n \in \mathbb{N}}$ is any Cauchy sequence of elements of (a, ∞), then $(f(x_n))_{n \in \mathbb{N}}$ must also be Cauchy.

2. If $f : [a, \infty) \to \mathbb{R}$ is a continuous function on an unbounded closed interval $[a, \infty)$, and $(x_n)_{n \in \mathbb{N}}$ is any Cauchy sequence of elements of $[a, \infty)$, then $(f(x_n))_{n \in \mathbb{N}}$ must also be Cauchy.

3. If $f : \mathbb{R} \to \mathbb{R}$ is a continuous function on the whole real line $\mathbb{R}$, and $(x_n)_{n \in \mathbb{N}}$ is any Cauchy sequence, then $(f(x_n))_{n \in \mathbb{N}}$ must also be Cauchy.

14 More about series

14.1 Absolute convergence

Almost all the series that we have worked on so far have consisted exclusively of non-negative terms (for the reasons set out in Chapter 7), the only significant exception being the (very special) case of alternating series, where a simple test applies. Our purpose now is to extend our knowledge to general series where a mixture of positively- and negatively-signed terms must be expected, and where the pattern of signs may well be unpredictable. Naturally, we try to do this by capitalising on the work we have already done, by turning a general series into a series of non-negative terms, and seeking to relate its behaviour to that of the original. There is, therefore, a kind of inevitability about the first definition:

14.1.1 Definition We call a series

$$\sum x_k$$

absolutely convergent when the series (of non-negative terms)

$$\sum |x_k|$$

is convergent.

14.1.2 Note

- Let us be clear immediately that convergence and absolute convergence are not the same thing. For instance, we know from the alternating series test that the 'alternating harmonic series'

$$\sum (-1)^k \frac{1}{k}$$

is convergent; and yet

$$\sum \left| (-1)^k \frac{1}{k} \right| = \sum \frac{1}{k}$$

is the (notoriously divergent) harmonic series. In the present terminology, the alternating harmonic series is convergent, but it is not absolutely convergent.

- A series that is convergent but not absolutely convergent is called *conditionally convergent*.

Undergraduate Analysis: A Working Textbook, Aisling McCluskey and Brian McMaster 2018.
© Aisling McCluskey and Brian McMaster 2018. Published 2018 by Oxford University Press

- On the other hand, you will never come across a series that is absolutely convergent but is not convergent. No such series can exist: and the way to see this important truth is to use the idea of Cauchy sequences that we encountered just a few pages back. Recall that Cauchy and convergent are equivalent for sequences, and so a (general) series converges if and only if its sequence of partial sums is Cauchy.

- We also remind you again about the triangle inequality: that $|a + b| \leq |a| + |b|$ for arbitrary numbers a and b. This basic form of the inequality extends immediately to the three-term version

$$|a + b + c| \ (\leq |a + b| + |c|) \ \leq |a| + |b| + |c|$$

and the four-term version

$$|a + b + c + d| \ (\leq |a + b + c| + |d|) \ \leq |a| + |b| + |c| + |d|$$

and, through an easy induction, to the general form

$$|a_1 + a_2 + a_3 + \cdots + a_n| \leq |a_1| + |a_2| + |a_3| + \cdots + |a_n|$$

which we have occasionally used, and which features again in the next demonstration:

14.1.3 Theorem   Every absolutely convergent series is convergent.

Proof

Let $\sum x_k$ be absolutely convergent, that is, let $\sum |x_k|$ be convergent. Our task is to show that $\sum x_k$ is convergent and, since that means studying two different partial-sum sequences (one for each series), we must take care to have different notations for the two of them. For instance, let us put

$$S_n = x_1 + x_2 + x_3 + \cdots + x_n = \sum_1^n x_k, \ \text{ and}$$

$$\overline{S}_n = |x_1| + |x_2| + |x_3| + \cdots + |x_n| = \sum_1^n |x_k| \ .$$

Convergence of $\sum |x_k|$ tells us that $(\overline{S}_n)$ is a (convergent, and therefore) Cauchy sequence so, given $\varepsilon > 0$, we can find n_0 such that

$$|\overline{S}_m - \overline{S}_n| < \varepsilon \ \text{ whenever } \ m, n \geq n_0.$$

Since the last line is empty of information when m equals n, we may as well assume $m \neq n$ here. Also, there is no harm in assuming that m is the larger and n is the smaller, since if we swop them over, the modulus signs will ensure that $|\overline{S}_m - \overline{S}_n|$ will remain unaltered. Thus we can write the previous display in slightly more convenient (but equivalent) forms:

$$|\overline{S}_m - \overline{S}_n| < \varepsilon \text{ whenever } m > n \geq n_0, \text{ that is,}$$

$$\left| \sum_1^m |x_k| - \sum_1^n |x_k| \right| < \varepsilon \text{ whenever } m > n \geq n_0, \text{ that is,}$$

$$\left| \sum_{n+1}^m |x_k| \right| < \varepsilon \text{ whenever } m > n \geq n_0, \text{ that is,}$$

$$\sum_{n+1}^m |x_k| < \varepsilon \text{ whenever } m > n \geq n_0.$$

This is where we get to use the 'enhanced' triangle inequality: whenever $m > n \geq n_0$

$$|S_m - S_n| = \left| \sum_1^m x_k - \sum_1^n x_k \right| = \left| \sum_{n+1}^m x_k \right| \leq \sum_{n+1}^m |x_k|$$

which, as we just saw, is $< \varepsilon$. In other words, (S_n) is also a Cauchy sequence, and therefore convergent. So $\sum x_k$ is, indeed, a convergent series.

14.1.4 Notes

1. What we now have is the basis of a strategy for deciding upon the convergence or divergence of a general series (as opposed to a series of non-negatives). If we are given such a series $\sum x_k$, we look instead at the modulussed series $\sum |x_k|$ and examine it by the techniques we acquired in Chapter 7. If they show that $\sum |x_k|$ is convergent, in other words, that $\sum x_k$ is absolutely convergent, then the theorem tells us that the original $\sum x_k$ is convergent also, and the task is completed. So far, so good.

2. However, what if our Chapter 7 skills tell us that $\sum |x_k|$ is NOT convergent? Then there is more and different work to do because the discovery, that $\sum x_k$ is not absolutely convergent, does not tell us whether it is convergent or not. (Look again at the alternating harmonic series: it is not absolutely convergent, but it is convergent; in contrast, a series such as $\sum (-1)^k \sqrt[k]{k}$ is not absolutely convergent, and it is not convergent either.) In brief: non-(absolute convergence) does not decide for us whether we do or do not have convergence.

3. However, some of the main tests in Chapter 7 were designed to help us get around this difficulty in many cases. Consider, for example, the nth root test (for non-negative terms, of course). *It did not simply say the following*:
   - if $\lim \sqrt[n]{x_n} < 1$ then $\sum x_n$ is convergent,
   - if $\lim \sqrt[n]{x_n} > 1$ then $\sum x_n$ is divergent.

Instead, it said something rather less symmetrical and slightly more awkward:

- if $\lim \sqrt[n]{x_n} < 1$ then $\sum x_n$ is convergent,
- if $\lim \sqrt[n]{x_n} > 1$ then x_n does not converge to zero and therefore $\sum x_n$ is divergent.

Perhaps you now see why it pays dividends to word it in this half-clumsy fashion? If we use the root test on $\sum |x_k|$, discover that $\lim \sqrt[n]{|x_n|} > 1$, and bring back merely the information that $\sum |x_k|$ is divergent, that leaves unanswered the question of whether $\sum x_k$ converges or diverges ... but that is only part of what the test (as we expressed it) is telling us: it actually says that $\sum |x_n|$ is divergent *because* $|x_n|$ does not converge to zero. Therefore x_n does not converge to zero either, and $\sum x_n$ cannot be convergent.

4. Consequently, the nth root test will deal with all the (general) series for which $\lim \sqrt[n]{|x_n|}$ can be calculated and is not equal to 1.

5. A very similar discussion will clarify that the ratio test will deal with all series (with no zero terms) for which the modulussed growth-rate limit $\lim \dfrac{|x_{n+1}|}{|x_n|}$ can be calculated and is not equal to 1.

By way of illustration, we'll now re-work a couple of Chapter 7's examples *but without the assumption that the parameter t or x is positive.*

14.1.5 **Example** For precisely which **real** values of t does the series

$$\sum \left(\frac{3n^2 - 1}{2n^2 - 1}\right)^n t^n$$

converge?

Solution

Put $x_n = $ the nth term here, and consider instead $\sum |x_n|$. All its terms are non-negative, and the nth root of $|x_n|$ is

$$\left(\frac{3n^2 - 1}{2n^2 - 1}\right) |t|$$

which has a limit of $\frac{3|t|}{2}$ so, by the root test:

1. for $|t| < 2/3$ the limit is < 1 and so the series $\sum |x_n|$ converges, that is, $\sum x_n$ is absolutely convergent, and therefore also convergent;

2. for $|t| > 2/3$ the limit is > 1 so $|x_n|$ cannot tend to zero, and neither can x_n, so the original series $\sum x_n$ must diverge.

It remains to ponder what happens when t is exactly $\pm 2/3$. Luckily, in that borderline case $|x_n|$ itself is

$$\left(\frac{3n^2-1}{2n^2-1}\right)^n \left(\frac{2}{3}\right)^n = \left(\frac{6n^2-2}{6n^2-3}\right)^n$$

which is (just) greater than 1 (in the final fraction, the top line exceeds the bottom line). This shows once again that neither $|x_n|$ nor x_n can tend to zero, and so the series $\sum x_n$ diverges.

We conclude that the given series $\sum x_n$ is (absolutely) convergent when $-2/3 < t < 2/3$, and divergent for every other value of t.

14.1.6 **Example**   For exactly which **real** values of x does the following series converge?

$$\sum w_n \quad \text{where} \quad w_n = \frac{(n+1)!(2n+2)!\,x^n}{(3n+3)!}$$

Solution

If $x=0$ then, although a ratio test would not be legal, the series definitely converges;[1] so from now on assume $x \neq 0$ and consider $\sum |w_n|$. Its growth rate $|w_{n+1}|/|w_n|$ cancels to

$$\frac{(n+2)(2n+3)(2n+4)|x|}{(3n+4)(3n+5)(3n+6)} = \frac{(2n+3)(2n+4)|x|}{3(3n+4)(3n+5)}$$

whose limit is $4|x|/27$. By the ratio test, therefore:

1. for $0 < |x| < 27/4$ the limit is less than 1 and $\sum |w_n|$ converges, that is, $\sum w_n$ converges absolutely, but

2. for $27/4 < |x|$ the limit exceeds 1, $|w_n|$ does not tend to zero, w_n equally does not tend to zero, and $\sum w_n$ diverges.

It remains unclear at first what will happen in the borderline cases $x = \pm 27/4$.
Notice, however, that when $x = \pm 27/4$, the growth rate for $|w_n|$ is actually

$$\frac{(2n+3)(2n+4)27}{12(3n+4)(3n+5)} = \frac{(6n+9)(6n+12)}{(6n+8)(6n+10)}$$

which is greater than 1 (look at the individual factors in the top and bottom lines). Thus $|w_{n+1}| > |w_n|$, the terms of $\sum |w_n|$ are increasing and cannot converge to zero, and thus w_n cannot tend to zero either and $\sum w_n$ again diverges.

We conclude that $\sum w_n$ converges (absolutely) when $-27/4 < x < 27/4$ but diverges in all other cases.

---

[1] Every term is zero, every partial sum is zero, and the limit of the partial sums is zero (and certainly exists)

14.1.7 **EXERCISE**   For which **real** values of t does the series

$$\sum \frac{(3n+1)^n}{n^{n-1}} t^n$$

converge, and for which does it diverge?

14.1.8 **EXERCISE**   Determine the range of values of the real number x for which the series

$$\sum \frac{(n!)^6 \, x^{3n}}{(6n)!}$$

converges.

Many textbooks present the ratio and root tests as tests upon general series, rather than as tests upon series of non-negative terms. Our preference is to proceed as above, that is, consciously to switch from $\sum x_k$ to $\sum |x_k|$, use the appropriate test there, and then switch back to see what we have learned about the original series. (For one thing, this forces awareness of the important fact that we are dealing with two distinct series, not one.) For the sake of completeness, however, here are the two tests as applicable to general series:

14.1.9 **The nth root test for general series**   Suppose that $\sum_{n=1}^{\infty} a_n$ is a series of real terms and that $\sqrt[n]{|a_n|}$ converges to a limit ℓ (as $n \to \infty$). Then:

1. if $\ell < 1$ then the series converges absolutely,

2. if $\ell > 1$ then a_n and $|a_n|$ do not tend to zero, and therefore the series diverges.

14.1.10 **D'Alembert's ratio test for general series**   Suppose that $\sum_{n=1}^{\infty} a_n$ is a series of non-zero terms and that the growth rate $|a_{n+1}|/|a_n|$ converges to a limit ℓ (as $n \to \infty$). Then:

1. if $\ell < 1$ then the series converges absolutely,

2. if $\ell > 1$ then a_n and $|a_n|$ do not tend to zero, and therefore the series diverges.

14.2 The 'robustness' of absolutely convergent series

For the learner, one of the most disturbing features of series is that it is sometimes possible to take a convergent series, 'add up' all its terms in a different order, and get a *different* sum to infinity (we shall see this soon). This outrageous behaviour flies in the face of what we all learned about basic arithmetic in elementary school, and one of the most reassuring features of *absolutely* convergent series is that they do not behave badly like this. Indeed, they can even be multiplied together in a more-or-less natural fashion. This section focusses on the 'good' behaviour of absolutely convergent series when subjected to arithmetical processes such as re-ordering and multiplying. Some of the proofs are relatively complicated and it will again be perfectly acceptable if you omit them on a first reading – but do take on board the results and the examples as soon as possible.

An easy topic to begin with (although not one that is really about absolute convergence) is that of insertion and removal of brackets. If $\sum x_n = x_1 + x_2 + x_3 + x_4 + \cdots$ is a convergent series, we can create many more series by imposing brackets in an arbitrary manner upon the stream of numbers, for instance:

$$(x_1 + x_2) + x_3 + (x_4 + x_5 + x_6) + (x_7 + x_8) + \cdots .$$

To explore the question of its convergence, we must examine its partial sums which, in the present example, begin with

$$x_1+x_2, x_1+x_2+x_3, x_1+x_2+x_3+x_4+x_5+x_6, x_1+x_2+x_3+x_4+x_5+x_6+x_7+x_8, \cdots .$$

Notice that these constitute a subsequence of the partial-sum sequence for the original series – indeed, this was inevitable, since the nth partial sum of the bracketed series is simply the $m(n)^{th}$ original partial sum where $m(n)$ is the label on the last term of the nth bracket (regarding each unbracketed term as sitting inside an invisible pair of brackets on its own). That is the only insight needed to establish:

14.2.1 **Theorem**   Any series arising by bracketing together blocks of terms in a convergent series is convergent, and to the same sum-to-infinity.

Proof

The partial-sum sequence for the bracketed series is a subsequence of the partial-sum sequence for the original series, and therefore converges to the same limit.

14.2.2 **Example**   On the other hand, removal of existing brackets can completely change the convergence status of a series. For a simple example, consider:

$$(1 - 1) + (2 - 2) + (3 - 3) + (4 - 4) + (5 - 5) + \cdots .$$

Clearly this converges, since every single term (every single bracket) is zero, and $\sum 0$ converges to 0. However, if we remove all[2] the brackets, it becomes

$$1 - 1 + 2 - 2 + 3 - 3 + 4 - 4 + 5 - 5 + \cdots$$

whose partial sums

$$1, 0, 2, 0, 3, 0, 4, 0, 5, 0, \cdots$$

---

[2] Indeed, a similar argument runs if we remove an infinite number of the brackets.

don't merely fail to converge to zero, they fail to converge at all since they are unbounded (the $(2n - 1)^{th}$ partial sum is n, for each $n \in \mathbb{N}$).

14.2.3 **EXERCISE**   In contrast, show that if a bracketed version of a series *of non-negative terms* converges, then so did the original series (and to the same sum).

Turning next to the question of 'robustness of convergence under rearrangement', we once again begin by tackling the easy special case of a series of *non-negative* terms.

14.2.4 **Roughwork**   Starting with a convergent series of non-negative terms

$$a_1 + a_2 + a_3 + a_4 + a_5 + \cdots$$

let us imagine a typical rearranged series consisting of exactly the same terms but in a different order, such as

$$a_9 + a_3 + a_{41} + a_2 + a_{17} + \cdots .$$

Look at the first few partial sums of the rearranged series (as indeed we must, if we seek its sum to infinity):

$$a_9, a_9 + a_3, a_9 + a_3 + a_{41}, a_9 + a_3 + a_{41} + a_2, a_9 + a_3 + a_{41} + a_2 + a_{17}, \cdots .$$

The bad news is that these are, of course, not partial sums of the original series. Give them a temporary name, say, *random handfuls*.

The better news is the observation that this is an increasing sequence, just as was the partial-sum sequence of the original series . . . and for increasing sequences, *limit* and *supremum* are the same thing. Furthermore, each of the random handfuls is *part of* a partial sum, and therefore *less than or equal to* a partial sum since all the terms are non-negative.[3] Each random handful is therefore $\leq$ the supremum of all the partial sums, that is, the limit of the original series, so the supremum of the random handfuls ($=$ the limit of the rearranged series) must be $\leq$ the limit of the original series. Presumably we now only have to reverse the argument to obtain the inequality the other way round?

14.2.5 **Theorem**   Any rearrangement of a convergent series of non-negative terms converges to the same sum.

---

[3] For instance, $a_9 + a_3$ is less than $\sum_1^9 a_k$, $a_9 + a_3 + a_{41}$ is less than $\sum_1^{41} a_k$ and so on.

Proof

Given that $\sum b_n$ is a rearrangement of a series $\sum a_n$ that converges to ℓ, and in which $a_n \geq 0$ for all n, recall that ℓ is the supremum of the partial sums of $\sum a_n$.

Each partial sum of $\sum b_n$ is the sum of finitely many a_ns scattered in some unpredictable pattern within the series $\sum a_n$, so *these* a_ns must all occur *before* some particular a_m in the original sequence and their total is therefore $\leq$ the mth partial sum of $\sum a_n$, and therefore also $\leq \ell$.

Since the partial-sum sequence for $\sum b_n$ is also increasing, it converges to some ℓ' where $\ell' \leq \ell$.

Now $\sum a_n$ is equally a rearrangement of $\sum b_n$ so, by the identical argument, $\ell \leq \ell'$.

Hence $\ell = \ell'$.

14.2.6 **Note**   In order to extend this conclusion to absolutely convergent series whose terms are a mixture of positives and negatives, we simply segregate out the positives from the negatives and think about the two streams separately. Given a series $\sum a_n$, we introduce the notation (for each $n \in \mathbb{N}$):

$$a_n^+ = a_n \text{ if } a_n \geq 0, \quad a_n^+ = 0 \text{ if } a_n < 0;$$
$$a_n^- = -a_n \text{ if } a_n < 0, \quad a_n^- = 0 \text{ if } a_n \geq 0.$$

Then

$$a_n = a_n^+ - a_n^-, \quad |a_n| = a_n^+ + a_n^-$$

in all cases (just check it out for non-negative a_n and for negative a_n : it works in both cases) but, importantly, every a_n^+ and every a_n^- is non-negative (and therefore we can use the preceding theorem on them separately).

Think what the remark $a_n = a_n^+ - a_n^-$ does to a typical partial sum of $\sum a_n$ and you will understand what we meant by 'segregating out the positives from the negatives'; for instance:

$$(3 - 5 - 2 + 1 + 6 - 4 + 2 - 7 - 3)$$
$$= (3 + 0 + 0 + 1 + 6 + 0 + 2 + 0 + 0) - (0 + 5 + 2 + 0 + 0 + 4 + 0 + 7 + 3).$$

In general, what we get is

$$\sum_1^n a_k = \sum_1^n a_k^+ - \sum_1^n a_k^-$$

and, likewise,

$$\sum_1^n |a_k| = \sum_1^n a_k^+ + \sum_1^n a_k^-.$$

The last display line shows that if $\sum_1^n a_k^+$ and $\sum_1^n a_k^-$ both converge, then so must $\sum_1^n |a_k|$; yet the converse is also true: if $\sum_1^n |a_k|$ converges then, because (for all n) $0 \le a_n^+ \le |a_n|$ and $0 \le a_n^- \le |a_n|$, the comparison test tells us that both $\sum_1^n a_k^+$ and $\sum_1^n a_k^-$ converge. Furthermore, their sums-to-infinity add in the obvious manner. We have proved:

14.2.7 **Lemma**   The series $\sum a_n$ is absolutely convergent if and only if both $\sum a_n^+$ and $\sum a_n^-$ converge. Furthermore, their sums then satisfy

$$\sum |a_n| = \sum a_n^+ + \sum a_n^- \text{ and}$$
$$\sum a_n = \sum a_n^+ - \sum a_n^- .$$

14.2.8 **Theorem**   Any rearrangement of an absolutely convergent series converges to the same sum.

Proof

Rearranging a given absolutely convergent $\sum a_n$ into a new order $\sum b_n$ will rearrange both $\sum a_n^+$ and $\sum a_n^-$ in exactly the same pattern. By the above, the two latter series are convergent, and by the previous theorem, this does not alter their sums, that is:

$$\sum a_n = \sum a_n^+ - \sum a_n^-$$
$$= \sum b_n^+ - \sum b_n^-$$
$$= \sum b_n .$$

14.2.9 **IMPORTANT EXERCISE**   Show that if a series $\sum a_n$ is conditionally convergent, that is, convergent but NOT absolutely convergent, then both $\sum a_n^+$ and $\sum a_n^-$ diverge to ∞.

For most learners, the surprise is not that absolute convergence is robust under rearrangement, but that non-absolute convergence isn't. It turns out that this may be demonstrated upon *any* convergent-but-not-absolutely-convergent series, but we shall demonstrate using the most obvious such object – the alternating harmonic series.

14.2.10 **Example**   The series $\sum_1^\infty (-1)^{k-1} k^{-1}$ is well known to converge but not absolutely. Furthermore, since in the expression

$$1 - \frac{1}{2} + \left(\frac{1}{3} - \frac{1}{4}\right) + \left(\frac{1}{5} - \frac{1}{6}\right) + \left(\frac{1}{7} - \frac{1}{8}\right) + \left(\frac{1}{9} - \frac{1}{10}\right) + \cdots$$

every bracket is positive, the sum-to-infinity (let us denote it by S), whatever it is, is more than 0.5. In particular, $S \ne 0$.

Suppose it were correct that every rearrangement of this series is also convergent to S (and now we shall *seek a contradiction*). In particular, the rearrangement

$$1 - \frac{1}{2} - \frac{1}{4} + \frac{1}{3} - \frac{1}{6} - \frac{1}{8} + \frac{1}{5} - \frac{1}{10} - \frac{1}{12} + \frac{1}{7} - \frac{1}{14} - \frac{1}{16} + \cdots$$

is a rearrangement since each original term appears once and once only (in the pattern of one positive term and two negative terms alternating) so it also converges to S.

From what we saw in 14.2.1 about imposing brackets, the modified series

$$\left(1 - \frac{1}{2}\right) - \frac{1}{4} + \left(\frac{1}{3} - \frac{1}{6}\right) - \frac{1}{8} + \left(\frac{1}{5} - \frac{1}{10}\right) - \frac{1}{12} + \left(\frac{1}{7} - \frac{1}{14}\right) - \frac{1}{16} + \cdots$$

that is,

$$\frac{1}{2} - \frac{1}{4} + \frac{1}{6} - \frac{1}{8} + \frac{1}{10} - \frac{1}{12} + \frac{1}{14} - \frac{1}{16} + \cdots$$

must also converge to S. Yet the last display is precisely one half of the original series so it converges to $S/2$. We deduce that $S = S/2$ which, since S is non-zero, is absurd.

Conclusion: rearrangement of at least some convergent (non-absolutely convergent) series can alter their convergence!

14.2.11 **EXERCISE**   Devise a rearrangement of the alternating harmonic series that diverges to ∞. Devise another that diverges to $-\infty$.

14.2.12 **HARDER EXERCISE**   Given a completely arbitrary series $\sum x_k$ that is conditionally convergent, and a completely arbitrary real number ℓ, think how you could devise a rearrangement of $\sum x_k$ that converges to ℓ.

Suggestion: the key ingredient in finding one is that both $\sum x_k^+$ and $\sum x_k^-$ diverge to infinity. You might begin by taking *just enough* of the non-negative terms of the series to make the running total greater than ℓ.

14.2.13 **Note**   Our last task in this section is to verify that absolutely convergent series are robust under multiplication and, prior to that, we must define what we mean by multiplying two series together. Tempting though it may at first appear to multiply them 'term by term' as we did successfully for sequences (that is, to define $\sum x_k$ times $\sum y_k$ to mean $\sum x_k y_k$), this definition completely fails to match how series are actually used in practice, so we had better begin by a forward glance at one of their key applications: power series representations of functions.

It is well known (and yes, we shall be checking this out in detail) that many important functions can be represented, or even optimally defined, by power series. For instance, you have probably encountered the following:

$$e^x = 1 + x + \frac{x^2}{2!} + \frac{x^3}{3!} + \cdots = \sum_0^\infty \frac{x^n}{n!},$$

$$\sin x = x - \frac{x^3}{3!} + \frac{x^5}{5!} - \frac{x^7}{7!} + \cdots = \sum_0^\infty (-1)^n \frac{x^{2n+1}}{(2n+1)!},$$

$$\cos x = 1 - \frac{x^2}{2!} + \frac{x^4}{4!} - \frac{x^6}{6!} + \cdots = \sum_0^\infty (-1)^n \frac{x^{2n}}{(2n)!}.$$

Much of the purpose of representing functions by power series is in order to be able to manipulate the power series instead of the functions. So, for instance, if two functions f and g are represented by power series $\sum a_k x^k$ and $\sum b_k x^k$, then we expect/need the sum $f + g$ to be represented by $\sum a_k x^k + \sum b_k x^k$ and the difference $f - g$ by $\sum a_k x^k - \sum b_k x^k$, and *somebody* should have taken care to *define* the sum and the difference of power series so that this does happen. Luckily, there is no problem here: we define

$$\sum a_k x^k + \sum b_k x^k \quad \text{to be} \quad \sum (a_k + b_k) x^k,$$

we define

$$\sum a_k x^k - \sum b_k x^k \quad \text{to be} \quad \sum (a_k - b_k) x^k,$$

and everything turns out to run so smoothly that there was really no need to make a fuss about it.

In the case of multiplication, it is rather less obvious what to do. We need the product function fg to be represented by $\sum a_k x^k \times \sum b_k x^k$, but how then ought that product to be defined?

We shall let our definition be guided by the simplest case: the case where f and g are polynomials. Then the power series that 'represent' them are just f and g themselves (or, to be really fussy, f and g with infinite strings of zero terms attached: so that, for instance, the function $f(x) = 2 - 5x + 3x^2 - 4x^3$ is represented by the power series

$$2 - 5x + 3x^2 - 4x^3 + 0x^4 + 0x^5 + 0x^6 + 0x^7 + \cdots$$

but, pragmatically, we are not going to waste paper and patience by writing out endless strings of zero terms).

Take the case of two cubics, say,

$$f(x) = a_0 + a_1 x + a_2 x^2 + a_3 x^3, \quad g(x) = b_0 + b_1 x + b_2 x^2 + b_3 x^3.$$

What power series $c_0 + c_1 x + c_2 x^2 + c_3 x^3 + \cdots$ shall represent their product fg? The question virtually answers itself, because their product *is* itself a polynomial, as a few tedious minutes with paper and pen will show you:

$$(fg)(x) = a_0 b_0 + (a_0 b_1 + a_1 b_0) x + (a_0 b_2 + a_1 b_1 + a_2 b_0) x^2 +$$
$$(a_0 b_3 + a_1 b_2 + a_2 b_1 + a_3 b_0) x^3 + (a_0 b_4 + a_1 b_3 + a_2 b_2 + a_3 b_1 + a_4 b_0) x^4 + \cdots$$

(and two more terms). Suddenly we are left with no freedom of action about how to multiply the series: the coefficient c_0 has to be $a_0 b_0$, c_1 has to be $a_0 b_1 + a_1 b_0$, c_2 has to be $a_0 b_2 + a_1 b_1 + a_2 b_0$ and so on. Any other decision we might consider making would create a definition that didn't even work correctly for polynomials, let alone for general (properly infinite) power series.

This is why the following definition,[4] complicated though it looks, is the right one for our purposes:

14.2.14 Definition The *Cauchy product* of two series[5] $\sum_0^\infty a_k$ and $\sum_0^\infty b_k$ is the series $\sum_0^\infty c_k$ defined by

$$c_0 = a_0 b_0, \quad c_1 = a_0 b_1 + a_1 b_0, \quad c_2 = a_0 b_2 + a_1 b_1 + a_2 b_0,$$

and, in general,

$$c_n = a_0 b_n + a_1 b_{n-1} + a_2 b_{n-2} + \cdots + a_n b_0 = \sum_{k=0}^{k=n} a_k b_{n-k}.$$

14.2.15 Theorem If the two series $\sum_0^\infty a_k$ and $\sum_0^\infty b_k$ are absolutely convergent, with sums A and B, say, then their Cauchy product is also absolutely convergent, and its sum is AB.

Roughwork

When we multiply a partial sum (call it $A_n = \sum_0^n a_k$) of the first series by a partial sum B_n of the second, the various fragments $a_i b_j$ do not naturally line up in a sequence but, rather, in a two-dimensional grid such as

$$
\begin{array}{cccc}
a_0 b_0 & a_1 b_0 & a_2 b_0 & a_3 b_0 & \cdots \\
a_0 b_1 & a_1 b_1 & a_2 b_1 & a_3 b_1 & \cdots \\
a_0 b_2 & a_1 b_2 & a_2 b_2 & a_3 b_2 & \cdots \\
a_0 b_3 & a_1 b_3 & a_2 b_3 & a_3 b_3 & \cdots \\
\vdots & \vdots & \vdots & \vdots &
\end{array}
$$

and so on. There are several different ways to string that array out into a sequence so that we can consider adding the terms up as a series. For one, we can create an expanding list of 'square shells' starting at the top left hand corner (follow these on the grid to see what we mean by that somewhat cryptic phrase):

---

[4] We have formulated it for arbitrary series, not just for power series, so all the x^ns have disappeared; but the main application we intend is still that of power series.

[5] We are starting the labelling at $k = 0$ instead of at $k = 1$, again mainly because of the focus on power series which do naturally begin with $a_0 x^0$ in order to accommodate a constant term.

$$a_0 b_0$$
$$+a_0 b_1 + a_1 b_1 + a_1 b_0$$
$$+a_0 b_2 + a_1 b_2 + a_2 b_2 + a_2 b_1 + a_2 b_0$$
$$+a_0 b_3 + a_1 b_3 + a_2 b_3 + a_3 b_3 + \quad \text{and so on} \cdots$$

Notice that, at the end of each line, the running totals are $A_0 B_0, A_1 B_1, A_2 B_2, A_3 B_3$ and so on – a sequence whose limit is easy to grasp.

On the other hand, if we sort out the array into a sequence/series by following 'diagonal sweeps' (again, please follow these on the grid to see what we mean), we get instead:

$$a_0 b_0$$
$$+a_0 b_1 + a_1 b_0$$
$$+a_0 b_2 + a_1 b_1 + a_2 b_0$$
$$+a_0 b_3 + a_1 b_2 + a_2 b_1 + a_3 b_0$$
$$+a_0 b_4 + a_1 b_3 + \quad \text{and so on} \cdots$$

and look: each line is now one of the Cauchy product coefficients – we are now building up $c_0 + c_1 + c_2 + c_3 + \cdots$ as, indeed, we must do if we want to address what this theorem claims.

What we now need is a guarantee that these quite different sorting processes will give ultimately the same sum to infinity, that is, we need to be able to rearrange and know that the sum is robust. For this, absolute convergence must be established first; and for that, we have to begin with the same array *but with modulus signs on every term*.

Proof

Let $\overline{A}_n, \overline{B}_n$ stand for the nth partial sums of $\sum_0^\infty |a_k|$ and $\sum_0^\infty |b_k|$ respectively (which we know to be convergent series) and, since the partial-sum sequences are bounded, find two positive constants P, Q such that (for all $n \in \mathbb{N}$)

$$\overline{A}_n \leq P \quad \text{and} \quad \overline{B}_n \leq Q.$$

The various numbers $|a_i b_j|$ that turn up when we multiply $\overline{A}_n$ and $\overline{B}_n$ together present themselves naturally in a two-dimensional (infinite) grid:

$|a_0 b_0|$	$|a_1 b_0|$	$|a_2 b_0|$	$|a_3 b_0|$	$\cdots$
$|a_0 b_1|$	$|a_1 b_1|$	$|a_2 b_1|$	$|a_3 b_1|$	$\cdots$
$|a_0 b_2|$	$|a_1 b_2|$	$|a_2 b_2|$	$|a_3 b_2|$	$\cdots$
$|a_0 b_3|$	$|a_1 b_3|$	$|a_2 b_3|$	$|a_3 b_3|$	$\cdots$
$\vdots$	$\vdots$	$\vdots$	$\vdots$	

To be precise, the items in the first $(n + 1)$ places of the first $(n + 1)$ rows add up to $\overline{A}_n\overline{B}_n$.

Any finite selection of terms from the grid will lie within 'the first $(n+1)$ places of the first $(n + 1)$ rows' if we choose n big enough, so the sum total of any finite selection is less than or equal to $\overline{A}_n\overline{B}_n$ for *that* n, and therefore cannot exceed PQ. That is, no matter how we string these items $|a_ib_j|$ together into a sequence, the resulting series (of non-negatives) has its partial sums bounded above (by PQ) and must therefore converge. In other words, if we strip out the modulus signs from the grid and organise its entries into a sequence *in any fashion whatsoever*, the resulting series is absolutely convergent to some sum S. Best of all: it is the same number S *no matter how* we chose to organise them: for rearranging an absolutely convergent series does not alter its sum.

So now consider the 'un-modulussed' grid:

$$
\begin{array}{llll}
a_0b_0 & a_1b_0 & a_2b_0 & a_3b_0 & \cdots \\
a_0b_1 & a_1b_1 & a_2b_1 & a_3b_1 & \cdots \\
a_0b_2 & a_1b_2 & a_2b_2 & a_3b_2 & \cdots \\
a_0b_3 & a_1b_3 & a_2b_3 & a_3b_3 & \cdots \\
\vdots & \vdots & \vdots & \vdots
\end{array}
$$

If, firstly, we choose to sort it into a sequence and then a series as follows:

$$a_0b_0$$
$$+a_0b_1 + a_1b_1 + a_1b_0$$
$$+a_0b_2 + a_1b_2 + a_2b_2 + a_2b_1 + a_2b_0$$
$$+a_0b_3 + a_1b_3 + a_2b_3 + a_3b_3 + \quad and\ so\ on \cdots$$

then its partial sums converge to S and, moreover, the subsequence comprising partial sums number $1, 4, 9, 16, 25, \cdots$ also converges to S. Yet partial sum number $(n + 1)^2$ is exactly

$$(a_0 + a_1 + a_2 + \cdots + a_n)(b_0 + b_1 + b_2 + \cdots + b_n)$$

which converges to AB. We now know that $S = AB$.

Secondly, let us sort the grid into a different sequence and series like this:

$$a_0b_0$$
$$+a_0b_1 + a_1b_0$$
$$+a_0b_2 + a_1b_1 + a_2b_0$$
$$+a_0b_3 + a_1b_2 + a_2b_1 + a_3b_0$$
$$+a_0b_4 + a_1b_3 + \quad and\ so\ on \cdots$$

This series also has to converge to $S = AB$, and so will the subsequence comprising items $1, 3, 6, 10, 15, \cdots$ of its partial sums (as indicated by the line-breaks here). Yet these are precisely the Cauchy product numbers

$$c_0, c_0 + c_1, c_0 + c_1 + c_2, c_0 + c_1 + c_2 + c_3$$

and so on.

We are (at last) able to conclude that the Cauchy product series converges to AB.

14.2.16 EXERCISE   We know that (for all $x \in (-1, 1)$) the series $1 + x + x^2 + x^3 + x^4 + \cdots$ converges to $\frac{1}{1-x}$ and the series $1 - x + x^2 - x^3 + x^4 - \cdots$ converges to $\frac{1}{1+x}$. Calculate (and simplify as necessary) the Cauchy product of these two series and confirm that, as predicted by the theorem, it converges to the product of the two functions.

14.2.17 EXERCISE   Assuming the correctness of

$$e^x = 1 + x + \frac{x^2}{2!} + \frac{x^3}{3!} + \cdots = \sum_0^\infty \frac{x^n}{n!},$$

calculate (and simplify as necessary) the Cauchy product of the power series representations of e^x and of e^y, and confirm that it converges to the product of the two functions.

14.3 Power series

A power series is a series of the form $\sum_0^\infty a_n(x - c)^n$ where c is a constant (the 'centre') and the 'coefficients' a_n are also constants. Most of the time, we change the variable by substituting, say, $y = x - c$ so that the appearance of the series simplifies to $\sum_0^\infty a_n y^n$ (and the centre becomes 0). Since this can always be done, most of the theory assumes that it has already taken place. That is, in practice, *a power series is a series of the form $\sum_0^\infty a_n x^n$.*

We need to be aware of which values of x make $\sum_0^\infty a_n x^n$ converge and which make it diverge, and for a series of this type there are just three possible scenarios: either it converges absolutely for all x, or only for $x = 0$, or (the most typical case) there is a number D such that the series converges absolutely whenever $|x| < D$ and diverges whenever $|x| > D$. The number D is known as the *radius of convergence*, mainly because all this theory can equally well be developed for the case in which x is a *complex* number, and then $|x| < D$ describes the inside of a circle centred at the origin and of radius D. Since we are discussing only real functions, we shall have to put up with the slightly odd use of the word 'radius' to describe half the length of an interval $(-D, D)$ (which is where we know that the real series converges). It is generally more difficult to determine whether it converges at $x = D$ or at $x = -D$.

The two extreme cases (when the series converges for all x, or only for $x = 0$) are conventionally represented by saying that the radius of convergence is infinite, or zero.

14.3.1 Lemma   Every power series has a radius of convergence.

Proof

If $\sum_0^\infty a_n x^n$ converges only at $x = 0$, then $D = 0$. If not, pick any non-zero t such that $\sum_0^\infty a_n t^n$ does converge. Then certainly $a_n t^n \to 0$ and is therefore bounded: there is $M > 0$ such that $|a_n t^n| < M$ for all n.

For any number u in the interval $(-|t|, +|t|)$, we have

$$|a_n u^n| \le \frac{M}{|t^n|}|u|^n = M\left|\frac{u}{t}\right|^n$$

and the last item belongs to a convergent geometric series so, by the comparison test, $\sum |a_n u^n|$ is also convergent and $\sum a_n u^n$ is absolutely convergent. In other words, whenever t is a point in the 'convergence zone' of the given power series, then every point in $(-|t|, +|t|)$ (every point that lies closer to zero than t does) is also in the convergence zone.

The only subsets of $\mathbb{R}$ that possess this property are $\mathbb{R}$ itself and the intervals that are centred on 0 (length $2D$, say). Hence either ∞ or D acts as radius of convergence for the series.

In many cases, the radius of convergence can be calculated quite easily:

14.3.2 Theorem   For a given power series $\sum a_n x^n$:

1. If $\sqrt[n]{|a_n|}$ converges to a limit ℓ then
   - $\ell > 0$ implies that the radius of convergence is $\dfrac{1}{\ell}$,
   - $\ell = 0$ implies that the radius of convergence is ∞;

2. If $\dfrac{|a_{n+1}|}{|a_n|}$ converges[6] to a limit ℓ then
   - $\ell > 0$ implies that the radius of convergence is $\dfrac{1}{\ell}$,
   - $\ell = 0$ implies that the radius of convergence is ∞.

Proof

Given that $\sqrt[n]{|a_n|}$ converges to ℓ, we see that $\sqrt[n]{|a_n x^n|} = \sqrt[n]{|a_n|}|x| \to \ell|x|$. The root test (applied to $\sum |a_n x^n|$) tells us that if $\ell|x| < 1$ then $\sum |a_n x^n|$ converges (so $\sum a_n x^n$ converges absolutely) whereas if $\ell|x| > 1$ then $|a_n x^n|$ does not tend to zero, $a_n x^n$ also does not tend to zero, and $\sum a_n x^n$ diverges. Separating out the cases $\ell > 0$ and $\ell = 0$, that proves the first part. The second emerges from the d'Alembert test in the same way.

---

[6] and, implicitly, a_n is non-zero for all sufficiently large values of n, of course.

14.3.3 **EXERCISE** Determine the radius of convergence of each of the following:

1. $(e^x =)\ 1 + x + \dfrac{x^2}{2!} + \dfrac{x^3}{3!} + \cdots = \sum_0^\infty \dfrac{x^n}{n!},$

2. $(\sin x =)\ x - \dfrac{x^3}{3!} + \dfrac{x^5}{5!} - \dfrac{x^7}{7!} + \cdots = \sum_0^\infty (-1)^n \dfrac{x^{2n+1}}{(2n+1)!},$

3. $(\cos x =)\ 1 - \dfrac{x^2}{2!} + \dfrac{x^4}{4!} - \dfrac{x^6}{6!} + \cdots = \sum_0^\infty (-1)^n \dfrac{x^{2n}}{(2n)!},$

4. $\sum (30n + n^{30})x^n,$

5. $\sum (n!)x^n,$

6. $\sum \left(\dfrac{n!(n+1)!(n+2)!}{(3n+1)!} \right) x^n,$

7. $\sum \dfrac{x^n}{\left(1 + \frac{3}{n}\right)^{2n^2}}.$

It is important to realise that when, for each $x \in (-D, D)$, a power series $\sum a_n x^n$ converges to a limit, then that limit will depend on the value of x, that is, it will itself be a function of x. The whole concept of a sequence or series of functions converging to a function is of great importance in analysis, but we can deal with only a few key aspects of it in this text. Most particularly, we need to discuss continuity and differentiability of the sum of a series of continuous or differentiable functions, but only in the special case of convergent power series. Continuity is fairly straightforward to check out:

14.3.4 **Theorem** When a power series $\sum a_n x^n$ has nonzero radius of convergence D, then its sum $f(x) = \sum a_n x^n$ is continuous at every point of $(-D, D)$. (This includes the special case $D = \infty$: then f is continuous on the whole of $\mathbb{R}$.)

Roughwork/preparation

Let us put $S_m(x) = \sum_0^m a_n x^n$, the mth partial sum. Since $\lim_{m \to \infty} S_m(x) = f(x)$, $S_m(x)$ will be a good approximation to $f(x)$ (provided m is taken big enough) for each individual x: but that is not strong enough to guarantee the approximation to be equally good for all values of x at once – indeed, the endpoints $\pm D$ of the interval could present serious problems since we do not even know whether the series converges at all there. Common sense therefore suggests that we stay away from $\pm D$ and work on a slightly smaller interval of the form $[-\rho, \rho]$ for some suitably chosen positive $\rho < D$.

Let $\varepsilon > 0$ be given. Since $\sum a_n \rho^n$ converges absolutely to $f(\rho)$, the modulussed series $\sum |a_n| \rho^n$ converges to some limit $\ell' = \sum_0^\infty |a_n| \rho^n$, and so we can find a positive integer m such that

$$\left| \ell' - \sum_0^m |a_n| \rho^n \right| = \left| \sum_{m+1}^\infty |a_n| \rho^n \right| < \frac{\varepsilon}{3}.$$

Now for any $x \in [-\rho, \rho]$, we have

$$\left| f(x) - \sum_0^m a_n x^n \right| = \left| \sum_{m+1}^{\infty} a_n x^n \right| \leq \sum_{m+1}^{\infty} |a_n||x|^n \leq \sum_{m+1}^{\infty} |a_n|\rho^n < \frac{\varepsilon}{3}$$

that is, the mth partial sum S_m is an $\frac{\varepsilon}{3}$-good approximation to f at every point of $[-\rho, \rho]$ simultaneously.

Proof

Given any point x_0 in $(-D, D)$ and any $\varepsilon > 0$, choose ρ so that $-\rho < x_0 < +\rho < D$, and choose $m \in \mathbb{N}$ as in the roughwork/preparation. Since $S_m(x)$ is continuous (being just a polynomial) we can find $\delta > 0$ so that

$$|S_m(x) - S_m(x_0)| < \frac{\varepsilon}{3} \text{ whenever } x \text{ lies in the interval } (x_0 - \delta, x_0 + \delta).$$

(We also take care that δ is small enough to fit $(x_0 - \delta, x_0 + \delta)$ inside $[-\rho, \rho]$.)
    Now for any $x \in (x_0 - \delta, x_0 + \delta)$:

$$|f(x) - f(x_0)| = |(f(x) - S_m(x)) + (S_m(x) - S_m(x_0)) + (S_m(x_0) - f(x_0))|$$
$$\leq |f(x) - S_m(x)| + |S_m(x) - S_m(x_0)| + |S_m(x_0) - f(x_0)|$$
$$< \frac{\varepsilon}{3} + \frac{\varepsilon}{3} + \frac{\varepsilon}{3} = \varepsilon.$$

That is, f is continuous at x_0. Since x_0 was an arbitrary element of $(-D, D)$, the proof is complete.

14.3.5 Notes

1. There are a couple of places in the above argument at which a conscientious student might perfectly justifiably feel anxious. For one thing, we used an infinite version of the triangle inequality although we have ever only proved finite versions. Yet this is legitimate: if (a_k) is a sequence then, for each $m \in \mathbb{N}$, we have already shown that

$$\left| \sum_1^m a_k \right| \leq \sum_1^m |a_k|.$$

Now provided that $\sum a_k$ is absolutely convergent, the second summation converges to its supremum ℓ, the number conventionally written as $\ell = \sum_1^{\infty} |a_k|$. Thus, for every m:

$$-\ell \leq \sum_1^m a_k \leq \ell.$$

However, absolute convergence implies convergence, so $\sum_1^m a_k$ also converges to its limit, the number conventionally written as $\sum_1^\infty a_k$. Taking limits across the previous display, we therefore obtain $-\ell \leq \sum_1^\infty a_k \leq \ell$, that is, $\left|\sum_1^\infty a_k\right| \leq \ell$, or

$$\left|\sum_1^\infty a_k\right| \leq \sum_1^\infty |a_k|,$$

as we desired.

2. Lines such as

$$f(x) = \sum_0^\infty a_n x^n \text{ therefore } f(x) - \sum_0^m a_n x^n = \sum_{m+1}^\infty a_n x^n,$$

plausible though they appear, should also create a pause for thought since two limiting processes are involved. Remember that, once the integer m has been fixed, $\sum_0^m a_n x^n$ is simply a real number, a constant.

If we take a convergent series $\sum_1^\infty a_k$, and add a constant K as a zeroeth term, every partial sum will increase by K and therefore so will the sum-to-infinity. In other notation,

$$\lim_{n \to \infty} (K + a_1 + a_2 + a_3 + \cdots + a_n) = K + \lim_{n \to \infty} (a_1 + a_2 + a_3 + \cdots + a_n).$$

In the case where $K = -\sum_1^m a_k$, and restricting the discussion to $n > m$ which will not disturb limiting behaviour, this says

$$\lim_{n \to \infty} (a_{m+1} + a_{m+2} + a_{m+3} + \cdots + a_n) = K + \lim_{n \to \infty} (a_1 + a_2 + a_3 + \cdots + a_n),$$

confirming the legitimacy of steps such as

$$\sum_{m+1}^\infty a_k = -\sum_1^m a_k + \sum_1^\infty a_k.$$

3. The idea of finding an approximation to a limit function $f(x)$ that is 'ε-good' simultaneously for an entire interval of values of x, instead of just for one x-value that interested us, is called *uniform convergence*. We mobilised it just once, in the roughwork preparation for the last theorem. The interested reader (with considerable time to spare) will find a great deal more about this in the literature.

The final result in the set, concerning how to differentiate the sum of a power series, looks like little more than common sense at first sight. The proof, however, is demanding, so we shall divide its burden across this chapter and Chapter 16. Given that

$$f(x) = a_0 + a_1 x + a_2 x^2 + a_3 x^3 + \cdots + a_n x^n + \cdots$$

converges on an open interval $(-D, D)$, the only reasonable guess that comes to mind is that the so-called *derived series*

$$a_1 + 2a_2 x + 3a_3 x^2 + \cdots + na_n x^{n-1} + \cdots$$

really ought to converge on the same interval, and its sum *really ought* to be the derivative of $f(x)$. This is, in fact, true. For the moment, we shall content ourselves with proving only the first 'really ought'.

14.3.6 Theorem If $\sum a_n x^n$ has radius of convergence $D > 0$, then the derived series $\sum na_n x^{n-1}$ also converges on $(-D, D)$.

Proof

For arbitrary x_0 in $(-D, D)$, again start by choosing positive ρ so that $-\rho < x_0 < +\rho < D$. Since $\sum a_n \rho^n$ converges, its terms tend to zero and are therefore bounded: there is a positive number M such that

$$|a_n \rho^n| < M, \text{ that is, } |a_n| < \frac{M}{\rho^n} \text{ for all } n \in \mathbb{N}.$$

Therefore

$$|na_n x_0^{n-1}| \le n\frac{M}{\rho^n}|x_0|^{n-1} = \frac{M}{\rho} \times n\left(\frac{|x_0|}{\rho}\right)^{n-1}.$$

Now $\frac{M}{\rho}$ is merely a constant, and the power series $\sum nx^{n-1}$ has radius of convergence 1, so the final term in the display belongs to a convergent series. The comparison test applies, and shows that $\sum na_n x_0^{n-1}$ is also absolutely convergent. Since x_0 was any element of $(-D, D)$, the proof is complete.

14.3.7 Remark Hence also the second derived series $\sum n(n-1)a_n x^{n-2}$, the third derived series $\sum n(n-1)(n-2)a_n x^{n-3}$, and so on, all converge absolutely on the interval $(-D, D)$, where D is the radius of convergence of the original power series.

It can also be shown that the radius of convergence of the derived series (and of the second, and of the third . . .) is exactly D.

15 Uniform continuity – continuity's global cousin

15.1 Introduction

Continuity is a local property of a function, not a global one.

It is all too easy to forget this, because most of the functions we meet in practice are continuous at every point in their domain; this tends to create a misleading impression that continuity is global in character. However, look at the definition of continuity on a set: a function $f : D \to \mathbb{R}$ is continuous on D if, *for every individual point p in D and every sequence (x_n) in D that converges to p, we find that the limit* of $f(x_n)$ equals $f(p)$. The phrase in italics reminds us that continuity upon a set needs to be assessed at each individual element of that set: that makes it essentially a 'local' property (because it is judged locally, one point at a time).

We can emphasise this further by reminding ourselves of the $\varepsilon - \delta$, challenge-response, input-output 'game' that we can play in order to confirm continuity, namely:

> *for each output challenge $\varepsilon > 0$, there is an input response $\delta > 0$*
>
> *such that $x \in D, |x - p| < \delta$ together imply $|f(x) - f(p)| < \varepsilon$.*

(Of course, we don't usually write most of the English words in that display – we are just reminding ourselves about the nature, the dynamics of the game.)

It goes without saying that the response depends on the challenge – that δ depends upon (is a function of) ε. We could have been hyper-fussy and written not δ but δ_ε or $\delta(\varepsilon)$ to make this point visible, but we don't: as was just remarked, it goes without saying.

What also goes without saying, and is about to become important, is that δ is also allowed to depend upon and vary with p. Because continuity is assessed by the $\varepsilon - \delta$ game one point at a time, there is no requirement that (for a particular ε) the δ you find at one point should equal the δ you find at another. It *might* happen, but for ordinary continuity it doesn't have to.

Undergraduate Analysis: A Working Textbook, Aisling McCluskey and Brian McMaster 2018.
© Aisling McCluskey and Brian McMaster 2018. Published 2018 by Oxford University Press

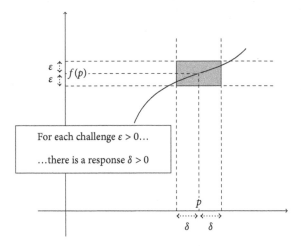

Ordinary continuity at p: δ may well depend on p

Straight line graph, positive gradient K

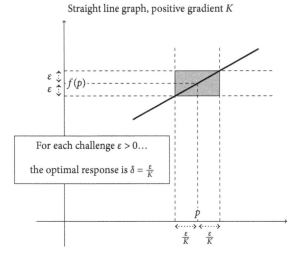

'Uniform' continuity: a straight-line function; optimal δ does not depend on p

We can throw some more light onto this issue by looking at a few simple illustrative examples. The straight-line function $s : \mathbb{R} \to \mathbb{R}$ given by $s(x) = Kx$ (where K is some positive constant) is just about the simplest of all functions upon which to play the $\varepsilon - \delta$ game because, for a given $\varepsilon > 0$, the choice $\delta = \varepsilon/K$ is optimal: it works, and it is the biggest possible choice of δ that does work.

Consider now a slightly more complicated function, say, the function $s' : [0,4] \to \mathbb{R}$ whose graph consists of the four pieces of straight line joining the points $(0,0)$ to $(1,1)$, $(1,1)$ to $(2,3)$, $(2,3)$ to $(3,6)$ and $(3,6)$ to $(4,10)$.

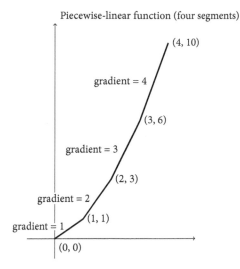

Uniform continuity: piecewise linear, four segments

Since their gradients are $1, 2, 3$ and 4, playing the $\varepsilon - \delta$ game (for a small value of ε) at 0.5 has an optimal choice of $\delta = \varepsilon/1$; but playing instead at 1.5, the optimal choice changes to $\delta = \varepsilon/2$, and playing instead at 2.5 and at 3.5 changes it to $\delta = \varepsilon/3$ and to $\delta = \varepsilon/4$. Of course, a general policy of choosing $\delta = \varepsilon/4$ would have dealt with all four of these points, since a choice of δ that is *smaller* than the optimal choice is still a perfectly valid choice – a winning move in the game, so to speak. Devoting a little extra care to the corners (because at the three junction points $(1, 1)$, $(2, 3)$ and $(3, 6)$, the graph actually doesn't have a gradient), you can readily check that the strategy $\delta = \varepsilon/4$ will win the $\varepsilon - \delta$ game at every point in $[0, 4]$.

Summary so far: for the function s', the natural choice of δ (for a given positive value of ε) varies from one point of its domain to another, but there is a uniform way to select δ that actually works irrespective of which point you focus on.

If we step up the last example by adding straight line fragments of gradient $5, 6, 7, \cdots, n$, then the discussion barely changes: the 'best' choice of δ at a point on the part of the graph that has gradient j is $\delta = \varepsilon/j$, and *that* varies from one point to another; but a systematic choice of $\delta = \varepsilon/n$ will work smoothly at every single point of the domain $[0, n]$.

Now take the final step in the direction in which we are travelling, by continuing to add endlessly many pieces of straight line of ever steeper gradient to the graph of s'.

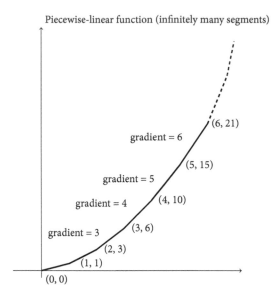

Piecewise-linear function (infinitely many segments)

Uniform continuity fails: piecewise linear, infinitely many segments

(If you wish, you can even obtain a formula for the function s'' whose graph this is:

$$s''(x) = nx + \frac{n - n^2}{2} \quad \text{while} \quad n - 1 \leq x < n \; (n \in \mathbb{N})$$

but, pragmatically, it is the shape of the graph that will help you to see what is happening, more than the formula does.)

For this function s'' it remains true that the optimal choice of δ at a point on the part of the graph that has gradient j is $\delta = \varepsilon/j$, but now there is *no* one-size-fits-all choice of δ that will work everywhere: if someone were to claim that (for a particular ε) a magical choice of $\delta = \delta'$ would work at all points of the domain $[0, \infty)$, then we could disprove that claim merely by finding an integer q greater than ε/δ', shifting our attention to a point on the graph at which the gradient is q, and observing that, at that point, the alleged all-purpose δ' is bigger than the optimal, the greatest acceptable value of δ (namely ε/q). Thus, for the function s'' (which certainly is continuous), *not only does δ vary naturally with the point at which you play the game, but IT HAS TO VARY*: there is no way to pick a δ that will work at all of the points in the domain of s''.

You may have noticed the word *uniform* sneaking into the discussion half a page back, and this is precisely what it means when it refers to continuity: a function is uniformly continuous on a set if not only can the $\varepsilon - \delta$ game be played and won at each point of the set, but also there is (for each $\varepsilon > 0$) a way to choose $\delta > 0$ that will work equally well at each and every point of the set; in other words, δ can be chosen in a way that is independent of the point at which the $\varepsilon - \delta$ game is to be played. So s and s' (as detailed above) were not just continuous but uniformly continuous, whereas s'' was continuous but *not* uniformly continuous. As George Orwell didn't quite get around to saying in *Animal Farm*, all (these) functions are continuous but some are more continuous than others.

15.2 Uniformly continuous functions

15.2.1 Definition Let $f : D \to \mathbb{R}$, and $A \subseteq D$. We say that f is *uniformly continuous* on A if

$$\text{for all } \varepsilon > 0 \text{ there exists } \delta > 0 \text{ such that}$$

$$x \in A, y \in A, |x - y| < \delta \text{ together imply } |f(x) - f(y)| < \varepsilon.$$

(We can usually assume that A is the *whole* of D since, if it were not, we could replace f by its restriction $f|_A$ to A and work with that function instead.)

15.2.2 Note Most people cannot, at first sight, see how this differs from 'ordinary' continuity on A but, hopefully, the introduction to this chapter will have clarified the distinction for you. In 'ordinary' continuity, you start with a y and an ε, and you go looking for a δ such that

$$|x - y| < \delta \text{ forces } |f(x) - f(y)| < \varepsilon \cdots\cdots (1).$$

In uniform continuity, on the other hand, you start only with an ε, and you go looking for a δ such that

$$|x - y| < \delta \text{ forces } |f(x) - f(y)| < \varepsilon \cdots\cdots (2).$$

In the first case, then, the δ needs to work only for a particular y and ε; in the second case, the δ has to work for a particular ε, but for all x and y that are δ-close no matter where in A they lie. This is asking considerably more: despite the apparent identity between (1) and (2), in (1) the y is fixed and only the x varies, whereas in (2) the x and the y are both free to vary.

15.2.3 Example To show that the function $f : [1, \infty) \to \mathbb{R}$ given by $f(x) = \sqrt{x}$ is uniformly continuous (on $[1, \infty)$).

Solution

We need to compare $|x - y|$ with $|f(x) - f(y)| = |\sqrt{x} - \sqrt{y}|$ and, fortunately, there is a simple algebraic connection between them:

$$(\sqrt{x} - \sqrt{y})(\sqrt{x} + \sqrt{y}) = x - y$$

from which it follows (keeping in mind that $x \geq 1$ and $y \geq 1$ in the present domain) that

$$|f(x) - f(y)| = |\sqrt{x} - \sqrt{y}| = \left| \frac{x-y}{\sqrt{x}+\sqrt{y}} \right| \leq \left| \frac{x-y}{2} \right|.$$

So, if $\varepsilon > 0$ is given, we may choose $\delta = 2\varepsilon > 0$ and see from the last display that $|x-y| < \delta$ will ensure that $|f(x) - f(y)| < \varepsilon$ no matter where x and y are in $[1, \infty)$: this is exactly what uniform continuity requires.

15.2.4 **EXAMPLE** Use a similar argument to prove that the function $g : [1, \infty) \to \mathbb{R}$ given by $g(x) = \sqrt[3]{x}$ is uniformly continuous (on $[1, \infty)$). You may find it helpful to use the factorisation $p^3 - q^3 = (p-q)(p^2 + pq + q^2)$.

Here is a small result that you would probably guess to be true, given what you know about 'ordinarily continuous' functions:

15.2.5 **Example**   To show that the composition of two uniformly continuous functions must be uniformly continuous.

Solution

To set up enough notation to discuss the posed question, let $f : A \to B$ and $g : B' \to C$ both be uniformly continuous, where $B \subseteq B'$. Then the composite map $g \circ f$ makes sense, and it is a function from A to C.

Let $\varepsilon > 0$ be given.

Since g is uniformly continuous, there is $\delta_1 > 0$ such that

$$p, q \in B', |p - q| < \delta_1 \text{ together imply } |g(p) - g(q)| < \varepsilon.$$

Since f is uniformly continuous and $\delta_1 > 0$, there is $\delta_2 > 0$ such that

$$x, y \in A, |x - y| < \delta_2 \text{ together imply } |f(x) - f(y)| < \delta_1.$$

Now for any x and y in A such that $|x - y| < \delta_2$, the second display tells us that $|f(x) - f(y)|$ is less than δ_1 so, putting $p = f(x)$ and $q = f(y)$ in the first display, we see that $|p - q|$ is less than δ_1, and therefore $|g(f(x)) - g(f(y))| < \varepsilon$. In other words, $|(g \circ f)(x) - (g \circ f)(y)|$ is less than the given ε.

Hence the result.

Recalling that the convergence of sequences gave us a particularly efficient way to describe continuity, it might have been expected that something similar would arise in uniform continuity. The following lemma provides just such a characterization, and it is often helpful in sorting out the subtle but important distinction between the two concepts.

15.2.6 **Lemma: the two-sequence characterization**   Let $f : A \to \mathbb{R}$. The following are equivalent:

1. f is uniformly continuous on A,

2. for every two sequences (x_n) and (y_n) in A such that $|x_n - y_n| \to 0$, we have that $|f(x_n) - f(y_n)| \to 0$ also.

SUGGESTION

Since it is clear that this lemma is a close relative of the corresponding ('one-sequence') result that connects the sequence-style definition and the epsilontics-style characterisation for ordinary continuity, you can reasonably expect its demonstration to follow the pattern of that result's proof. Try constructing such a demonstration *before* you read the account that we present next.

Proof

(I): (1) implies (2).
- Suppose that condition (1) is satisfied.
- Let $(x_n)_{n \in \mathbb{N}}$ and $(y_n)_{n \in \mathbb{N}}$ be any two sequences of elements of A for which $|x_n - y_n| \to 0$.
- Given $\varepsilon > 0$, use condition (1) to obtain $\delta > 0$ such that whenever $x \in A$ and $y \in A$ and $|x - y| < \delta$, we have $|f(x) - f(y)| < \varepsilon$.
- Since $|x_n - y_n| \to 0$, there is $n_0 \in \mathbb{N}$ such that $n \geq n_0$ guarantees $|x_n - y_n| < \delta$.
- Therefore $n \geq n_0 \Rightarrow |x_n - y_n| < \delta \Rightarrow |f(x_n) - f(y_n)| < \varepsilon$.
- That is, $|f(x_n) - f(y_n)| \to 0$. This proves (2).

(II): (2) implies (1).
- Suppose that condition (1) is not satisfied.
- That is, there exists a value of $\varepsilon > 0$ for which *no* suitable $\delta > 0$ can be found.
- In particular, for each $n \in \mathbb{N}$, $\delta = 1/n$ is not suitable...
- ... and so there are points $x_n, y_n \in A$ such that $|x_n - y_n| < 1/n$ and yet $|f(x_n) - f(y_n)| \geq \varepsilon$.
- Therefore $|x_n - y_n| \to 0$, and yet $|f(x_n) - f(y_n)|$ does not converge to 0.
- In other words, condition (2) is not satisfied.
- By contraposition, (2) implies (1).

15.2.7 **Corollary**   Any uniformly continuous function on A is continuous on A.

Proof

Suppose $f : A \to \mathbb{R}$ to be uniformly continuous. Given a convergent sequence $x_n \to \ell$ in A, let (y_n) be the constant sequence $(\ell, \ell, \ell, \ell, \cdots)$. Obviously $|x_n - y_n| \to 0$, so the two-sequence lemma gives $|f(x_n) - f(y_n)| \to 0$, that is, $|f(x_n) - f(\ell)| \to 0$. This is the same as saying $f(x_n) \to f(\ell)$. Since f therefore preserves limits of sequences, f is continuous at each point of A, as required.

Alternative proof

In the definition of uniform continuity, take the special case where y is held constant, and you immediately get f continuous at y (for each y).

15.2.8 Example Of course $f(x) = x^2$ defines a continuous function on $\mathbb{R}$. We show that it is NOT uniformly continuous.

Solution:

The two-sequence lemma suggests we look for a pair of sequences that get close to one another but whose squares do not, and the Introduction suggests we try to get out into the high-gradient parts of the graph. So take $x_n = n$, $y_n = n + \frac{1}{n}$ for each $n \in \mathbb{N}$. That decision creates two sequences in $[0, \infty)$ and, since $|x_n - y_n| = \frac{1}{n}$, we certainly have $|x_n - y_n| \to 0$. However,

$$\left| n^2 - \left(n + \frac{1}{n} \right)^2 \right| = \left| n^2 - n^2 - 2 - \frac{1}{n^2} \right| = 2 + \frac{1}{n^2}$$

which does not converge to zero. By the two-sequence lemma, x^2 cannot be uniformly continuous.

15.2.9 Remark Almost exactly the same proof will show that x^2 is not uniformly continuous on any interval of the form $[a, \infty)$, nor on any interval of the form (a, ∞).

15.2.10 EXERCISE

- Show that $f(x) = x^3$ fails to define a uniformly continuous function on $\mathbb{R}$.
- Show that $f(x) = x\sqrt{x}$ fails to define a uniformly continuous function on $[0, \infty)$.

The relationship between uniformly continuous functions and Cauchy sequences is close and very useful, but it is also a little complicated:

15.2.11 Theorem Uniformly continuous functions preserve Cauchyness. That is, if f is uniformly continuous on a set A of real numbers and $(x_n)_{n \in \mathbb{N}}$ is a Cauchy sequence in A, then $(f(x_n))_{n \in \mathbb{N}}$ is also a Cauchy sequence.

Proof

Given $\varepsilon > 0$, we use uniform continuity to find $\delta > 0$ so that $|x - y| < \delta, x \in A$, $y \in A$ together imply $|f(x) - f(y)| < \varepsilon$. Since $(x_n)_{n \in \mathbb{N}}$ is Cauchy, now choose $n_0 \in \mathbb{N}$ so that $m, n \geq n_0$ will force $|x_m - x_n| < \delta$. Then $m, n \geq n_0$ also forces $|f(x_m) - f(x_n)| < \varepsilon$ and we have what we wanted.

The interesting question now is: is the converse true? And the irritating answer is: *sometimes. . .*

15.2.12 **Theorem**   Let $f : A \to \mathbb{R}$ be a function defined on a **bounded** set A and suppose that it preserves Cauchyness. Then f is uniformly continuous on A.

Proof

If not, then there is a positive number ε such that, no matter how we choose $\delta > 0$, there will be points of A within δ of one another whose f-values are at least ε apart. In particular, for each positive integer n (and choosing $\delta = \frac{1}{n}$), there exist x_n, y_n in A for which $|x_n - y_n| < \frac{1}{n}$ and yet $|f(x_n) - f(y_n)| \geq \varepsilon$.

Since A is bounded, Bolzano-Weierstrass tells us that $(x_n)_{n \in \mathbb{N}}$ has a convergent subsequence $(x_{n_k})_{k \in \mathbb{N}}$ converging to a limit ℓ. Since, for all $k \geq 1$:

$$|y_{n_k} - \ell| \leq |y_{n_k} - x_{n_k}| + |x_{n_k} - \ell| < \frac{1}{n_k} + |x_{n_k} - \ell| \to 0,$$

we see that $(y_{n_k})_{k \in \mathbb{N}}$ also converges to the same limit ℓ. If we now 'interleave' these two sequences, we get

$$(x_{n_1}, y_{n_1}, x_{n_2}, y_{n_2}, x_{n_3}, y_{n_3}, x_{n_4}, y_{n_4}, \cdots)$$

again converging to ℓ (see Example 5.2.5), and therefore Cauchy, and yet

$$(f(x_{n_1}), f(y_{n_1}), f(x_{n_2}), f(y_{n_2}), f(x_{n_3}), f(y_{n_3}), f(x_{n_4}), f(y_{n_4}), \cdots)$$

fails to be Cauchy since endlessly many pairs of its terms are at least ε apart: we have thus achieved a *contradiction*.

15.2.13 **Note**   If we were to drop the word 'bounded', this result would cease to be true: for instance, we know that the x^2 function on (the unbounded set) $\mathbb{R}$ is not uniformly continuous, and yet it is easy to check (do so) that it preserves Cauchyness.

The next theorem is generally viewed as the most important basic result about uniform continuity. We'll offer two different proofs of it.

15.2.14 **Key theorem**   Any continuous function f on a **bounded closed** interval $[a, b]$ is uniformly continuous there.

First proof — only using the definition

If not, then there exists $\varepsilon > 0$ for which NO choice of $\delta > 0$ will work. In particular, for each $n \in \mathbb{N}$, the condition

$$x \in [a, b], y \in [a, b], |x - y| < \frac{1}{n}$$

fails to force $|f(x) - f(y)| < \varepsilon$.

That is, there exist (for each $n \in \mathbb{N}$) points $x_n, y_n \in [a, b]$ such that

$$|x_n - y_n| < \frac{1}{n} \text{ and yet } |f(x_n) - f(y_n)| \geq \varepsilon \cdots \cdots (1)$$

By Bolzano-Weierstrass, *some* subsequence (x_{n_k}) of (x_n) converges (and, since $a \leq x_{n_k} \leq b$ for each k, its limit must satisfy the same inequality):

$$x_{n_k} \to x \in [a, b] \text{ as } k \to \infty \cdots \cdots (2)$$

Now

$$|x - y_{n_k}| \leq |x - x_{n_k}| + |x_{n_k} - y_{n_k}| \to 0 \text{ as } k \to \infty$$

that is, $y_{n_k} \to x$ also[1] (as $k \to \infty$). By ordinary continuity, $f(y_{n_k}) \to f(x)$ as $k \to \infty$, and also (using (2)) $f(x_{n_k}) \to f(x)$ as $k \to \infty$.

Subtract, and we get $f(x_{n_k}) - f(y_{n_k}) \to 0$ as $k \to \infty$.

This contradicts (1), and completes the proof.

Second proof – using Cauchy sequences

Let $(x_n)_{n \in \mathbb{N}}$ be any Cauchy sequence in $[a, b]$.

Because it is Cauchy, $x_n \to \ell$ for some $\ell \in \mathbb{R}$.

Since $a \leq x_n \leq b$ for all n, also $a \leq \ell \leq b$.

Since f is continuous on $[a, b]$, and $\ell \in [a, b]$, f is *in particular* continuous at ℓ, so $f(x_n) \to f(\ell)$.

Because it is convergent, $(f(x_n))_{n \in \mathbb{N}}$ is Cauchy.

Now f is Cauchy-preserving on a bounded set, so (by 15.2.12 above) it is uniformly continuous.

15.2.15 **Example**   Let interval I and uniformly continuous $f : I \to \mathbb{R}$ be given.

1. If I is closed and bounded, we show that f^2 is uniformly continuous.

2. We show by example that (1) can fail to be true if we omit the words 'and bounded'.

Solution to 1: first method

(*Roughwork:* knowing that we can make $|f(x) - f(y)|$ as small as we wish just by taking x and y close together, here we need to make $|f^2(x) - f^2(y)|$ small as well. What connection can we see between them? Well, the second one factorises:

---

[1] Refer back to Exercise 2.7.14 if this is not clear enough.

$$|f^2(x) - f^2(y)| = |f(x) + f(y)| \, |f(x) - f(y)|$$

and this will be less than $M|f(x) - f(y)|$ provided that we can find a constant M that is always bigger than $|f(x) + f(y)|$. Luckily, we *can* find some such constant since the *continuous* function f will be bounded on the closed, bounded interval I. Then forcing $|f(x) - f(y)|$ to be smaller than ε/M will guarantee that $|f^2(x) - f^2(y)|$ is smaller than ε. Now let's write that out properly...)

Since f is uniformly continuous, it is certainly continuous on the closed, bounded interval I, and therefore f itself is bounded: there exists $K > 0$ so that $|f(x)| \le K$ for all $x \in I$. Notice that (for any p and q)

$$|p^2 - q^2| = |(p - q)(p + q)| \le (|p| + |q|)|p - q|.$$

Now f^2 is the function defined by $f^2(x) = (f(x))^2$, $x \in I$, so

$$|f^2(x) - f^2(y)| \le \big(|f(x)| + |f(y)|\big) \, |f(x) - f(y)| \le 2K|f(x) - f(y)|.$$

Given $\varepsilon > 0$ choose $\delta > 0$ such that

$$(x \in I, y \in I, |x - y| < \delta) \Rightarrow |f(x) - f(y)| < \frac{\varepsilon}{2K}$$

$$\Rightarrow |f^2(x) - f^2(y)| < \varepsilon \text{ as required.}$$

Solution to 1: second method

Since f is uniformly continuous, it is continuous. By standard algebra of continuous functions, f^2 is also continuous, and upon a closed, bounded interval. By the key theorem, f^2 is uniformly continuous there.

Solution to 2:

With $g : \mathbb{R} \to \mathbb{R}$ defined by $g(x) = x$, it is really trivial that g is uniformly continuous. Yet g^2 is now the x^2 function that we have shown not to be uniformly continuous. So the boundedness of the interval I cannot be thrown away in part 1.

Of course the key theorem is not able to prove a function to be uniformly continuous on an *unbounded* interval; nevertheless, it can sometimes be employed to carry out a *significant part* of that task; look first at the following:

15.2.16 **EXERCISE** Let I and J be two intervals that share an endpoint from opposite sides: that is, I is either $(-\infty, b]$ or $(a, b]$ or $[a, b]$, while J is either $[b, \infty)$ or $[b, c)$ or $[b, c]$. Let $f : I \cup J \to \mathbb{R}$ be uniformly continuous on I, and also uniformly continuous on J. Show that f is uniformly continuous on $I \cup J$.

Remarks

- This turns out to be valuable more often than you might expect, because the most obvious reason why some function is uniformly continuous can vary from one part of its domain to another. The above result allows us to 'glue together' uniform continuity that has been 'separately evidenced' in different parts of its domain. (Consider, as an illustration, the next example.)
- The only non-routine part of the proof is to show that nearby points $x \in I$ and $y \in J$ have f-values that are suitably close together. To help with this, notice that both x and b, and also b and y, will be nearby, and that
$$|f(x) - f(y)| \le |f(x) - f(b)| + |f(b) - f(y)|.$$

15.2.17 Example To prove that the formula $f(x) = \sqrt{x}, x \in [0, \infty)$ defines a uniformly continuous function on $[0, \infty)$.

Solution

Since f is continuous on the closed, bounded interval $[0, 1]$, the key theorem provides evidence of its uniform continuity there. Also, an earlier example (15.2.3) showed its uniform continuity on $[1, \infty)$. Now we can appeal to 15.2.16 to deduce that it is uniformly continuous on the union $[0, 1] \cup [1, \infty) = [0, \infty)$ as was required.

15.2.18 EXERCISE Prove that the function f given by $f(x) = \sqrt[3]{x}, x \in [0, \infty)$ is uniformly continuous on $[0, \infty)$.

15.2.19 Examples To determine whether the following real functions are uniformly continuous on the intervals indicated.

1. On $(0, \infty)$ we define f by the formula $f(x) = \frac{1}{x}$.
2. On $[1, \infty)$ we define f by the formula $f(x) = \frac{1}{x}$.
3. On $[0, 10]$ we define f by the formula $f(x) = \sin(\cos^2(\sqrt{1 + x^3} + e^{x + x^4}))$.
4. On $[0, 2]$ we define $f(x) = \lfloor x \rfloor$, the floor of x.
5. On $(0, 1]$ we define $f(x) = \dfrac{\sin x}{x}$ (you should assume that the limit of $f(x)$ as $x \to 0$ is 1).
6. On $(0, \infty]$ we define $f(x) = e^x$ (you should assume basic facts about the function ln, including that it is continuous on positive numbers).

Solution

1. The sequence $\left(\frac{1}{n}\right)_{n \ge 1}$ in $(0, 1)$ is Cauchy because it converges (to 0). Yet $\left(f\left(\frac{1}{n}\right)\right)_{n \ge 1}$ is the sequence (n) and that is not Cauchy: indeed, it is not even bounded. So f, since it does not preserve Cauchyness, cannot be uniformly continuous.

2. When $x \geq 1$ and $y \geq 1$, we see that

$$|f(x) - f(y)| = \left| \frac{1}{x} - \frac{1}{y} \right| = \frac{|y - x|}{xy} < |x - y|.$$

So, given $\varepsilon > 0$, if we choose $\delta = \varepsilon$, we shall have $|x - y| < \delta$ implying $|f(x) - f(y)| < \varepsilon$. Therefore f is uniformly continuous on this set.

3. Despite its complicated formula, this expression has been built up from components that we know to be continuous; so f is therefore continuous. Because $[0, 10]$ is closed and bounded, the key theorem assures us that f is uniformly continuous here.

4. As $x \to 1$ this function does not have a limit, so it is not continuous and therefore cannot possibly be uniformly continuous.

5. The question is more awkward than (3) because the domain is not closed. However, f is again composed from continuous components at every point of its domain and without division by zero, and is therefore continuous there. Also, as $x \to 0$, $\dfrac{\sin x}{x} \to 1$. Therefore, if we now define a 'new' function F on the *closed* interval $[0, 1]$ by the formula

$$F(x) = f(x) \text{ if } 0 < x \leq 1, F(x) = 1 \text{ if } x = 0$$

then F will be continuous not only on $(0, 1]$ but at 0 as well, that is, it is continuous on closed, bounded $[0, 1]$. By the key theorem, F is uniformly continuous on $[0, 1]$ and, in particular, on its subset $(0, 1]$. Yet on this interval, F and f are the same function – so f is uniformly continuous on its given domain.

6. (After some trial-and-error along the lines of the roughwork thinking we did towards showing that x^2 was not uniformly continuous), for each positive integer n we try $x_n = \ln(n)$ and $y_n = \ln(n + 1)$. Then $|x_n - y_n| = \ln(n + 1) - \ln(n) = \ln\left(1 + \frac{1}{n}\right)$ and, as $n \to \infty$, this expression $\to \ln(1) = 0$ since $\ln$ is continuous. On the other hand, $|f(x_n) - f(y_n)| = |n - (n + 1)| = 1$ which does not converge to 0. By the two-sequence lemma, f is not uniformly continuous.

15.2.20 Example Suppose that $f : I \to \mathbb{R}$ is uniformly continuous and never takes the value 0 (where I is an interval).

1. If I is both bounded and closed, we show that $\frac{1}{f}$ is also uniformly continuous on I.

2. We show by example that if I is closed but not bounded, then $\frac{1}{f}$ could fail to be uniformly continuous on I.

3. We show by example that if I is bounded but not closed, then $\frac{1}{f}$ could fail to be uniformly continuous on I.

Solution

1. Being uniformly continuous, f is certainly continuous. Since dividing among continuous functions (but scrupulously avoiding division by 0) always gives continuous functions, $\frac{1}{f}$ is continuous (on I). Since the interval I is bounded and closed, $\frac{1}{f}$ is also uniformly continuous there by the key theorem.

2. For instance, on the closed unbounded interval $I = [1, \infty)$, $f : I \to \mathbb{R}$ described by $f(x) = \frac{1}{x^2}$ is non-zero and uniformly continuous (as a proof very like that of Example 15.2.19, part 2, will show). But here, $\frac{1}{f}$ is the x^2 function which we know how to prove to be *not* uniformly continuous.

3. For instance, on the bounded, non-closed interval $(0, 1]$, it is very easy to check that the function $f(x) = x$ is uniformly continuous (and never exactly zero). Yet, very much as we saw in an earlier example, its reciprocal, the $\frac{1}{x}$ function, is not.

15.3 The bounded derivative test

Just as differentiability sometimes provides us with a quick way to confirm continuity, the first mean value theorem sometimes gives us a quick way to confirm *uniform* continuity. To add a little perspective, we first formulate a related definition:

15.3.1 Definition: Lipschitz functions   A function $f : I \to \mathbb{R}$ is said to be *Lipschitz*, or to *satisfy the Lipschitz condition* on I, if there is a positive constant K such that
$$|f(x) - f(y)| \leq K|x - y| \text{ for all } x, y \in I.$$

15.3.2 Lemma   Lipschitz functions are uniformly continuous.

Proof

Given $\varepsilon > 0$, define $\delta = \frac{\varepsilon}{K}$ where K is the constant in the Lipschitz definition. Then
$$x, y \in I, |x - y| < \delta \Rightarrow |f(x) - f(y)| \leq K|x - y| < K\delta = \varepsilon.$$

15.3.3 EXERCISE   (Perhaps unfortunately,) not all uniformly continuous functions are Lipschitz. For example, the function given by
$$f : [-1, 1] \to \mathbb{R}, \ f(x) = \sqrt{1 - x^2}$$

is uniformly continuous by the key theorem, because it is continuous on the closed bounded interval $[-1, 1]$. Prove that it is not Lipschitz.

15.3.4 Theorem: the bounded derivative test for uniform continuity   Suppose that the real function f is continuous on an interval I and differentiable at each interior point of I, and that its derivative $f'(x)$ is bounded. Then f is uniformly continuous on I.

Proof

For any $a < b$ in I the conditions of the first mean value theorem are met on the subinterval $[a, b]$, so there is a point $c \in (a, b)$ such that

$$\frac{f(b) - f(a)}{b - a} = f'(c).$$

There is also a positive constant K such that $|f'| < K$ at every point of the interval (a, b), so

$$|f(b) - f(a)| \le K|b - a|.$$

Hence f is a Lipschitz function, and is therefore uniformly continuous.

15.3.5 Notes

- The converse of this theorem is not true: for instance, the uniformly continuous function $f(x) = \sqrt{1 - x^2}$ has an unbounded derivative on $(-1, 1)$.
- This may be a good moment at which to review a variety of uniformly continuous functions, and observe how the bounded derivative test can make it easier for us to see why they are so.

1. On the interval $[1, \infty), f(x) = x^{-1}$ is uniformly continuous because the modulus of its derivative $|-x^{-2}|$ never exceeds 1.

2. On the interval $[1, \infty), f(x) = x^{-2}$ is uniformly continuous because the modulus of its derivative $|-2x^{-3}|$ never exceeds 2.

3. If $f(x) = ax + b$ then $|f'(x)| = |a|$ always, and is bounded, so f is uniformly continuous on any interval.

4. If

$$f(x) = e^{\sin x + \cos x}$$

then (assuming basic results about the trig functions)

$$f'(x) = (\cos x - \sin x)e^{\sin x + \cos x}$$

which cannot exceed $2e^2$ in modulus, so this function is uniformly continuous on any interval.

5. Here is another way to show that the function $f(x) = \sqrt{x}$ is uniformly continuous on $[0, \infty)$. Firstly, it is continuous on $[0, 1]$ and therefore, by the key theorem, uniformly continuous there. Secondly, on $[1, \infty)$ its derivative is $\dfrac{1}{2\sqrt{x}}$ and therefore (in modulus) never more than $\frac{1}{2}$, so by the bounded derivative test it is uniformly continuous there also. Now we can invoke Exercise 15.2.16 to see that it is uniformly continuous on the union $[0, 1] \cup [1, \infty) = [0, \infty)$.

15.3.6 **EXERCISE** Show that the function specified by $f(x) = \sqrt[3]{x^2}$ is uniformly continuous on $\mathbb{R}$. Suggestion: break up the domain into $(-\infty, -1] \cup [-1, 1] \cup [1, \infty)$.

15.3.7 **EXERCISE** Suppose that $f : (0, \infty) \to \mathbb{R}$ is everywhere differentiable, and that the derivative $f'(x) \to \infty$ as $x \to \infty$ (so: the derivative is unbounded in a rather extreme manner.) Show that f cannot be uniformly continuous on $(0, \infty)$.

15.3.8 **Note** In many areas of mathematics, an important question to ask about a continuous function is whether it can be extended over a bigger domain without destroying its continuity. Indeed, we recently saw a use of this idea in part 5 of Example 15.2.19, where we proved uniform continuity of a function on the domain $(0, 1]$ by extending it to become a continuous function on the domain $[0, 1]$: even that tiny augmentation of domain turned out to be strategically beneficial. There are – as that quoted example already indicates – strong connections between this kind of 'continuous extensibility' on the one hand, and uniform continuity on the other. We shall round off the chapter by taking a brief look at this issue.

15.3.9 **Proposition** Let $f : (a, b] \to \mathbb{R}$. Then f can be extended to a continuous real function on $[a, b]$ if and only if f is uniformly continuous.

Proof

(I) Suppose firstly that f can be so extended: that is, there is a continuous function $F : [a, b] \to \mathbb{R}$ whose restriction to $(a, b]$ is exactly f. By the key theorem, F is uniformly continuous on $[a, b]$ and therefore, in particular, uniformly continuous on $(a, b]$. Yet F and f are identical on $(a, b]$, so f is uniformly continuous.

(II) Conversely, suppose that f is uniformly continuous (on $(a, b]$). Choose a sequence[2] (y_n) in $(a, b]$ whose limit is a. Since (convergent) (y_n) is Cauchy, we know from 15.2.11 that $(f(y_n))$ is also Cauchy, and consequently converges to some limit – let us call it ℓ.

Now consider *any* sequence (x_n) in $(a, b]$ that converges to a. If (as we did before) we interleave the two sequences thus: $(x_1, y_1, x_2, y_2, x_3, y_3, \cdots)$, we create a new sequence converging to a, and therefore Cauchy, and we see from 15.2.11 that

$$(f(x_1), f(y_1), f(x_2), f(y_2), f(x_3), f(y_3), \cdots)$$

is Cauchy and therefore has to tend to some limit (let us be cautious and call it ℓ' for the moment). However, since the subsequence $(f(y_1), f(y_2), f(y_3) \cdots)$ must also converge to ℓ' but actually does converge to ℓ, the two numbers ℓ and ℓ' are, in fact, identical. Hence the 'complementary' subsequence $(f(x_1), f(x_2), f(x_3) \cdots)$ converges to ℓ also.

---

[2] for example, $y_n = a + \frac{b-a}{n+1}$ would give one suitable choice.

What the last paragraph shows us is that the function $F : [a, b] \to \mathbb{R}$ defined by the formula

$$F(x) = f(x) \text{ for } x \in (a, b]; \quad F(a) = \ell$$

possesses a limit as $x \to a$, and that this limit is ℓ which equals $F(a)$. Therefore F is continuous at a, as well as (trivially) continuous everywhere else in $[a, b]$. Thus we have managed to find a continuous extension of f over $[a, b]$.[3]

15.3.10 **EXERCISE** Think how much[4] you would need to modify that argument in order to show that a real function defined on a bounded open interval (a, b) or, indeed, on a finite union of bounded open intervals $(a_1, b_1) \cup (a_2, b_2) \cup \cdots \cup (a_n, b_n)$ can be continuously extended over the corresponding closed interval(s) if and only if it is uniformly continuous.

---

[3] Incidentally, the same 'last paragraph' also shows that F is *unambiguously* defined in the sense that the number ℓ we selected to be the value of $F(a)$ does not depend on *how* we chose the sequence (y_n): any different choice of (y_n) would have resulted in exactly the same number ℓ.

[4] The short answer is: *not very much!*

16 Differentiation – mean value theorems, power series

16.1 Introduction

Recall *Rolle's theorem*: a function that is continuous on a bounded closed interval, differentiable on the corresponding open interval, and of equal value at the endpoints must have zero derivative at one point (at least) in between.

Recall the *first mean value theorem*: a function that is continuous on a bounded closed interval and differentiable on the corresponding open interval must, somewhere between the endpoints, have derivative equal to the average (the mean) gradient of its graph across the entire interval.

Given the use of the word *first* in the title, it will hardly surprise anyone to learn that there are other 'mean value' theorems. This chapter is going to look at some of the others: what they say, why they are true and what use can be made of them. This study will lead us into questions of how to represent 'highly differentiable' functions by power series and thus, inevitably, back to the foreshadowed theorem on the differentiation of power series themselves.

16.2 Cauchy and l'Hôpital

As we commented in an earlier chapter, Rolle can be seen as a special case of the FMVT and, furthermore, the FMVT is most readily proved by re-engineering its hypotheses into a form to which Rolle can apply. Analogous remarks apply to the next result, which is a kind of 'double' FMVT that deals with two functions at once.

16.2.1 Cauchy's Mean Value Theorem ('CMVT')   Let f and g both be continuous on $[a, b]$ and differentiable on (a, b), with $g'(x)$ non-zero at every point of (a, b). Then there is (at least) one point $c \in (a, b)$ such that

$$\frac{f(b) - f(a)}{g(b) - g(a)} = \frac{f'(c)}{g'(c)}.$$

Proof

We define a new function h by the formula $h(x) = f(x) - \lambda g(x)$ where the constant λ will be chosen in such a way that RT can be applied to h.

First, how to choose λ? Certainly h will be continuous and differentiable where f and g were, no matter how we decide to pick λ, so all we need to arrange is that $h(a) = h(b)$, that is

$$f(a) - \lambda g(a) = f(b) - \lambda g(b)$$

which solves easily to give

$$\lambda = \frac{f(b) - f(a)}{g(b) - g(a)}$$

provided that the bottom line is non-zero. Fortunately, $g(b) - g(a)$ cannot be zero because, if it were, RT applied to g would tell us that g' was zero somewhere, which is explicitly not the case.

Now that λ has been thus chosen, RT applied to h gives us the existence of $c \in (a, b)$ for which $0 = h'(c) = f'(c) - \lambda g'(c)$ and so, again because g' cannot go zero,

$$\frac{f'(c)}{g'(c)} = \lambda,$$

which is the declared result.

16.2.2 Remarks

1. In the special case where $g(x) = x$ for all relevant x (and therefore $g'(x) = 1$ which is certainly non-zero) we get, from CMVT,

$$\frac{f(b) - f(a)}{b - a} = \frac{f'(c)}{1},$$

   that is, the FMVT is a particular case of the CMVT.

2. Occasionally useful is a slightly different version of CMVT in which we do not assume that g' is non-zero on the relevant interval. It says:
   'Let f and g both be continuous on $[a, b]$ and differentiable on (a, b). Then there is (at least) one point $c \in (a, b)$ such that

$$(f(b) - f(a))g'(c) = (g(b) - g(a))f'(c).\text{''}$$

   Proof
   Case 1: if $g(b) - g(a) \neq 0$ then essentially the proof we gave before still runs.
   Case 2: if $g(b) - g(a) = 0$ then RT says there is a point c such that $g'(c) = 0$ and then the result is immediate.

16.2.3 Example Let f be any function that is continuous on $[0, 1]$ and differentiable on $(0, 1)$, and n be any positive integer. We show that there must be a number c in $(0, 1)$ such that $f'(c) = nc^{n-1}(f(1) - f(0))$.

Roughwork

Only one function f is visible here, but the group of symbols nc^{n-1} should make us suspect that 'the other function' is x^n. Certainly $g(x) = x^n$ is continuous and differentiable wherever we need it to be, and its derivative goes to zero only at $x = 0$, that is, *not* anywhere in $(0, 1)$ ...

Solution

The CMVT does apply to the two functions f and g where $g(x) = x^n$, and it says there is $c \in (0, 1)$ such that

$$\frac{f(1) - f(0)}{1^n - 0^n} = \frac{f'(c)}{nc^{n-1}}.$$

This quickly rearranges into the form required.

16.2.4 EXERCISE The following *alleged proof* of CMVT is *incorrect*. Find out precisely why.

(If necessary, try running the argument of the purported proof on a couple of simple functions such as $f(x) = x^2$ and $g(x) = x^3$ over $[0, 1]$.)

'Since f satisfies the conditions of FMVT over the interval $[a, b]$, we know that

$$\text{there exists } c \in (a, b) \text{ such that } f'(c) = \frac{f(b) - f(a)}{b - a}.$$

'By exactly the same argument on g:

$$\text{there exists } c \in (a, b) \text{ such that } g'(c) = \frac{g(b) - g(a)}{b - a}.$$

'Dividing one by the other (and remembering that $g(b) - g(a)$ cannot be zero, else RT would give $g' = 0$ somewhere, contradiction) we get

$$\frac{f'(c)}{g'(c)} = \frac{f(b) - f(a)}{g(b) - g(a)}$$

as desired.'

One of the most immediate and useful applications of CMVT is a result that you may very well have used already, called l'Hôpital's Rule, for determining function limits in the most awkward case where unaided common sense hits the nonsense barrier of zero divided by zero.

16.2.5 **L'Hôpital's Rule**   Suppose that f and g are both differentiable on an open interval $(p - h, p + h)$ centred on a real number p, that both $f(p)$ and $g(p)$ are zero, that g' is non-zero here except possibly at p,[1] and that $f'(x)/g'(x)$ tends to a limit ℓ as $x \to p$. Then also

$$\frac{f(x)}{g(x)} \to \ell \ \text{ as } x \to p.$$

Proof

(I) For k positive and less than h, the two functions f, g satisfy the conditions of CMVT on the interval $[p, p + k]$. Therefore there is a number c such that $p < c < p + k$ and

$$\frac{f'(c)}{g'(c)} = \frac{f(p + k) - f(p)}{g(p + k) - g(p)} = \frac{f(p + k)}{g(p + k)}.$$

As $k \to 0$ (but through positive values) the fact that $p < c < p + k$ gives us $c \to p$ also, so (by hypothesis) $\dfrac{f'(c)}{g'(c)} \to \ell$. Therefore $\dfrac{f(p + k)}{g(p + k)} \to \ell$ also.

Let x stand for $p + k$ here. In the language of one-sided limits (which is what we are presently speaking, since we have so far only considered points just to the right of p), what this establishes is that

$$\lim_{x \to p^+} \frac{f(x)}{g(x)} = \ell.$$

(II) For k *negative* and lying between $-h$ and 0, a virtually identical argument upon the interval $[p + k, p]$ shows that

$$\lim_{x \to p^-} \frac{f(x)}{g(x)} = \ell.$$

Putting the two one-sided limits together, we get (as we wanted)

$$\lim_{x \to p} \frac{f(x)}{g(x)} = \ell.$$

There are many different versions of this Rule, of which the two most obvious are what we actually proved above for one-sided limits:

---

[1] Actually, we do not even need $g'(p)$ to exist, so long as g is continuous at p.

16.2.6 L'Hôpital's Rule: right–hand limits   Suppose that f and g are both differentiable on an open interval $(p, p + h)$ and continuous on $[p, p + h)$, that $f(p)$ and $g(p)$ are zero, that g' is non-zero on $(p, p + h)$, and that $f'(x)/g'(x)$ tends to a limit ℓ as $x \to p^+$. Then also

$$\frac{f(x)}{g(x)} \to \ell \text{ as } x \to p^+.$$

Now the left-handed variety of this resembles it so closely that it is barely worth stating:

16.2.7 L'Hôpital's Rule: left–hand limits   Suppose that f and g are both differentiable on an open interval $(p - h, p)$ and continuous on $(p - h, p]$, that $f(p)$ and $g(p)$ are zero, that g' is non-zero on $(p - h, p)$, and that $f'(x)/g'(x)$ tends to a limit ℓ as $x \to p^-$. Then also

$$\frac{f(x)}{g(x)} \to \ell \text{ as } x \to p^-.$$

Here is another in which the control variable tends to infinity:

16.2.8 L'Hôpital's Rule: as $x \to \infty$ Suppose that f and g are both differentiable on an open interval (a, ∞), that both $f(x)$ and $g(x)$ converge to 0 as $x \to \infty$, that g' is non-zero on (a, ∞), and that $\dfrac{f'(x)}{g'(x)}$ tends to a limit ℓ as $x \to \infty$. Then also

$$\frac{f(x)}{g(x)} \to \ell \text{ as } x \to \infty.$$

Proof

We shall use 11.1.14 to switch between limits at infinity and limits at zero (and there is no loss of generality in assuming $a > 0$).

Define two 'new' functions F, G on the interval $\left(0, \dfrac{1}{a}\right)$ by setting $F(x) = f\left(\dfrac{1}{x}\right)$, $G(x) = g\left(\dfrac{1}{x}\right)$. Then F, G are differentiable and, if we additionally define $F(0) = 0, G(0) = 0$, they also become continuous on $\left[0, \dfrac{1}{a}\right)$ (because their limiting values, as $x \to 0^+$, are 0 and thus coincide with the values that we attributed to them *at* 0). Furthermore, $G'(x) = -g'\left(\dfrac{1}{x}\right) x^{-2}$ is non-zero, and

$$\lim_{x \to 0^+} \frac{F'(x)}{G'(x)} = \lim_{x \to 0^+} \frac{-f'\left(\frac{1}{x}\right) x^{-2}}{-g'\left(\frac{1}{x}\right) x^{-2}} = \lim_{x \to 0^+} \frac{f'\left(\frac{1}{x}\right)}{g'\left(\frac{1}{x}\right)}$$

$$= \lim_{t \to \infty} \frac{f'(t)}{g'(t)} = \ell.$$

By a one-sided version of the Rule (16.2.6) that we have already established, that gives

$$\lim_{x \to 0^+} \frac{F(x)}{G(x)} = \ell,$$

which is equivalent to

$$\lim_{x \to \infty} \frac{f(x)}{g(x)} = \ell.$$

16.2.9 Notes

1. Resist the temptation to think (or write) that the essence of the Rule is that (in the zero-over-zero case)

$$\lim \frac{f(x)}{g(x)} = \lim \frac{f'(x)}{g'(x)}.$$

   While it is true that l'Hôpital does say this, the main point is that *if* the second limit exists, *then* so must the first. Their numerical equality is secondary to that.

2. One more procedural detail before we settle down to a batch of examples. The Rule is written as if you begin with knowledge of the limit of f'/g' and proceed from there to knowledge of the limit of f/g, but that is not exactly what happens in practice. We actually begin with curiosity about the limit of f/g, turn it into curiosity about f'/g', solve that question if we can, and feed it back into an answer for the limit of f/g.

3. The additional point is that if our first attack on the limit of f'/g' *also* hits the nonsense wall of zero divided by zero, we ought not to give up the struggle: we should, instead, drive the process further into curiosity about the limit of f''/g''. If that is answerable, then the answer we get pans back to one for f'/g' and, in turn, for f/g. If not, consider f'''/g''', and so on. (Of course, if the derivatives are becoming unmanageable, this process should not be continued past the point of reasonable hopes.) Take care to check that the conditions of the Rule are fully satisfied each time you invoke it.

16.2.10 Example To determine, if it exists, the limit of $\dfrac{x^4 - 16}{x^6 - 64}$ as $x \to 2$.

Solution

An initial, common-sense attempt of replacing x by 2 gives the meaningless (but encouraging) response of zero divided by zero, so an application of the Rule is indicated.

Putting $f(x) = x^4 - 16$ and $g(x) = x^6 - 64$ we see that $g'(x) = 6x^5$ is zero only at $x = 0$ so, if we operate over (say) the interval $(1, 3)$ then that derivative is non-zero and

$$\frac{f'(x)}{g'(x)} = \frac{4x^3}{6x^5} = \frac{2}{3x^2}$$

whose limit is obviously $\frac{1}{6}$. Therefore the limit also of $\dfrac{f(x)}{g(x)}$ exists, and equals $\frac{1}{6}$.

16.2.11 **Example**   (Assuming knowledge of how to differentiate basic exponential and trigonometric functions), to determine whether or not the limit exists of

$$\frac{x(e^x - 1)}{\sin^2 x}$$

as $x \to 0$.

Roughwork

Putting $x = 0$ gives us zero divided by zero, which is not an answer, but suggests we should try l'Hôpital.
  Let
$$f(x) = x(e^x - 1), \quad g(x) = \sin^2 x.$$
Then (using product rule and chain rule)

$$f'(x) = xe^x + (e^x - 1), \quad g'(x) = 2 \sin x \cos x = \sin(2x).$$

Now $x = 0$ in this *still* gives zero over zero, so try again:

$$f''(x) = (1 + x)e^x + e^x, \quad g''(x) = 2\cos(2x).$$

This time the limits (as $x \to 0$) are 2 and 2, so we have 'broken through the nonsense wall'.

Solution

Putting $f(x) = x(e^x - 1), g(x) = \sin^2 x$ we see that $f'(x) = xe^x + e^x - 1$, $g'(x) = \sin(2x)$ and $f''(x) = (2 + x)e^x, g''(x) = 2\cos(2x)$. All are visibly differentiable (and continuous).
  Now on $(-\frac{\pi}{4}, \frac{\pi}{4})$ we have g'' non-zero, f' and g' are zero at 0, and $\dfrac{f''(x)}{g''(x)} \to \dfrac{2}{2} = 1$ as $x \to 0$. By the Rule, $\dfrac{f'(x)}{g'(x)} \to 1$ also.
  Secondly, on $(-\frac{\pi}{2}, \frac{\pi}{2})$ we have g' non-zero except at $0, f$ and g are zero at 0, and $\dfrac{f'(x)}{g'(x)} \to 1$ as $x \to 0$. By the Rule, $\dfrac{f(x)}{g(x)} \to 1$ also.
  The desired limit does exist (and equals 1).

16.2.12 **EXERCISE**   Evaluate

$$\lim_{x \to 4} \frac{x - \sqrt{x}}{4 - x}.$$

16.2.13 EXERCISE   Assuming that the derivative of e^x is e^x, evaluate

$$\lim_{x \to 3} \frac{e^x + (2 - x)e^3}{(x - 3)^2}.$$

16.2.14 EXERCISE

1. Use l'Hôpital's Rule to investigate

$$\lim_{x \to \infty} x \left(\frac{\pi}{2} - \arctan x \right).$$

(You can assume that the derivative of $\arctan x$ is $(1 + x^2)^{-1}$.)
It may be useful to express the desired function in the form

$$\frac{\frac{\pi}{2} - \arctan x}{\frac{1}{x}}.$$

2. Re-work the problem by putting $t = \frac{1}{x}$ and so converting it into

$$\lim_{t \to 0^+} \frac{\frac{\pi}{2} - \arctan\left(\frac{1}{t}\right)}{t}.$$

16.3 Taylor series

If f is a function that is differentiable several times on an open interval including a then we can easily write down a list of polynomials that – in some sense at least – give better and better approximations to f itself. For, consider the following:

$$p_1(x) = f(a) + f'(a)(x - a)$$
$$p_2(x) = f(a) + f'(a)(x - a) + \frac{f''(a)}{2!}(x - a)^2$$
$$p_3(x) = f(a) + f'(a)(x - a) + \frac{f''(a)}{2!}(x - a)^2 + \frac{f'''(a)}{3!}(x - a)^3$$
$$p_4(x) = f(a) + f'(a)(x - a) + \frac{f''(a)}{2!}(x - a)^2 + \frac{f'''(a)}{3!}(x - a)^3 + \frac{f^{iv}(a)}{4!}(x - a)^4$$

... and so on. It is routine to check that the first of these, p_1, has the same value and the same derivative as f had at a, the second p_2 has the same value, the same first derivative and the same second derivative as f had at a, the third has additionally the same third derivative as f at a, and so on.

It would be nice to believe that these so-called *Taylor polynomials* were approximating f better and better **not only at a but on the interval as a whole.**

Unfortunately, this is not always true. However, it is true in many important cases. This is what Taylor's theorem is about: it sets out to examine how well the list of polynomials that we just described approximates f over an interval of values of x around $x = a$. There are many slightly different versions of it, but we'll focus on just one of them:

16.3.1 Taylor's theorem   Suppose that f is differentiable at least $k + 1$ times on an open interval including a and x. Let p_k denote the kth Taylor polynomial

$$p_k(x) = f(a) + f'(a)(x - a) + \frac{f''(a)}{2!}(x - a)^2 + \cdots + \frac{f^k(a)}{k!}(x - a)^k.$$

Then

$$f(x) = p_k(x) + \frac{f^{k+1}(\xi)}{(k + 1)!}(x - a)^{k+1}$$

for some number ξ lying between a and x.

Proof

Consider the function

$$F(t) = f(t) + f'(t)(x - t) + \frac{f''(t)}{2!}(x - t)^2 + \cdots + \frac{f^k(t)}{k!}(x - t)^k$$

where t varies over the interval in question. The first nice thing about this function (please check this out) is just how its derivative simplifies:

$$F'(t) = \frac{f^{k+1}(t)}{k!}(x - t)^k$$

and the second nice thing (check this one also) is that $F(x) - F(a)$ simplifies to $f(x) - p_k(x)$.

Now with $G(t) = (x - t)^{k+1}$, use the Cauchy mean value theorem on the functions F, G on the interval joining a and x. It tells us that

$$\frac{F(x) - F(a)}{G(x) - G(a)} = \frac{F'(\xi)}{G'(\xi)}$$

for some number ξ between a and x, that is,

$$\frac{f(x) - p_k(x)}{-(x - a)^{k+1}} = \frac{\frac{f^{k+1}(\xi)}{k!}(x - \xi)^k}{-(k + 1)(x - \xi)^k} = -\frac{f^{k+1}(\xi)}{(k + 1)!}$$

which, cancelling the minuses, gives the declared result.

16.3.2 Notes

1. The series (of functions) whose partial sums are these Taylor polynomials is known as the *Taylor series* of the function f at the point a.

2. For historical reasons, the special case $a = 0$ is named after Maclaurin as well as after Taylor: Maclaurin's theorem, Maclaurin polynomials, and so on.

3. Think about Taylor's theorem as saying 'original function = approximating polynomial + error term', where all three are functions of x, of course, and both the polynomial and the error depend on k (on how far we have gone with the approximation process). So the main point of the theorem above is to give us a usable formula for the kth-stage error. Typical questions then are: does the error always tend to zero, at least for each value of x across a range? Can we make the error as small as we please over a range of x values simultaneously? Determine a value of f to so-many decimal places (etc.)

4. The kth-stage error term is often called the *remainder after k terms* and denoted by $R_k(x)$:

$$f(x) = p_k(x) + R_k(x).$$

16.3.3 Example

Assuming that e^x is its own derivative, we use the theorem to show that (for every real number x) e^x is the limit (as $k \to \infty$) of

$$1 + x + \frac{x^2}{2!} + \frac{x^3}{3!} + \frac{x^4}{4!} + \cdots + \frac{x^k}{k!}.$$

Solution

Since the exponential function equals all of its own derivatives, and since they all take the value 1 at 0, it is easy to see that the formula displayed here is just $p_k(x)$. Our task, then, is only to show that the remainder tends to zero. Also, via Taylor's theorem,

$$R_k(x) = \frac{e^\xi}{(k+1)!} x^{k+1}$$

(for some ξ between 0 and x) whose modulus is at most $\dfrac{e^{|x|}|x|^{k+1}}{(k+1)!}$, which does indeed tend to zero (see paragraph 6.2.9).

16.3.4 Example

We use Taylor's theorem to show that, for the interval $J = [-1000, 1000]$, we can find a polynomial that differs from $\sin x$ at each point of J by less than 0.000001. (Assume standard facts about the trig functions.)

Solution

In the theorem, take $f(x) = \sin x$, $a = 0$. All the derivatives of f are either $\pm \sin x$ or $\pm \cos x$ so they never exceed 1 in modulus. The remainder term

$$|R_k(x)| = \left| \pm(\sin, \ \cos)(\xi) \frac{x^{k+1}}{(k+1)!} \right|$$

cannot exceed $1000^{k+1}/(k+1)!$, which tends to zero as before. Choosing a value k_0 of k that makes the latter expression less than 0.000001, we then see that $|\sin x - p_{k_0}(x)| = |R_{k_0}(x)|$ is less than 0.000001 at every point of J, that is, the Taylor polynomial $p_{k_0}(x)$ is as good an approximation to $\sin x$ as the question wanted.

16.3.5 **Example**   Estimate $\ln(1.12)$ to four significant figures.

Solution

Take $f(x) = \ln x, a = 1$. The derivatives fall into an obvious pattern $f'(x) = x^{-1}$, $f''(x) = -x^{-2}, f'''(x) = +2x^{-3}, f^{iv}(x) = -3! x^{-4}$ and, in general, we see that $f^k(x) = (-1)^{(k-1)}(k-1)! x^{-k}$. Setting $a = 1$ in these and appealing to Taylor's theorem, we see that the Taylor polynomials take the form

$$0 + 1(x-1) - \frac{1}{2}(x-1)^2 + \frac{1}{3}(x-1)^3 - \frac{1}{4}(x-1)^4 + \cdots$$

to the appropriate number of terms, and that the remainder is

$$R_k(x) = \frac{f^{k+1}(\xi)}{(k+1)!}(x-1)^{k+1} = \pm \frac{\xi^{-k-1}}{k+1}(x-1)^{k+1}$$

where we are about to replace x by 1.12 and know, therefore, that ξ lies between 1 and 1.12. Thus the modulus of the remainder after k terms cannot be more than $\frac{1}{k+1}(0.12)^{k+1}$. Experimenting now with a calculator, we soon find that $k = 4$ gives $\frac{1}{k+1}(0.12)^{k+1} = 0.000005$ approximately, which ought to be good enough for four significant figures, and the 4th Taylor polynomial approximation gives

$$p_4(1.12) = 0 + 1(1.12 - 1) - \frac{1}{2}(1.12 - 1)^2 + \frac{1}{3}(1.12 - 1)^3 - \frac{1}{4}(1.12 - 1)^4$$

which calculates out as 0.113324...

We find an answer of 0.1133 to four significant figures.

16.4 Differentiating a power series

What the previous section established can be expressed as follows: every infinitely differentiable function[2] for which it can be shown that the Taylor remainder after k terms tends to zero in a satisfactory manner *is* the sum of a power series. (The inclusion of the proviso about the remainder after k terms breaks up the symmetry of what we are trying to say, but it cannot be avoided: there are infinitely differentiable functions whose Taylor remainders totally fail to tend to zero and which, therefore, are very different from the sum of their Taylor series!)

[2] that is, every function that can be repeatedly differentiated as often as you wish

What we now seek is a converse, to the effect that every convergent power series gives, as its limit, an infinitely differentiable function. (The only proviso needed this time is that the radius of convergence shall be greater than zero, and this is obviously necessary since a series whose radius of convergence was exactly zero would define a sum function *only* at $x = 0$, so the issue of differentiability would simply not arise.) We started this investigation back in Chapter 14, from which the following revision material is taken:

Given that the power series

$$a_0 + a_1 x + a_2 x^2 + a_3 x^3 + \cdots + a_n x^n + \cdots$$

has radius of convergence $D > 0$ and therefore converges (absolutely) to a sum $f(x)$ at every point x of the open interval $(-D, D)$, we can create another series, called its *derived series*, by mindlessly differentiating each individual term in the preceding one:

$$a_1 + 2a_2 x + 3a_3 x^2 + \cdots + n a_n x^{n-1} + \cdots$$

and, continuing, another *second derived series*

$$2a_2 + 6a_3 x + \cdots + n(n-1) a_n x^{n-2} + \cdots$$

(and more such series, if the need arises) by exactly the same mechanical process. We showed in Chapter 14 that all of these series also converge (absolutely) at every point of $(-D, D)$, and we expressed the aspiration that the sum function of the derived series 'ought' to be the derivative $f'(x)$ of $f(x)$. It is time to show that this is not mere wishful thinking.

16.4.1 Theorem: power series can be differentiated term–by–term

Proof

This is, by a good margin, the biggest and most demanding proof in the entire text, so

- we shall split it down into (hopefully) more comprehensible chunks,
- do not worry unduly if you don't fully understand it,
- do pay careful attention to what the theorem says, which is both useful and natural, even if you decide to shelve the proof till later,
- do read through it carefully at some stage, because understanding *why* a result works gives a deeper understanding of what it can do.

STEP 1: CLARIFY THE TASK

We fix our attention at a typical point x_0 in $(-D, D)$ and we choose r so that $|x_0| < r < D$. (See the comments *after* the proof as to why it is important to do this.) With

$$f(x) = a_0 + a_1 x + a_2 x^2 + a_3 x^3 + \cdots + a_n x^n + \cdots$$

and, say,

$$g(x) = a_1 + 2a_2 x + 3a_3 x^2 + \cdots + na_n x^{n-1} + \cdots$$

we need to prove that the limit, as $h \to 0$, of the expression

$$E(h) = \frac{f(x_0 + h) - f(x_0)}{h} - g(x_0)$$

is zero.

Let $\varepsilon > 0$ be given. We need to show that, for all sufficiently small (non-zero) values of h, $|E(h)| < \varepsilon$.

STEP 2: BREAK THE TASK UP INTO MANAGEABLE PIECES

For each positive integer N we need to think about the Nth partial sum of what is happening here, so let

$$S_N(x) = \sum_0^N a_k x^k, \quad T_N(x) = \sum_{N+1}^\infty a_k x^k$$

and use this notation to decompose $E(h)$ into three fragments (with the intention of handling each separately):

$$E(h) = E_1(h) + E_2 + E_3(h)$$

where

$$E_1(h) = \frac{S_N(x_0 + h) - S_N(x_0)}{h} - S_N'(x_0), \quad E_2 = S_N'(x_0) - g(x_0), \quad \text{and}$$

$$E_3(h) = \frac{T_N(x_0 + h) - T_N(x_0)}{h}.$$

Fortunately, two of these three pieces are pretty easy to deal with. After all, S_N is just a polynomial, and polynomials surely are differentiable, so the limit (as $h \to 0$) of E_1 is zero; continuing, S_N' is the Nth partial sum of the series that defines g, so the limit of E_2 is also zero. The sole abiding difficulty is E_3.

STEP 3: DEAL WITH THE REMAINING DIFFICULT CASE

At this point it will pay us to recall a (rather forgettable) factorisation trick from basic algebra:

$$x^k - y^k = (x - y)(x^{k-1} + x^{k-2}y + x^{k-3}y^2 + \cdots + y^{k-1}).$$

If, in this, you replace x by $x_0 + h$ and y by x_0, it yields

$$\frac{(x_0 + h)^k - x_0^k}{h} = (x_0 + h)^{k-1} + (x_0 + h)^{k-2}x_0 + (x_0 + h)^{k-3}x_0^2 + \cdots + x_0^{k-1}$$

so, if we can arrange that both x_0 and $x_0 + h$ have modulus less than r:

$$\left| \frac{(x_0 + h)^k - x_0^k}{h} \right| < r^{k-1} + r^{k-1} + r^{k-1} + \cdots + r^{k-1} = kr^{k-1}$$

(using the triangle inequality yet again). *Now* we are fully prepared to look more closely at E_3:

$$E_3(h) = \frac{1}{h} \left\{ \sum_{N+1}^{\infty} a_k(x_0 + h)^k - \sum_{N+1}^{\infty} a_k x_0^k \right\}$$

$$= \frac{1}{h} \left\{ \sum_{N+1}^{\infty} a_k \{ (x_0 + h)^k - x_0^k \} \right\},$$

therefore

$$|E_3(h)| = \left| \frac{1}{h} \left\{ \sum_{N+1}^{\infty} a_k \{ (x_0 + h)^k - x_0^k \} \right\} \right| \leq \sum_{N+1}^{\infty} |a_k| \times kr^{k-1}.$$

The reason why *this* is good news is that $\sum_0^{\infty} k a_k r^{k-1}$ converges absolutely (to $g(r)$) so the last summation on the line above is a tail of a convergent series, and therefore must tend to zero (see 7.3.20) as $N \to \infty$.

STEP 4: PUT THE PIECES TOGETHER (IN THE RIGHT ORDER)

Remember that $\varepsilon > 0$ was given some time ago. *How* small must we make h in order that all the pieces of this jigsaw-puzzle of a proof shall come together?

1. Because $S_N'(x_0)$ is the Nth partial sum of the series whose limit (by definition) is $g(x_0)$, there is a positive integer N_0 such that, for every $N \geq N_0$, we will get

$$|E_2| = |S_N'(x_0) - g(x_0)| < \frac{\varepsilon}{3}.$$

2. Because $\sum_0^\infty k a_k r^{k-1}$ converges absolutely, there is a positive integer N_2 such that, for every $N \geq N_2$, the 'remainder' of the (modulussed) series $\sum_{N+1}^\infty |a_k| \times k r^{k-1}$ will be less than $\frac{\varepsilon}{3}$.

3. Choose now and fix a value of N that is bigger than each of N_1, N_2.

4. Because S_N is merely a differentiable polynomial, the limit of E_1 is zero. Hence there is $\delta_1 > 0$ such that $0 < |h| < \delta_1$ will guarantee that

$$|E_1(h)| < \frac{\varepsilon}{3}.$$

5. Because $|x_0| < r < D$, if we pick $\delta_2 = r - |x_0|$ then $0 < |h| < \delta_2$ will guarantee that (not only $|x_0|$ but also) $|x_0 + h|$ will be less than r: in consequence of which, the first round of estimates in STEP 3 will work.

6. So now the second round of estimates in STEP 3 gives

$$|E_3(h)| \leq \sum_{N+1}^\infty |a_k| \times k r^{k-1}$$

from which (2) gives

$$|E_3(h)| < \frac{\varepsilon}{3}.$$

Define $\delta = \min\{\delta_1, \delta_2\} > 0$. Then, provided only that $0 < |h| < \delta$, we get

$$|E(h)| = |E_1(h) + E_2 + E_3(h)| \leq |E_1(h)| + |E_2| + |E_3(h)|$$
$$< \frac{\varepsilon}{3} + \frac{\varepsilon}{3} + \frac{\varepsilon}{3} = \varepsilon.$$

16.4.2 Comments   Why did we select a number r for which $|x_0| < r < D$?

- The superficial reason is that bad things can happen to power series at $\pm D$, that is, at the endpoints of the interval whose 'radius' is the radius of convergence of the series. By selecting such an r and then only working inside $[-r, r]$, we were keeping these dangers at a small but safe distance away from our calculations.

- More precisely, the estimates taking place in STEP 3 were all rounded up, so to speak, to the behaviour of the derived power series *at the single point r*. If the power series had not converged at r, these estimates would have collapsed. There was no guarantee that the derived series would converge at D itself: we needed a point *less than D* upon which to hang these estimates.

- Furthermore, that same rounding-up process by which we estimated our various power series by the *absolutely* convergent derived series at r shows that they are all absolutely convergent also: and this we needed in order to be able to rearrange them, which we did extensively in breaking up $E(h)$ into individually-handled fragments. Such a radical reorganisation of the terms of those infinite series would have been illegal without a guarantee of their *absolute* convergence.

17 Riemann integration – area under a graph

17.1 Introduction

The words *integrate, integral, integration* occur very often in later school mathematics with what seem to be, at first sight, two entirely different meanings. Let us begin by reviewing the sorts of pre-university questions in this area that you have certainly encountered many times before.

17.1.1 Example A To find an (indefinite) integral of the function $f(x) = x \cos x$ (with respect to x).

Interpretation

This means: find, by any means whatsoever (not excluding trial and error) another function $g(x)$ whose derivative $g'(x)$ is exactly the given $f(x)$. (This is sometimes described informally as *un-differentiation* of $f(x)$.)

Solution

You probably know tricks such as 'integration by parts' for tackling this sort of question (and if so, just use them), but don't undervalue trial and error either. The presence of cos suggests that sin ought to be part of the answer, so $x \sin x$ is a reasonable first guess. Differentiate (via the product rule) and see if we are close:

$$\frac{d}{dx}\{x \sin x\} = x(\sin x)' + (\sin x)(x)' = x \cos x + \sin x.$$

Not bad, but we need to get rid of that last $\sin x$. Of course, the derivative of $\cos x$ is $-\sin x$ which would cancel it out, so ... a second guess:

$$\frac{d}{dx}\{x \sin x + \cos x\} = x(\sin x)' + (\sin x)(x)' - \sin x$$

$$= x \cos x + \sin x - \sin x = x \cos x.$$

Success. An answer is $g(x) = x \sin x + \cos x$. Indeed, for any constant C that you care to select, another answer is $g(x) = x \sin x + \cos x + C$ since added-in constants disappear under differentiation.

Undergraduate Analysis: A Working Textbook, Aisling McCluskey and Brian McMaster 2018.
© Aisling McCluskey and Brian McMaster 2018. Published 2018 by Oxford University Press

17.1.2 **Example B**   To find the (definite) integral from $x=1$ to $x=6$ of the function f described by

$$f(x) = x - 1 \text{ while } 1 \le x \le 3; \quad f(x) = 2 \text{ while } 3 < x \le 6.$$

Interpretation

This means: find the area of the region of the coordinate plane lying under the graph of f, above the horizontal axis, and between the vertical lines $x = 1$ and $x = 6$.

Solution

Even a very rough sketch graph will clarify what this region is: it consists of a right-angled triangle with vertices at $(1, 0)$, $(3, 0)$, $(3, 2)$ and a rectangle with vertices at $(3, 0)$, $(6, 0)$, $(6, 2)$, $(3, 2)$. Primary school mathematics is perfectly good enough to identify its area as $2 + 6 = 8$.

(Incidentally, the terms *definite* and *indefinite* are often left out, and so may be the phrase *with respect to x* provided that there are no other variables in play.)

Now for an exercise that brings the two ideas together:

17.1.3 **Example C**   To determine the area of the region lying under the graph of $f(x) = x \cos x$, above the x-axis and between $x = 0$ and $x = \frac{\pi}{4}$.

Procedure

- First, find an indefinite integral g of f. We'll lift our earlier answer from Example A: $g(x) = x \sin x + \cos x$.
- Second, calculate the change in value of g from $x = 0$ to $x = \frac{\pi}{4}$; this is often denoted by $[g(x)]_{x=0}^{x=\frac{\pi}{4}}$:

$$[g(x)]_{x=0}^{x=\frac{\pi}{4}} = g\left(\frac{\pi}{4}\right) - g(0) = \frac{\pi}{4}\sin\left(\frac{\pi}{4}\right) + \cos\left(\frac{\pi}{4}\right) - 0\sin(0) - \cos(0)$$
$$= \frac{\pi\sqrt{2} + 4\sqrt{2} - 8}{8}.$$

- Answer: the area is

$$\frac{\pi\sqrt{2} + 4\sqrt{2} - 8}{8}.$$

17.1.4 **Remarks**   Why we should expect that procedure to give the right answer is very hard to explain in a school classroom, no matter how easy to apply and immediately useful the result itself is. Part of our brief in this chapter is to present evidence of its reliability. Before that, however, there are several other issues to address. What exactly does 'area under a graph' mean? Does it have a logically watertight meaning – a definition that agrees with common sense but does not depend upon it – for the graph of every function? If not, for which ones? Can we expect to be able to un-differentiate any given formula? How, if at all, can we proceed with a function if it is not possible to un-differentiate it? Questions such as these are normally side-stepped in school mathematics; over the next dozen or so pages you will see why.

17.2 Riemann integrability – how closely can rectangles approximate areas under graphs?

Throughout the next several paragraphs, $f : [a, b] \to \mathbb{R}$ will be a given *bounded* function on a closed bounded interval. We imagine the graph of a typical such function:

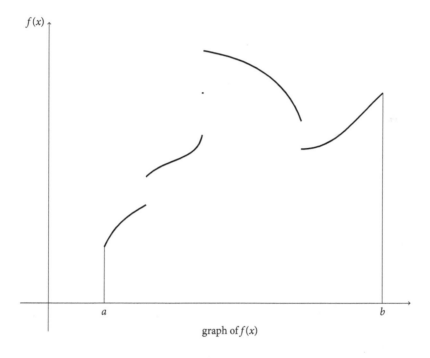

graph of $f(x)$

17.2.1 **Definition: partition**

1. A *partition* of $[a, b]$ is a finite set Δ of elements in $[a, b]$ that includes both the endpoints a and b. Since Δ is finite, we can always label its elements as $x_0, x_1, x_2 \cdots, x_n$ in such a way that

$$a = x_0 < x_1 < x_2 < \cdots < x_{n-1} < x_n = b$$

and by default we shall always assume that this has been done. The effect of the partition is to divide up $[a, b]$ into a list of subintervals

$$[x_0, x_1], [x_1, x_2], [x_2, x_3], \cdots, [x_k, x_{k+1}], \cdots, [x_{n-1}, x_n]$$

that overlap only at the shared endpoints.

2. If two partitions Δ_1, Δ_2 (of the same interval $[a, b]$) satisfy $\Delta_1 \subseteq \Delta_2$, we call Δ_2 a *refinement* of Δ_1, and say that Δ_2 is *finer* than Δ_1. Since both of these sets are finite, it is always legitimate to regard a refinement of Δ_1 as having been created by adding in the extra points one at a time, breaking up just one of the subintervals into two sub-subintervals each time.

3. Of two arbitrary partitions Δ_1, Δ_2 it is by no means guaranteed that one of them is a refinement of the other. For instance, consider the partitions $\{0, 0.5, 0.75, 1\}$ and $\{0, 0.25, 0.5, 1\}$ of $[0, 1]$.

4. If Δ_1, Δ_2 are any two partitions of $[a, b]$ then their union $\Delta_3 = \Delta_1 \cup \Delta_2$ is a partition also, and it is a refinement of each of them.

17.2.2 Definition: upper and lower Riemann sums

1. Given a partition $\Delta = \{a = x_0 < x_1 < x_2 < \cdots < x_{n-1} < x_n = b\}$ of $[a, b]$, since f is bounded on the whole of $[a, b]$, it is certainly bounded on each of the subintervals $[x_0, x_1], [x_1, x_2], [x_2, x_3], \cdots, [x_k, x_{k+1}], \cdots, [x_{n-1}, x_n]$ and possesses an infimum and a supremum on each of them. Put

$$m_k = \inf\{f(x) : x \in [x_k, x_{k+1}]\}, \quad M_k = \sup\{f(x) : x \in [x_k, x_{k+1}]\}$$

for each $k = 0, 1, 2, \cdots, n - 1$.

2. Notice that *if f happens to have a smallest and/or a biggest value over $[x_k, x_{k+1}]$* then m_k and/or M_k will be these values. In particular, this will happen in cases where f is continuous and in cases where f is either increasing or decreasing, and *when it happens* it simplifies the rest of the argument. However, it fails to happen with many of the functions that we need to consider.

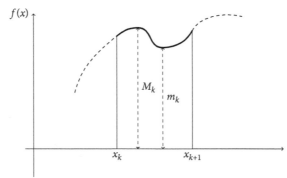

The sup and inf of f over a partition subinterval

3. The *lower Riemann sum* and the *upper Riemann sum* (for this function, this interval and this partition) are

$$L(f, [a, b], \Delta) = \sum_{k=0}^{n-1} m_k(x_{k+1} - x_k);$$

$$U(f, [a, b], \Delta) = \sum_{k=0}^{n-1} M_k(x_{k+1} - x_k).$$

Notice that $L(f, [a, b], \Delta) \leq U(f, [a, b], \Delta)$ just because $m_k \leq M_k$ for all relevant k.

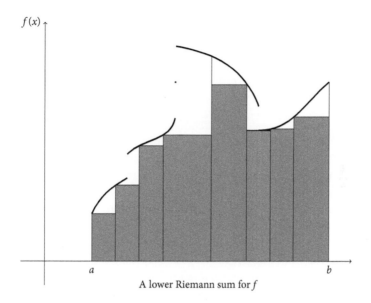

A lower Riemann sum for f

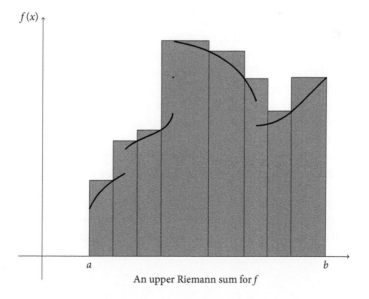

An upper Riemann sum for f

These sums have natural geometric interpretations (at least, in the case where $f(x) > 0$ always) as the area of the largest 'histogram' figure (placed upon the subintervals of the partition) that definitely fits under the graph, and the area of the smallest 'histogram' figure (placed upon the subintervals of the partition) that definitely fits over the graph. At this point, common sense suggests that, whatever we eventually define the area under the graph to mean, it should lie somewhere between these under- and overestimates.

4. In any extended argument in which f or $[a, b]$ doesn't change, it is normal practice to stop labelling the lower and upper sums with them: so that

$$L([a, b], \Delta), U([a, b], \Delta) \ or \ L(f, \Delta), U(f, \Delta) \ or \ L(\Delta), U(\Delta)$$

become legitimate abbreviations. Likewise, in a discussion involving several functions but only one interval and only one partition, shorthand symbols such as
$$L(f), L(g), L(h), \cdots U(f), U(g), U(h) \cdots$$
are perfectly acceptable.

5. Informally, our overall position is this. There may be (indeed, there are) both practical and philosophical difficulties involved in being clear about the area of a region that has one or more curved or discontinuous edges . . . but there certainly aren't any about rectangles, nor about unions of rectangles that don't overlap except along their edges. The 'histogram' trick allows us to find such a figure that approximates the region (whose area we would like to define) from beneath, and another that does so from above, using the underlying partition as a control of how fine the approximation is. As we move through successively finer and finer underlying partitions, we can expect that the approximations ought to get better and better (based on the rather imprecise insight that opting for narrower and narrower rectangles should allow us to reduce the size of the errors). This suggests that the area could be defined as the limit of one, or perhaps both, of these streams of approximations as the partition tends to . . . *and there is the difficulty:* the partition Δ is not an integer (so we are not grappling with the limit of a *sequence* of approximations) nor is it a real number that we can allow to tend to infinity and calculate the limit of an approximation function $A(x)$ as $x \to \infty$. Though this argument clearly points to a limit of some kind, it is not a type of limit that we have met previously, and we must take care not to *assume* that it works in precisely the same way as our experience of sequence and real-function limits says it should.

17.2.3 **Lemma**   If we add one extra point to a partition, it does not decrease the lower Riemann sum nor increase the upper Riemann sum. That is, if Δ^+ is Δ together with one extra point y where, say, $x_k < y < x_{k+1}$, then

$$L(f, [a, b], \Delta) \leq L(f, [a, b], \Delta^+) \quad \text{and} \quad U(f, [a, b], \Delta) \geq U(f, [a, b], \Delta^+).$$

Proof

The change from $L(f, [a, b], \Delta)$ to $L(f, [a, b], \Delta^+)$ only alters the individual summand $m_k(x_{k+1} - x_k)$ to $m'(y - x_k) + m''(x_{k+1} - y)$ where m', m'' are the infima of f on the sub-subintervals $[x_k, y]$ and $[y, x_{k+1}]$. Since, clearly, $m' \geq m_k$ and $m'' \geq m_k$, this is a nett increase or, more precisely, cannot be a nett decrease.

A similar analysis deals with the upper sum.

17.2.4 Improved lemma   If Δ' is a refinement of Δ, then

$$L(f, [a, b], \Delta) \leq L(f, [a, b], \Delta') \quad \text{and} \quad U(f, [a, b], \Delta) \geq U(f, [a, b], \Delta').$$

Proof

We can evolve from Δ to Δ' in stages, adding one new point at a time. At each stage, the first lemma tells us that the lower sum increases (or stays still) and that the upper sum decreases (or stays still). Hence the final result.

17.2.5 Proposition   Any lower sum is $\leq$ any upper sum.

Proof

What this says is that if Δ_1 and Δ_2 are any partitions at all of $[a, b]$, then $L(\Delta_1) \leq U(\Delta_2)$. The proof consists of recalling that $\Delta_3 = \Delta_1 \cup \Delta_2$ is a refinement of each of the given partitions, and then applying the Improved Lemma:

$$L(\Delta_1) \leq L(\Delta_3) \leq U(\Delta_3) \leq U(\Delta_2).$$

17.2.6 Definition

1. Let us put $A = \{L(f, [a, b], \Delta) : \text{all partitions } \Delta \text{ of } [a, b]\}$. From the proposition, the non-empty set A of real numbers is bounded above by *any* $U(f, [a, b], \Delta)$ and therefore has a supremum, a least upper bound (write it temporarily as J^-) such that $J^- \leq U(f, [a, b], \Delta)$ for every partition Δ.

   In turn, the set $B = \{U(f, [a, b], \Delta) : \text{all partitions } \Delta \text{ of } [a, b]\}$ is a non-empty set of real numbers bounded below by J^-, so it has an infimum, a greatest lower bound J^+ such that $J^- \leq J^+$.

2. The numbers J^- and J^+ are called the *lower Riemann integral* and the *upper Riemann integral* of f over $[a, b]$, and are more usually denoted by $\underline{\int} f$ and $\overline{\int} f$.

3. What we know so far is that, for any partition Δ:

$$L(f,[a,b],\Delta) \leq \underline{\int} f \leq \overline{\int} f \leq U(f,[a,b],\Delta).$$

4. We call f *Riemann integrable* over $[a,b]$ if $\underline{\int} f = \overline{\int} f$, and in that case, their common value is called the *Riemann integral* of f over $[a,b]$ and written as $\int f$. Sometimes, a more elaborated symbol such as $\int_a^b f$ or $R\int f$ or $\int_a^b f(x)\,dx$ will be employed if we feel a need to stress which interval we are operating over, or which procedure (for there are others beyond Riemann's) we are using, or which variable is associated with the horizontal axis.

5. Since Riemann's is the only integration procedure (apart from naïve un-differentiation) being discussed in this chapter, we shall feel free to abbreviate *Riemann integable* to *integrable*, *Riemann integral* to *integral*, and *upper* or *lower Riemann sum* to *upper* or *lower sum* when it makes the text easier to read.

We have by now achieved a logically reliable definition of what the integral of a bounded function over a closed bounded interval is, and when it exists, that does not depend on intuiting areas. Unfortunately, this definition is cumbersome and time-consuming to use, even for quite simple functions.

17.2.7 **Example**   To show that a constant function is Riemann integrable over any closed bounded interval, and to evaluate the integral.

Solution

We consider $f : [a,b] \to \mathbb{R}$ given by $f(x) = C, x \in [a,b]$ where C is a constant. In this case, for any partition

$$\Delta = \{a = x_0 < x_1 < x_2 < \cdots < x_{n-1} < x_n = b\},$$

we see that $m_k = M_k = C$ for every k, so

$$L(f,[a,b],\Delta) = \sum_0^{n-1} C(x_{k+1} - x_k) = C\sum_0^{n-1}(x_{k+1} - x_k) = C(b-a)$$

and, likewise, $U(f,[a,b],\Delta) = C(b-a)$. The sets we denoted by A and B in the definition paragraph above each consists of the single number $C(b-a)$, so it is entirely trivial to determine inf and sup for them: $\underline{\int} f = C(b-a)$ and $\overline{\int} f = C(b-a)$. Therefore f is integrable, and its integral over $[a,\overline{b}]$ is $C(b-a)$.

17.2.8 Example Given $b > 0$, and f defined by $f(x) = x$ on the interval $[0, b]$, to show that $R \int f$ exists and equals $\dfrac{b^2}{2}$.

Solution

For any partition

$$\Delta = \{0 = x_0 < x_1 < x_2 < \cdots < x_{n-1} < x_n = b\}$$

of $[0, b]$, the fact that f is increasing tells us that, on the typical subinterval $[x_k, x_{k+1}], f(x_k) = x_k$ is the least value of f and $f(x_{k+1}) = x_{k+1}$ is the greatest value of f: that is, $m_k = x_k$ and $M_k = x_{k+1}$. Thus, typical lower and upper sums take the form

$$\sum_0^{n-1} x_k(x_{k+1} - x_k), \quad \sum_0^{n-1} x_{k+1}(x_{k+1} - x_k).$$

Consider now the special-case partition Δ_n all of whose subintervals have the same length b/n (where n is a particular positive integer). In this case, x_k is simply kb/n for each k, so the lower and upper sums become much more accessible to calculation:

$$L(f, [0, b], \Delta_n) = \sum_0^{n-1} x_k(x_{k+1} - x_k) = \sum_0^{n-1} (kb/n)((k+1)b/n - kb/n)$$

$$= \sum_0^{n-1} (kb/n)(b/n) = (b^2/n^2) \sum_0^{n-1} k = (b^2/n^2)(n(n-1))/2$$

$$= b^2 \left(\frac{n-1}{2n} \right) = \frac{b^2}{2} \left(1 - \frac{1}{n} \right)$$

and this is an increasing sequence with limit (and therefore supremum) $\dfrac{b^2}{2}$. Of course, the Δ_ns are only *some* of the possible partitions, so it is imaginable that the supremum of *all* their lower sums might be different from the supremum of this sample. Yet it certainly cannot be smaller:[1] so we learn that $\underline{\int} f \geq \dfrac{b^2}{2}$.

---

[1] If $\emptyset \neq A_1 \subseteq A_2$ where A_2 is a bounded-above set of real numbers, then $\sup A_1 \leq \sup A_2$.

Likewise for the upper sums:

$$U(f, [0, b], \Delta_n) = \sum_0^{n-1} x_{k+1}(x_{k+1} - x_k) = \sum_0^{n-1} ((k+1)b/n)((k+1)b/n - kb/n)$$

$$= \sum_0^{n-1} ((k+1)b/n)(b/n) = (b^2/n^2) \sum_0^{n-1} (k+1) = (b^2/n^2)(n(n+1))/2$$

$$= \frac{b^2}{2} \left(\frac{n+1}{n} \right) = \frac{b^2}{2} \left(1 + \frac{1}{n} \right)$$

which is a decreasing sequence with limit (and therefore infimum) $\dfrac{b^2}{2}$. The infimum of *all* the upper sums might conceivably differ from the infimum of this special sample, but it cannot be greater: therefore $\overline{\int} f \leq \dfrac{b^2}{2}$.

Bearing in mind that $\underline{\int} f \leq \overline{\int} f$ in all cases, we now deduce that $\underline{\int} f = \overline{\int} f = \dfrac{b^2}{2}$, as expected.

17.2.9 Remark   It took us a full typed page of calculations to establish the integral of the absurdly simple function $f(x) = x$. Your rational, entirely legitimate response to that observation should be one of bitter disappointment, combined with a determination to get access to better methods as soon as possible. The first step in that direction is the following lemma, strongly reminiscent of the Cauchy condition's ability to detect the existence of a limit (for a sequence) without any need to know what number that limit might turn out to be.

17.2.10 Darboux's (or Riemann's) integrability criterion   The function f is Riemann integrable over $[a, b]$ if and only if, for each $\varepsilon > 0$, there is a partition Δ_ε of $[a, b]$ for which

$$U(f, [a, b], \Delta_\varepsilon) - L(f, [a, b], \Delta_\varepsilon) < \varepsilon.$$

Proof

If, for each $\varepsilon > 0$, such a partition exists, then our inequality

$$L(f, [a, b], \Delta) \leq \underline{\int} f \leq \overline{\int} f \leq U(f, [a, b], \Delta) \quad \text{for all partitions } \Delta$$

says that $\underline{\int} f$ and $\overline{\int} f$ differ, if at all, by less than ε. Since the lower and upper integrals do not depend on ε, that can only happen if they are equal. Hence f is integrable.

Conversely, suppose that f is integrable. Then $J = \underline{\int} f = \overline{\int} f$ is both the supremum of the lower sums and the infimum of the upper sums. Given $\varepsilon > 0$, it is therefore possible to find a partition Δ_1 with $L(\Delta_1) > J - \frac{\varepsilon}{2}$, and a partition Δ_2 with $U(\Delta_2) < J + \frac{\varepsilon}{2}$. Put $\Delta_3 = \Delta_1 \cup \Delta_2$:

$$J - \frac{\varepsilon}{2} < L(\Delta_1) \le L(\Delta_3) \le U(\Delta_3) \le U(\Delta_2) < J + \frac{\varepsilon}{2}.$$

Therefore the gap between $L(\Delta_3)$ and $U(\Delta_3)$ is smaller than ε, and the Darboux criterion holds.

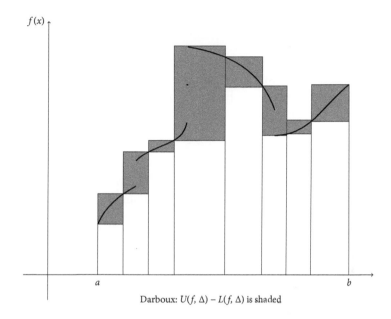

Darboux: $U(f, \Delta) - L(f, \Delta)$ is shaded

Although the criterion seems to concern only the *existence* of the integral of f, its numerical value quite often emerges from the same calculations. Note that:

17.2.11 Corollary If f is Riemann integrable over $[a, b]$, then its integral is the *unique* number J such that

$$L(f, [a, b], \Delta) \le J \le U(f, [a, b], \Delta) \quad \text{for all partitions } \Delta.$$

Proof

The integral certainly does lie between every lower sum and every upper sum, by its definition. If it were not unique in this respect, there would have to be two distinct numbers J_1 and J_2 with

$$L(f, [a, b], \Delta) \le J_1 < J_2 \le U(f, [a, b], \Delta) \quad \text{for all partitions } \Delta.$$

Yet this implies $U(\Delta) - L(\Delta) \ge \varepsilon = J_2 - J_1 > 0$ for every partition Δ, in contradiction to the Darboux criterion.

As a first indication of the usefulness of the Darboux test, here is a worked example of a function that is *not* Riemann integrable. Given how labour intensive it seems to be to show that a simple function *is* integrable, you may be surprised (or even irritated) to see how straightforward this is.

17.2.12 **Example**   We define a function $f : [0, 5] \to \mathbb{R}$ by $f(x) = 2$ if x is rational, $f(x) = 3$ if x is irrational. Show that f is not Riemann integrable over $[0, 5]$.

Solution

Let Δ be absolutely any partition of $[0, 5]$. Since, in every interval, there are both rationals and irrationals, f will take both the value 2 and the value 3 somewhere in each of the subintervals $[x_k, x_{k+1}]$ into which Δ carves up $[0, 5]$. So (for each k) $m_k = 2$ and $M_k = 3$. Hence[2]

$$L(f, \Delta) = \sum 2(x_{k+1} - x_k) = 2 \sum (x_{k+1} - x_k) = 2(5) = 10.$$

Likewise, $U(f, \Delta) = 15$. So the supremum of 'all' the lower sums is just the sup of the one single number 10, namely 10. Likewise, the infimum of 'all' the upper sums is 15.

We conclude that *no* partition can force $U(f, \Delta) - L(f, \Delta)$ to be smaller than $15 - 10 = 5$, so Darboux alerts us that this function is not integrable. (Alternatively: we have just shown that $\underline{\int} f = 10 \neq 15 = \overline{\int} f$, so $\int f$ does not exist via the definition.)

Clearly, the choice of the numbers $0, 5, 2$ and 3 has no real bearing on the outcome: a function that takes a constant value on the rationals and a different constant value on the irrationals is not integrable over any non-degenerate interval.

Here is another worked example illustrating how we can use Darboux plus its corollary (17.2.11) both to guarantee the existence of a Riemann integral and to determine its numerical value.

17.2.13 **Example**   Determine the Riemann integral of the function $f : [1, 4] \to \mathbb{R}$ described by

$$f(x) = \begin{cases} 0 & \text{if } 1 \leq x < 2, \\ 7 & \text{if } x = 2, \\ 0 & \text{if } 2 < x \leq 4. \end{cases}$$

Solution

For any positive $h < 1$ let us consider the partition $\Delta_h = \{1, 2 - h, 2 + h, 4\}$ (whose intention is to isolate the somewhat anomalous value $x = 2$ of the domain). We see that $L(f, \Delta_h) = 0$ and that $U(f, \Delta_h) = 14h$.

Firstly, given $\varepsilon > 0$, if we choose h to be, say, $\min\left\{\frac{1}{2}, \frac{\varepsilon}{15}\right\}$, then $L(f, \Delta_h)$ and $U(f, \Delta_h)$ differ by less than ε, so Darboux guarantees that the Riemann integral $\int_1^4 f$ exists.

---

[2] Note that $\sum(x_{k+1} - x_k)$ is always the total length of all the subintervals, that is, the length of the whole interval in question.

Secondly, now that the integral is known to exist, Corollary 17.2.11 observes that it is the unique number that lies between $L(f, \Delta)$ and $U(f, \Delta)$ for *every* partition Δ. Since 0 is the only number lying between $L(f, \Delta_h) = 0$ and $U(f, \Delta_h) = 14h$ (for every h between 0 and 1) even for the *particular* partitions Δ_h that we have examined, $\int_1^4 f$ can only be zero.

17.3 The integral theorems we ought to expect

As a second (and much larger) illustration of how we can use Darboux's criterion, here are a few of the several integration theorems that your pre-university experience probably leads you to expect to be true for the Riemann integral:

17.3.1 Theorem: integral addition over back-to-back intervals　　If f is (Riemann) integrable over $[a, b]$ and over $[b, c]$, where $a < b < c$, then it is also integrable over $[a, c]$, and

$$\int_a^c f = \int_a^b f + \int_b^c f.$$

Proof

Given $\varepsilon > 0$, use Darboux to find a partition Δ_1 of $[a, b]$ such that

$$U([a, b], \Delta_1) - L([a, b], \Delta_1) < \frac{\varepsilon}{2}.$$

Likewise, use Darboux again to find a partition Δ_2 of $[b, c]$ such that

$$U([b, c], \Delta_2) - L([b, c], \Delta_2) < \frac{\varepsilon}{2}.$$

Now $\Delta_3 = \Delta_1 \cup \Delta_2$ is a partition of $[a, c]$ and, purely by the definitions,

$$L([a, c], \Delta_3) = L([a, b], \Delta_1) + L([b, c], \Delta_2)$$

and

$$U([a, c], \Delta_3) = U([a, b], \Delta_1) + U([b, c], \Delta_2).$$

It follows that

$$U([a, c], \Delta_3) - L([a, c], \Delta_3) =$$
$$= U([a, b], \Delta_1) - L([a, b], \Delta_1) + U([b, c], \Delta_2) - L([b, c], \Delta_2)$$
$$< \frac{\varepsilon}{2} + \frac{\varepsilon}{2} = \varepsilon.$$

At this point, the Darboux test tells us that $\int_a^c f$ exists.

Secondly, we know that

$$L([a,b], \Delta_1) \leq \int_a^b f \leq U([a,b], \Delta_1), \quad L([b,c], \Delta_2) \leq \int_b^c f \leq U([b,c], \Delta_2).$$

Adding these, we get

$$L([a,b], \Delta_1) + L([b,c], \Delta_2) \leq \int_a^b f + \int_b^c f \leq U([a,b], \Delta_1) + U([b,c], \Delta_2)$$

that is, in the light of our comments above,

$$L([a,c], \Delta_3) \leq \int_a^b f + \int_b^c f \leq U([a,c], \Delta_3).$$

Also, of course,

$$L([a,c], \Delta_3) \leq \int_a^c f \leq U([a,c], \Delta_3).$$

Comparing these two displays, we see that $\int_a^b f + \int_b^c f$ and $\int_a^c f$ differ, if at all, by no more than $L([a,c], \Delta_3)$ and $U([a,c], \Delta_3)$ differ: that is, by less than ε. Since they are independent of ε, which is arbitrary – and could therefore be made arbitrarily small – that can be true only if they are exactly equal.

It is very convenient to be able to jettison the requirement $a < b < c$ from this result, and a simple notational convention allows this to happen:

17.3.2 **Definition** We define (for arbitrary $a \in \mathbb{R}$)

$$\int_a^a f = 0$$

and, whenever $b < a$, we define

$$\int_a^b f = - \int_b^a f.$$

The effect of this (rather artificial seeming) convention is that the equality

$$\int_a^c f = \int_a^b f + \int_b^c f.$$

now becomes true no matter how the three numbers a, b, c are arranged on the real line (always provided, of course, that the two integrals on the right-hand side do exist). For instance, if $a = c$, the equality merely says that $0 = \int_a^b + \int_b^a$, that is, that $\int_a^b = - \int_b^a$, which is correct by the convention. Again, if $a < c < b$ then the equality $\int_a^c = \int_a^b + \int_b^c$ decodes under the convention as $\int_a^c = \int_a^b - \int_c^b$, that is, $\int_a^b = \int_a^c + \int_c^b$ which is the originally established version of the theorem when the limits of integration occur in that order.

There is also a valid converse to 17.3.1: if $a < b < c$ and f is integrable over $[a, c]$, then it is necessarily integrable also over $[a, b]$ and over $[b, c]$. In fact, with really no additional work we can prove something slightly more general:

17.3.3 Theorem: integrability over a subinterval   Suppose that f is integrable over $[a, b]$ and that $a \le c < d \le b$; then f is integrable over $[c, d]$.

Proof

Given $\varepsilon > 0$, first use Darboux to find a partition Δ of $[a, b]$ for which $U(f, [a, b], \Delta) - L(f, [a, b], \Delta) < \varepsilon$. Of course, it is perfectly possible that Δ will not include the points c and d ... but if we refine Δ by adding them in, the lower sum will increase or stay still, and the upper sum will decrease or stay still: so, after the refinement (if necessary) the gap between $U(\ldots)$ and $L(\ldots)$ will still be less than ε. For that reason, we may as well assume that this has been done already, and that Δ does include both c and d.

With that understanding, $\Delta' = \Delta \cap [c, d]$ is now a partition of $[c, d]$, and $U(f, [c, d], \Delta') - L(f, [c, d], \Delta')$ is just the sum of those expressions

$$(M_k - m_k)(x_{k+1} - x_k)$$

for which the subinterval $[x_k, x_{k+1}]$ happens to lie within $[c, d]$.

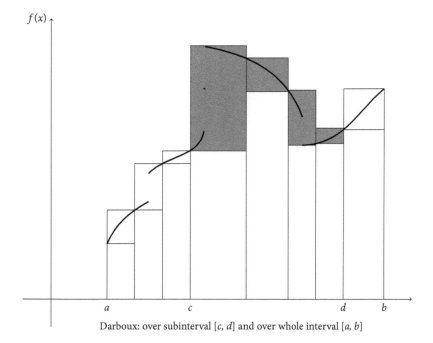

Darboux: over subinterval $[c, d]$ and over whole interval $[a, b]$

It is therefore $\leq$ the total of all such expressions across the whole of $[a, b]$; that is:

$$U(f, [c, d], \Delta') - L(f, [c, d], \Delta') \leq U(f, [a, b], \Delta) - L(f, [a, b], \Delta) < \varepsilon.$$

By Darboux, f is integrable over $[c, d]$.

17.3.4 Theorem: integral addition of functions If f and g are each (Riemann) integrable over $[a, b]$ then so is $f + g$, and

$$\int_a^b (f + g) = \int_a^b f + \int_a^b g.$$

Proof

Given $\varepsilon > 0$ we use Darboux (twice) to find partitions Δ_1, Δ_2 of $[a, b]$ such that

$$U(f, \Delta_1) - L(f, \Delta_1) < \frac{\varepsilon}{2}, \quad U(g, \Delta_2) - L(g, \Delta_2) < \frac{\varepsilon}{2}.$$

Putting $\Delta_3 = \Delta_1 \cup \Delta_2$, it follows that

$$U(f, \Delta_3) - L(f, \Delta_3) < \frac{\varepsilon}{2}, \quad U(g, \Delta_3) - L(g, \Delta_3) < \frac{\varepsilon}{2}.$$

Thinking back to the definitions of m_k and M_k, and necessarily now enhancing that notation to refer explicitly to the function it concerns, we have

$$
\begin{aligned}
m_k(f + g) &= \inf\{f(x) + g(x) : x_k \leq x \leq x_{k+1}\} \\
&\geq \inf\{f(x)) : x_k \leq x \leq x_{k+1}\} + \inf\{g(x)) : x_k \leq x \leq x_{k+1}\} \\
&= m_k(f) + m_k(g)
\end{aligned}
$$

for each $k = 0, 1, 2, \cdots, n - 1$ and, likewise

$$M_k(f + g) \leq M_k(f) + M_k(g).$$

Therefore

$$
\begin{aligned}
L(f + g, \Delta_3) &= \sum_0^{n-1} (m_k(f + g))(x_{k+1} - x_k) \\
&\geq \sum_0^{n-1} (m_k(f) + m_k(g))(x_{k+1} - x_k) \\
&= \sum_0^{n-1} (m_k(f))(x_{k+1} - x_k) + \sum_0^{n-1} (m_k(g))(x_{k+1} - x_k) \\
&= L(f, \Delta_3) + L(g, \Delta_3),
\end{aligned}
$$

and

$$U(f+g,\Delta_3) = \sum_0^{n-1}(M_k(f+g))(x_{k+1}-x_k)$$

$$\leq \sum_0^{n-1}(M_k(f)+M_k(g))(x_{k+1}-x_k)$$

$$= \sum_0^{n-1}(M_k(f))(x_{k+1}-x_k) + \sum_0^{n-1}(M_k(g))(x_{k+1}-x_k)$$

$$= U(f,\Delta_3) + U(g,\Delta_3).$$

From these, we get

$$U(f+g,\Delta_3) - L(f+g,\Delta_3) \leq U(f,\Delta_3) + U(g,\Delta_3) - L(f,\Delta_3) - L(g,\Delta_3)$$

$$= U(f,\Delta_3) - L(f,\Delta_3) + U(g,\Delta_3) - L(g,\Delta_3) < \frac{\varepsilon}{2} + \frac{\varepsilon}{2} = \varepsilon,$$

and it follows from Darboux that $\int(f+g)$ exists.

Secondly, we know that

$$L(f,\Delta_3) \leq \int f \leq U(f,\Delta_3) \text{ and } L(g,\Delta_3) \leq \int g \leq U(g,\Delta_3)$$

so, adding,

$$L(f,\Delta_3) + L(g,\Delta_3) \leq \int f + \int g \leq U(f,\Delta_3) + U(g,\Delta_3)........(1)$$

On the other hand, we learned above that

$$L(f,\Delta_3) + L(g,\Delta_3) \leq L(f+g,\Delta_3) \leq \int(f+g)$$

$$\leq U(f+g,\Delta_3) \leq U(f,\Delta_3) + U(g,\Delta_3)..............(2)$$

Since the gap between $L(f,\Delta_3)+L(g,\Delta_3)$ and $U(f,\Delta_3)+U(g,\Delta_3)$ is smaller than ε, we see from (1) and (2) that the difference (if there is one) between $\int f + \int g$ and $\int(f+g)$ must also be smaller than ε.

Because the integral expressions are independent of ε, which is arbitrary, this can only be true if they are exactly equal.

17.3.5 Theorem: integral of a scaled function For any constant C, if f is integrable over $[a,b]$ then so is Cf, and

$$\int Cf = C\int f.$$

Proof

- Case 1: $C > 0$. Given $\varepsilon > 0$, apply Darboux to f with the tolerance $\frac{\varepsilon}{C}$: there is a partition Δ of $[a, b]$ such that $U(f, \Delta) - L(f, \Delta) < \frac{\varepsilon}{C}$. Now (for this same partition)

$$m_k(Cf) = \inf\{Cf(x) : x_k \leq x \leq x_{k+1}\}$$
$$= C\inf\{f(x) : x_k \leq x \leq x_{k+1}\} = Cm_k(f)$$
$$\text{and likewise } M_k(Cf) = CM_k(f),$$

so

$$L(Cf, \Delta) = \sum m_k(Cf)(x_{k+1} - x_k) = C\sum m_k(f)(x_{k+1} - x_k) = CL(f, \Delta)$$
$$\text{and likewise } U(Cf, \Delta) = CU(f, \Delta).$$

Therefore $U(Cf, \Delta) - L(Cf, \Delta) = C(U(f, \Delta) - L(f, \Delta)) < C\left(\frac{\varepsilon}{C}\right) = \varepsilon$, showing via Darboux that Cf is integrable.

Secondly, now that we know $\int Cf$ exists,

$$L(Cf, \Delta) \leq \int Cf \leq U(Cf, \Delta); \ldots\ldots\ldots(3)$$

but also $L(f, \Delta) \leq \int f \leq U(f, \Delta)$ so, multiplying across by C, we find that $CL(f, \Delta) \leq C\int f \leq CU(f, \Delta)$, that is,

$$L(Cf, \Delta) \leq C\int f \leq U(Cf, \Delta)\ldots\ldots(4)$$

Comparing (3) and (4), we again find that the difference between $C\int f$ and $\int Cf$ is less than arbitrary ε, so the two expressions must coincide.
- Case 2: $C = 0$. Here, Cf is a constantly zero function, whose integral is zero, so the result is entirely trivial.
- EXERCISE: check out the details of Case 3: $C < 0$. Be aware that scaling by a negative number will swop over sups and infs: this time we have to anticipate $m_k(Cf) = CM_k(f), L(Cf, \Delta) = CU(f, \Delta)$, and so on.

Here is perhaps the easiest of this set of theorems to believe and to prove:

17.3.6 Theorem: integrating across an inequality If $f(x) \leq g(x)$ for every $x \in [a, b]$, and f and g are integrable over $[a, b]$, then

$$\int f \leq \int g.$$

Proof

For any partition Δ, any resulting subinterval $[x_k, x_{k+1}]$ and any x in that subinterval, we know $f(x) \leq g(x)$. That feeds through the sups and infs to give us $m_k(f) \leq m_k(g)$ and $M_k(f) \leq M_k(g)$, feeds through the formation of sums to give us $L(f, [a, b], \Delta) \leq L(g, [a, b], \Delta)$ and $U(f, [a, b], \Delta) \leq U(g, [a, b], \Delta)$, and feeds through more sups and infs to provide $\underline{\int} f \leq \underline{\int} g$ and $\overline{\int} f \leq \overline{\int} g$. Since the functions are integrable, that is equivalent to what we had to prove.

17.3.7 Corollary If $K \leq f(x) \leq L$ for all $x \in [a, b]$, where K and L are constants, then

$$K(b - a) \leq \int f \leq L(b - a).$$

Proof

Immediate upon applying the theorem to f together with each of the constant functions K and L (whose integrals we determined some time ago).

17.3.8 Theorem: integral of the modulus If $f : [a, b] \to \mathbb{R}$ is integrable then so is $|f|$, and

$$\int |f| \geq \left| \int f \right|.$$

Proof

All we have to show is that $|f|$ can be integrated: because then

$$\text{(for all } x) \quad f(x) \leq |f(x)| \Rightarrow \int f \leq \int |f|, \quad \text{and}$$

$$\text{(for all } x) \quad -f(x) \leq |f(x)| \Rightarrow -\int f = \int(-f) \leq \int |f|$$

by the previous theorem. Then $|\int f|$ is either $\int f$ or $-\int f$ and, whichever of the two it is, it is $\leq \int |f|$.

Using Darboux again (and given $\varepsilon > 0$), the integrability of f says there is a partition Δ for which (in the usual notation)

$$U(f, \Delta) - L(f, \Delta) = \sum (M_k(f) - m_k(f))(x_{k+1} - x_k) < \varepsilon.$$

Consider the embedded expression $M_k(f) - m_k(f)$. If f happened to have largest and smallest values $f(x')$ and $f(x'')$ over the subinterval $[x_k, x_{k+1}]$ then that expression would have been $f(x') - f(x'')$, that is, the biggest difference-in-value that two values of f could achieve in the subinterval. In general, of course, 'biggest' values may fail to happen, and we shall need to settle for the second-best option, the supremum. What that indicates (and you can prove it formally without much difficulty) is that $M_k(f) - m_k(f)$ is the supremum of the differences-in-value $|f(t) - f(u)|$ as t and u vary across the subinterval $[x_k, x_{k+1}]$.

Although that is a slightly awkward way to think about $M_k(f) - m_k(f)$ most of the time, it is the optimal approach to take to it in this particular proof, because the inverse triangle inequality:

$$\big|\, |p| - |q|\, \big| \le |p - q|, \quad p, q \in \mathbb{R}$$

gives us a neat way to connect these values for f and for $|f|$; look:

$$\big|\, |f(t)| - |f(u)|\, \big| \le |f(t) - f(u)|, \quad \text{all } t, u \text{ in the subinterval.}$$

Taking sups across that last line gives $M_k(|f|) - m_k(|f|) \le M_k(f) - m_k(f)$, and therefore

$$\sum (M_k(|f|) - m_k(|f|))(x_{k+1} - x_k) \le \sum (M_k(f) - m_k(f))(x_{k+1} - x_k),$$

that is,

$$U(|f|, \Delta) - L(|f|, \Delta) \le U(f, \Delta) - L(f, \Delta) < \varepsilon.$$

Therefore, via Darboux once more, $|f|$ is indeed integrable.

The last instalment in this catalogue of expected theorems is the one that says that the product of two integrable functions is integrable. It is quite intricate to prove this directly, so we shall instead sneak up on it from behind using the following lemma as cover:

17.3.9 **Lemma: integrating the square of a function**   If $f : [a, b] \to \mathbb{R}$ is integrable (over $[a, b]$) then so is f^2.

Proof

Think again about the quantity we denote by $M_k - m_k$ as being the supremum of all the differences $|f(t) - f(u)|$ as t and u vary over the kth subinterval of the partition. We need to compare this quantity as calculated for f with the same quantity as calculated for f^2, and this – keeping in mind that f is bounded, so $|f(x)| < K$ for some constant K and for all x – turns out to be rather easy:

$$|f^2(t) - f^2(u)| = |(f(t) + f(u))(f(t) - f(u))| = |(f(t) + f(u))|\,|(f(t) - f(u))|$$
$$\le (|(f(t)| + |f(u))|)\,|(f(t) - f(u))| \le 2K|(f(t) - f(u))|.$$

Taking sups across that line, we find that (for each relevant k):

$$M_k(f^2) - m_k(f^2) \le 2K(M_k(f) - m_k(f))$$

and, in consequence,

$$U(f^2, \Delta) - L(f^2, \Delta) \le 2K(U(f, \Delta) - L(f, \Delta)).$$

Now, given $\varepsilon > 0$, Darboux says that (when f is integrable) there is a partition Δ for which $U(f, \Delta) - L(f, \Delta) < \dfrac{\varepsilon}{2K}$. Feed that into the previous line, and we get

$$U(f^2, \Delta) - L(f^2, \Delta) \le 2K \left(\frac{\varepsilon}{2K}\right) = \varepsilon.$$

Now a second mobilisation of Darboux shows that f^2 is also integrable.

17.3.10 Theorem: multiplying two integrable functions   Suppose that f and g are both integrable over $[a, b]$. Then so is their product fg.

Proof

Forget Darboux for once: this is just basic school algebra. For any $x \in [a, b]$, and abbreviating $f(x)$ to f and $g(x)$ to g in order to minimise the clutter:

$$\frac{(f+g)^2 - (f-g)^2}{4} = fg$$

and, from previous theorems, all of the following are integrable:

$$f + g, -g = (-1)g, f - g = f + (-g), (f+g)^2, (f-g)^2,$$
$$-(f-g)^2 = (-1)(f-g)^2, (f+g)^2 - (f-g)^2 = (f+g)^2 + (-(f-g)^2)$$

and, finally, so is

$$\frac{(f+g)^2 - (f-g)^2}{4},$$

whence the result.

17.4 The fundamental theorem of calculus

Here we are, some twenty pages into the theory of the Riemann integral and, so far, $f(x) = x$ is the *most complicated function* that we have actually integrated. How bad is that?[3]

The reality is that the Riemann definition of integral and integrability, successful though it is in putting the intuitive idea of area-under-graph on a logically sound basis, is extremely unwieldy on its own as a tool for actually calculating integrals.

---

[3] Very.

However, the theorems you have seen developing from it over the last several pages now allow us to construct a much more efficient and easier calculation method, not for all integrable functions, but for a huge range of commonly occurring ones. More specifically, we now set out to investigate whether the method shown in Example C in the introduction to this chapter is valid for the Riemann integral. The label 'the fundamental theorem of calculus' is used to refer to both of the theorems in this section (and, on occasions, to other similar results).

17.4.1 **Theorem**   Given that $f : [a, b] \to \mathbb{R}$ is integrable, we can[4] define another function $F : [a, b] \to \mathbb{R}$ as follows:

$$F(x) = \int_a^x f, \quad x \in [a, b].$$

Then:

1. F is continuous on $[a, b]$.
2. If f is continuous at a point p of (a, b), then F is differentiable at p, and $F'(p) = f(p)$.
3. If f is continuous on $[a, b]$, then $F' = f$ everywhere in (a, b).

Proof

Since f is integrable, it must be bounded. Choose therefore a positive constant K such that $|f(x)| \le K$ always.

1. For any $x \in [a, b)$ let positive h be small enough to ensure that $y = x + h$ also belongs to $[a, b]$. Then

$$|F(x + h) - F(x)| = \left| \int_a^{x+h} f - \int_a^x f \right| = \left| \int_x^{x+h} f \right|$$

$$\le \int_x^{x+h} |f|$$

$$\le \int_x^{x+h} K = K((x + h) - x) = Kh.$$

Since the final item tends to zero as $h \to 0^+$, we get the one-sided limit $\lim_{y \to x^+} F(y) = F(x)$.

Similarly, for any $x \in (a, b]$, $\lim_{y \to x^-} F(y) = F(x)$.

The agreement of the two one-sided limits and of the value of F tells us that F is continuous at each point of (a, b) (and we also got the correct one-sided limits at a and at b, where it is *only* one-sided limits that are relevant). This proves (1).

---

[4] thanks to the theorem on integrability over subintervals

2. Again, let positive h be small enough to ensure that $y = p + h$ also belongs to $[a, b]$, Then

$$\frac{F(p + h) - F(p)}{h} - f(p) = \frac{F(p + h) - F(p) - hf(p)}{h}$$

$$= \frac{\int_a^{p+h} f - \int_a^p f - hf(p)}{h}$$

$$= \frac{\int_p^{p+h} f - hf(p)}{h}$$

$$= \frac{\int_p^{p+h} f - \int_p^{p+h} f(p)}{h}$$

(Remember that $f(p)$ is just a constant.)

$$= \frac{\int_p^{p+h} \{f(x) - f(p)\}}{h}.$$

(We're writing in the (x) here just to stress that f varies across the interval of integration, whereas $f(p)$ does not.)

Next, take the modulus:

$$\left| \frac{F(p + h) - F(p)}{h} - f(p) \right| = \left| \frac{\int_p^{p+h} \{f(x) - f(p)\}}{h} \right|$$

$$\leq \frac{\int_p^{p+h} |f(x) - f(p)|}{h}$$

(Consider now the supremum of all the values of $|f(x) - f(p)|$ as x varies over $[p, p + h]$:)

$$\leq \frac{\int_p^{p+h} \sup |f(x) - f(p)|}{h} = \frac{h \sup |f(x) - f(p)|}{h}$$

$$= \sup_{p \leq x \leq p+h} |f(x) - f(p)|.$$

Now continuity of f tells us that this final supremum tends to zero as $h \to 0^+$. Thus we have a one-sided limit

$$\lim_{h \to 0^+} \left| \frac{F(p + h) - F(p)}{h} - f(p) \right| = 0,$$

which is equivalent to

$$\lim_{h \to 0^+} \frac{F(p+h) - F(p)}{h} = f(p).$$

A very similar argument establishes the *other* one-sided limit:

$$\lim_{h \to 0^-} \frac{F(p+h) - F(p)}{h} = f(p)$$

and completes the proof of part(2) since the two one-sided limits agree upon the number $f(p)$.

3. This is immediate from part (2).

17.4.2 Comments

- We actually proved a bit more than we claimed in part 2 above. Provided that you interpret the phrase 'F is differentiable on $[a, b]$' to mean differentiability on (a, b) *plus* existence of both

$$\lim_{h \to 0^+} \frac{F(a+h) - F(a)}{h} \quad \text{and} \quad \lim_{h \to 0^-} \frac{F(b+h) - F(b)}{h}$$

then our proof established that F was differentiable on the closed interval $[a, b]$ rather than just on the open interval (a, b).

- Note that we just now showed that every continuous function *is* the derivative of *something*. However, that is no guarantee that we can come up with a simple, explicit *formula* for that *something*. Furthermore, the converse is false: there are plenty of discontinuous functions that are derivatives of something.[5]

17.4.3 Theorem   Given that $f : [a, b] \to \mathbb{R}$ is continuous and integrable, suppose that we can find a function $G : [a, b] \to \mathbb{R}$ such that

- G is continuous on $[a, b]$,
- G is differentiable on (a, b), and
- $G' = f$ on (a, b).

Then

$$\int f = [G(x)]_{x=a}^{x=b} = G(b) - G(a).$$

---

[5] For instance, differentiate $f(x) = x^2 \sin(x^{-1})$ for $x \neq 0, f(0) = 0$, paying careful attention to exactly what happens at the difficult point $x = 0$, and you should find that the derivative exists at every point (including 0) but doesn't even have a limit as $x \to 0$.

Proof

With F as defined in the previous theorem, notice that $F - G$ is continuous on $[a, b]$ and has zero derivative everywhere in (a, b). Therefore[6] it is constant. In particular, $F(b) - G(b) = F(a) - G(a)$, which we rearrange into

$$F(b) - F(a) = G(b) - G(a).$$

However, $F(b) = \int_a^b f$ and $F(a) = \int_a^a f = 0$ by definition, so the previous display is exactly what we had to prove.

17.4.4 Comment   If T S Eliot is right when he says that ... *the end of all our exploring will be to arrive where we started and know the place for the first time*, then that is just about the point we have reached in our brief exploration of the integral as defined by the Riemann process: for the previous result now justifies the way in which you have been calculating integrals up to now. In particular, the phrase 'suppose that we can find' is, in a way, liberating: when you are looking for an expression whose derivative is the given function f, you are free to use any sixth-form tricks, any trial-and-error or sheer guesswork process, even mindlessly looking up a cookbook table of standard derivatives and integrals, so long as you check that the derivative of the thing you 'found' actually is f – and this check is almost always a routine process since differentiating, unlike un-differentiating, is normally a pretty algorithmic business. After that, the actual calculation of the integral is just arithmetic.

17.4.5 Examples   Assuming that the following expressions are integrable over the indicated intervals, calculate their integrals. (You should assume, where appropriate, basic properties of trig, logarithmic and exponential functions.)

1. $f(x) = x^n$ on $[a, b]$ assuming that $n \neq -1$;
2. $f(x) = x^2(5 + 2x^3)^{3/4}$ over the interval $[0, 1]$;
3. $f(x) = \sin^2 x \cos^3 x$ over $[0, \pi/6]$;
4. $f(x) = x^2 e^{-x}$ on $[0, \ln 2]$.

Solution

1. Since the derivative of $\frac{1}{n+1}x^{n+1}$ is exactly x^n, the answer is

$$\left[\frac{1}{n+1}x^{n+1} \right]_a^b = \frac{1}{n+1}(b^{n+1} - a^{n+1}).$$

2. By some process (perhaps the method called 'change of variable' or 'substitution') we find that $G(x) = \frac{2}{21}(5 + 2x^3)^{7/4}$ does actually have $G' = f$.

---

[6] This is an easy consequence of the first mean value theorem.

Then the integral is

$$[G(x)]_0^1 = G(1) - G(0) = \frac{2}{21}(5+2)^{7/4} - \frac{2}{21}(5)^{7/4} = \frac{2}{21}\left(7^{\frac{7}{4}} - 5^{\frac{7}{4}}\right).$$

3. Change of variable and the black arts of trigonometry will provide one way to stumble into the function $G(x) = \frac{1}{3}\sin^3 x - \frac{1}{5}\sin^5 x$, whose derivative is easily seen to be equal to $f(x)$. Thus the answer is

$$[G(x)]_0^{\pi/6} = G(\pi/6) - G(0)$$
$$= \frac{1}{3}\sin^3(\pi/6) - \frac{1}{5}\sin^5(\pi/6) - 0 = \frac{1}{3}\frac{1}{8} - \frac{1}{5}\frac{1}{32} = \frac{17}{480}.$$

4. (Perhaps using integration by parts twice?) we come up with the function $G(x) = -e^{-x}(x^2 + 2x + 2)$ whose derivative is readily checked to equal $f(x)$. Therefore the answer is

$$[G(x)]_0^{\ln 2} = G(\ln 2) - G(0)$$
$$= -e^{-\ln 2}((\ln 2)^2 + 2(\ln 2) + 2) - \{-e^{-0}(0^2 + 2(0) + 2)\}$$
$$= -\frac{1}{2}((\ln 2)^2 + 2(\ln 2) + 2) + 2$$
$$= 1 - \frac{1}{2}((\ln 2)^2 + 2\ln 2).$$

17.4.6 **Another comment**   There is a strong temptation to ignore or omit the phrase 'assuming that the following expressions are integrable' from that group of examples, probably due to all our pre-university experiences that, once you can un-differentiate $f(x)$, there is nothing left to stop you evaluating the integral of f. This is, however, not true in general: there are functions for which you can successfully un-differentiate in the way we have been describing, and which are, nevertheless, not Riemann integrable!

A relatively easy example to show the truth of this (rather annoying) reality is as follows. Consider the function $g : [0, 1] \to \mathbb{R}$ defined by

$$g(x) = x^2 \sin(x^{-2}) \text{ if } x \neq 0, \quad g(0) = 1.$$

It is easy to see that g is differentiable at each point of $(0, 1]$ – just use the various rules of differentiation to find $g'(x)$ there. With a bit more caution, you can also verify that $g'(0)$ exists (and equals 0, if you're curious about it). So the function g' upon the interval $[0, 1]$ can certainly be un-differentiated. Yet, look at the formula you get for g' and you will see that it is unbounded, and therefore cannot be Riemann integrated.

In fact, much stranger things than that can happen. It is possible to define a function that is differentiable everywhere in $\mathbb{R}$, *and whose derivative is bounded everywhere in* $\mathbb{R}$, and yet that derivative is not Riemann integrable over a closed interval. The construction used in the definition is seriously sophisticated: search for *Volterra's function* if you have a lot of time and patience to spare.

For our purposes, the main point here is that un-differentiability is not a good enough reason to assert that a function can be integrated. Our final major task for this chapter is to identify a good range of functions (but by no means all) that definitely can.

17.4.7 Theorem   Every continuous function on a closed bounded interval is Riemann integrable.

Proof

If $f : [a, b] \to \mathbb{R}$ is continuous then, by the key theorem on uniform continuity, it is also uniformly continuous there. Let $\varepsilon > 0$ be given.

By definition, we can find $\delta > 0$ such that any two points of $[a, b]$ that are less than δ apart have f-values that are less than ε apart. Choose any partition Δ of $[a, b]$ whose subintervals are each shorter than δ. On each subinterval $[x_k, x_{k+1}]$ the *continuous* function f has biggest and smallest values (which will be the numbers M_k and m_k used in the Riemann integral's construction) and they will necessarily differ by less than ε, that is:

$$\text{(for each } k\text{) } M_k - m_k < \varepsilon, \quad \text{so} \quad U(f, \Delta) - L(f, \Delta) = \sum (M_k - m_k)(x_{k+1} - x_k)$$

$$< \varepsilon \sum (x_{k+1} - x_k) = \varepsilon(b - a).$$

Go back over the last paragraph and re-run it with ε replaced by $\dfrac{\varepsilon}{b - a}$ (this kind of in-flight course correction in 'epsilontics' should be almost second nature to you by now) and the revised conclusion $U(f, \Delta) - L(f, \Delta) < \varepsilon$ shows via Darboux that f is integrable.

Now we know that it is OK to dump the phrase 'assuming that the following expressions are integrable' out of the last batch of examples, because all the expressions there presented actually were (fairly obviously) continuous.

There are also many functions that are not continuous but are nevertheless Riemann integrable. Here is one source:

17.4.8 Theorem   Every monotonic function on a closed bounded interval is Riemann integrable.

Proof

Suppose that $f : [a, b] \to \mathbb{R}$ is monotonically increasing (the decreasing case is very similar or, in that scenario, you could choose to look at $-f$ which is then increasing ...). If f is actually constant then we already know it is integrable, so suppose not: that is, suppose $f(a) < f(b)$. Also let $\varepsilon > 0$ be given, so that we are ready to try Darboux again.

On each subinterval $[x_k, x_{k+1}]$ created by an arbitrary partition Δ, the *increasing* function f has biggest value $f(x_{k+1})$ and smallest value $f(x_k)$ (which will be the numbers M_k and m_k). So

$$U(f, \Delta) - L(f, \Delta) = \sum (M_k - m_k)(x_{k+1} - x_k) = \sum (f(x_{k+1}) - f(x_k))(x_{k+1} - x_k).$$

Pick a partition whose subintervals all have the same length (say, h) and we now get

$$U(f, \Delta) - L(f, \Delta) = \sum (f(x_{k+1}) - f(x_k))(x_{k+1} - x_k)$$
$$= h \sum (f(x_{k+1}) - f(x_k)) = h(f(b) - f(a)).$$

Now reverse-engineer the last paragraph by choosing $h = \dfrac{\varepsilon}{2(f(b) - f(a))}$, and we find that

$$U(f, \Delta) - L(f, \Delta) = h(f(b) - f(a))$$
$$= \left(\frac{\varepsilon}{2(f(b) - f(a))} \right)(f(b) - f(a)) = \frac{\varepsilon}{2} < \varepsilon.$$

(Again, the in-flight correction of retro-fitting h to the developing requirements is the sort of thing we all need to do when working through an unfamiliar question.) Darboux is now content to deliver the conclusion we asked for.

There are also plenty of integrable functions that are neither continuous nor monotonic. Here is an exercise to help you find some of them for yourself.

17.4.9 **EXERCISE** Let $f : [a, b] \to \mathbb{R}$ be given by

$$f(x) = 0 \ \text{ if } \ x \neq c, \ \ f(c) = 1$$

where c is some particular point in $[a, b]$. Use Darboux to show that f is integrable over $[a, b]$ and check that its integral is zero.

(Paragraph 17.2.13 offers a useful approach. The cases where c equals either a or b need a little extra attention.)

Extend this result (using whichever theorems help you) to show that a function that is zero on a closed interval except at a *finite* number of particular points is integrable.

Extend it further to verify that if two bounded functions f, g on $[a, b]$ are equal in value at all but a finite number of points, and one of them is integrable, then so is the other one, and their integrals are equal.

17.4.10 EXERCISE

1. Given that f is bounded on $[0, 2]$ and that, for each positive integer n, it is integrable over $[0, 2 - \frac{1}{n}]$, show that f is also integrable over $[0, 2]$, and that

$$\int_0^2 f = \lim_{n \to \infty} \int_0^{2 - \frac{1}{n}} f.$$

2. Give an example of an integrable function on a bounded closed interval I that is not monotonic and is discontinuous at an infinite number of points in I.

17.4.11 EXERCISE

Show by example that the following assertion is false: if f, g are integrable over a closed bounded interval I and $g(x) > 0$ at every point x of I, then $\dfrac{f(x)}{g(x)}$ must be integrable over I also.

Think how you might slightly modify this false statement to make (and then prove) a true one about integrability of the quotient of two integrable functions.

17.4.12 EXERCISE

Calculate the integrals of each of the following expressions over the interval indicated (assuming, where appropriate, basic properties of trig, logarithmic and exponential functions).

1. $f(x) = x \ln x$ over $[1, e]$;
2. $f(x) = (0.5 + xe^{x^2})(x + e^{x^2})^6$ over the interval $[0, 1]$;
3. $f(x) = \sin^2 x \cos^2 x$ over $[0, \frac{\pi}{2}]$;
4. $f(x) = e^x \sin x$ over $[0, \pi]$.

A final idea to examine in this chapter is a test that, in some sense, really belongs in Chapter 14 since it concerns convergence of series, but which had to wait until we had developed the idea of integration. There are, in fact, not very many series problems for which it is useful; however, *for those few*, it is usually the only reasonably obvious test that will work at all.

17.4.13 The integral test for series

Suppose that $f : [1, \infty) \to \mathbb{R}$ is continuous, positive and decreasing.[7] Then the following statements are equivalent:

1. the series $\sum_{k=1}^{\infty} f(k)$ is convergent;
2. the sequence $\left(\int_{x=1}^n f(x) \, dx \right)_{n \in \mathbb{N}}$ is convergent.

---

[7] That is, $1 \leq x < y$ implies that $f(x) \geq f(y)$.

Proof

For the integral just mentioned, $f(2)+f(3)+f(4)+\cdots+f(n)$ is a lower Riemann sum, and $f(1)+f(2)+f(3)+\cdots+f(n-1)$ is an upper Riemann sum (where $n \geq 1$ is an integer). If, as usual, we denote by S_n the nth partial sum of the series $\sum_{k=1}^{\infty} f(k)$, we therefore have:

$$S_n - f(1) \leq \int_{x=1}^{n} f(x)\ dx \leq S_n - f(n).$$

This shows that if either of the sequences $\left(\int_{x=1}^{n} f(x)\ dx \right), (S_n)$ is *bounded above*, then so must the other be. Since both sequences are increasing, this is the same as saying that if either of them is *convergent*, then so must the other be.

<u>17.4.14 **Illustration**</u>   An alternative proof that the harmonic series diverges: the function $f(x) = \frac{1}{x}$ certainly is positive, continuous and decreasing on $[1, \infty)$, and (as is well known) $\int_{1}^{n} f = [\ln x]_{1}^{n} = \ln n - \ln 1 = \ln n$. Since the sequence $(\ln n)_{n \geq 1}$ is unbounded and therefore divergent, the integral test informs us that the series

$$\sum_{k=1}^{\infty} f(k) = \sum_{k=1}^{\infty} \frac{1}{k}$$

(which is the harmonic series) diverges also.

<u>17.4.15 **Example**</u>   Does the series $\sum_{2}^{\infty} a_k$, given by

$$a_k = \frac{1}{k \ln k}$$

converge or diverge?

Solution

In order to use the integral test, we need to consider the corresponding real function f specified by

$$f(x) = \frac{1}{x \ln x}.$$

Notice that this formula goes bad at $x = 1$ since $\ln 1 = 0$, but that this does not impede our progress since the series started at $n = 2$ (for the same necessitating reason). We'll therefore regard f as being defined on $[2, \infty)$ (and we also need a slightly modified version of the test, in which $k = 1$ and $x = 1$ are replaced by $k = 2$ and $x = 2$).

Now it needs a little bit of insight or experience (or luck) to notice that $f(x)$ is precisely the derivative of the function $g(x) = \ln(\ln x), x \in [2, \infty)$. The rest of the argument is routine:

$$\int_2^n f(x)\, dx = [g(x)]_2^n = \ln(\ln n) - \ln(\ln 2)$$

which is unbounded[8] and therefore divergent. By the integral test, the given series is also divergent.

17.4.16 **EXERCISE** Show that[9]

$$\sum_{n=3}^{\infty} \frac{1}{n \ln n \ln(\ln n)}$$

diverges.

17.4.17 **EXERCISE**

1. Extend the argument of the integral test to show that, for each (fixed) positive integer n_0, if $\int_{n_0}^n f \to \ell$ as $n \to \infty$, then

$$\ell \le \sum_{k=n_0}^{\infty} f(k) \le \ell + f(n_0).$$

2. Estimate the sum of the (convergent) series $\sum_1^\infty \frac{1}{k^5}$ with an error less than 0.001.

17.4.18 **Note** It's occasionally useful to notice that, in the context and notation of the integral test, when the two (equivalent) conditions hold then $f(x)$ has a limit of 0 as $x \to \infty$. For suppose that $f : [0, \infty) \to \mathbb{R}$ is continuous,[10] positive and decreasing and that $\sum_1^\infty f(n)$ converges. Then certainly $f(n) \to 0$ as $n \to \infty$. For any given $\varepsilon > 0$ we can therefore find $n_0 \in \mathbb{N}$ such that $f(n) < \varepsilon$. Thinking now of $f(x)$ decreasing (and also positive) as the real variable x increases, we see that $x \in \mathbb{R}, x \ge n_0$ together imply that $0 < f(x) < \varepsilon$. Hence $f(x) \to 0$ as $x \to \infty$, as claimed.

---

[8] (for instance) because, for any positive constant K, if we choose a value of n greater than e^{e^K}, then we shall get $g(n) > K - \ln(\ln 2) > K$.

[9] Notice that the function $(x \ln x \ln(\ln x))^{-1}$ is positive and decreasing on $[3, \infty]$ but not on $[2, \infty]$: indeed, it is not even defined at $x = e$.

[10] actually, continuity does not play any role in this part of the argument.

18 The elementary functions revisited

18.1 Introduction

One of the benefits of now having a logically watertight definition of integral is that we can at last provide reliable definitions of the so-called elementary functions such as $\ln x$, e^x and $\sin x$, and prove that the basic properties of these entities that we have been cheerfully using throughout – in order to enrich our library of examples and exercises – do actually hold good in all circumstances. If it seems surprising and even counter-intuitive that integration theory should be required for this task, pause and think about the useful facts that the area under the graph of $f(x) = \frac{1}{x}$ between $x = 0$ and $x = a$ is $\ln a$, that the area under the graph of $f(x) = e^x$ between $x = a$ and $x = b$ is $e^b - e^a$ (assuming that $a < b$), and that the number π that is so critical to trigonometry is the area of a circle of unit radius. *Area* is evidently quite central to how these functions operate, so perhaps it is more natural than it initially seems that area (interpreted as integral) should provide a means for defining them in a way that is consistent with intuition and common sense, but not dependent on either.

18.2 Logarithms and exponentials

18.2.1 Definition Consider the function $f : (0, \infty) \to \mathbb{R}$ specified by $f(x) = \frac{1}{x}$. We notice that f is positive, strictly decreasing, continuous and differentiable on its domain, that $f(x) \to \infty$ as $x \to 0^+$, and that $f(x) \to 0$ as $x \to \infty$. This will let us sketch its graph with decent accuracy.

Now for each $t > 0$ we define

$$\ln t = \int_1^t f.$$

The fact that f is continuous on the interval from 1 to t guarantees that the integral exists, and the following details about the so-called *natural logarithm* function thus created are immediate:

Undergraduate Analysis: A Working Textbook, Aisling McCluskey and Brian McMaster 2018.
© Aisling McCluskey and Brian McMaster 2018. Published 2018 by Oxford University Press

1. $\ln 1 = 0$;
2. if $t > 1$ then $\ln t > 0$;
3. if $t < 1$ then $\ln t = \int_1^t f = -\int_t^1 f < 0$.

18.2.2 **Lemma** The function ln is differentiable at the point t (for each $t > 0$) and its derivative there is $\dfrac{1}{t}$.

Proof

Immediate from the fundamental theorem of calculus.

18.2.3 **Lemma** If $n \geq 2$ is a positive integer, then

$$\left(\sum_1^n \frac{1}{k}\right) - 1 < \ln n < \left(\sum_1^n \frac{1}{k}\right) - \frac{1}{n}.$$

Proof

It is easy to 'see' this from a sketch graph:

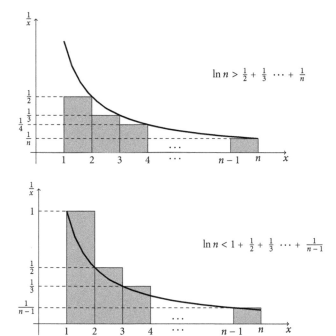

but a more logically robust reason is the fact that the first and third items in that display are exactly the lower and upper Riemann sums for f using the partition $1 < 2 < 3 < 4 < \cdots < n$ of the interval $[1, n]$.

18.2.4 Corollary $\ln x \to \infty$ as $x \to \infty$.

Proof

From the Lemma, and the fact that the harmonic series diverges, we get[1] $\ln n \to \infty$ as $n \to \infty$. Now for any real x we have $x \geq \lfloor x \rfloor$ so, since $\ln$ is increasing,[2] $\ln x \geq \ln\lfloor x \rfloor$ and, letting $x \to \infty$ (and consequently $\lfloor x \rfloor \to \infty$ also) we get $\ln x \to \infty$.

Mildly reassuring though these details are, we are still missing the essential point of *what logarithms are for*: their prime purpose, whether for calculation, for algebraic simplification or for theoretical arguments, is to convert multiplication into addition: '$\ln(xy) = \ln x + \ln y$'. We next need to establish this fundamental 'law'.

18.2.5 Theorem   For any positive numbers x and y, $\ln(xy) = \ln x + \ln y$.

Proof

Consider any real constant $a > 0$, and define a real function $g : (0, \infty) \to \mathbb{R}$ by the formula

$$g(x) = \ln(ax) - \ln x.$$

Using the chain rule (and our known derivative of $\ln$), it is easy to differentiate this at any positive x:

$$g'(x) = \left(\frac{1}{ax}\right) a - \frac{1}{x} = 0.$$

It follows that g must be constant, and that its constant value is

$$g(1) = \ln a - \ln 1 = \ln a,$$

in other words, that $\ln(ax) = \ln x + \ln a$. Since a and x are arbitrary members of $(0, \infty)$, the proof is concluded.

18.2.6 Corollary 1   For any $y > 0$, $\ln \dfrac{1}{y} = -\ln y$.

Proof

(Using the theorem):

$$0 = \ln 1 = \ln \left(y \left(\frac{1}{y} \right) \right) = \ln y + \ln \left(\frac{1}{y} \right).$$

---

[1] See 2.9.9.
[2] If $0 < a < b$ then $\ln b - \ln a = \int_1^b \frac{1}{x} dx - \int_1^a \frac{1}{x} dx = \int_a^b \frac{1}{x} dx$ which is positive.

18.2.7 Corollary 2   For any $x > 0$ and $y > 0$, $\ln \dfrac{x}{y} = \ln x - \ln y$.

Proof

(Using the theorem and its first corollary):

$$\ln\left(\frac{x}{y}\right) = \ln\left(x \times \frac{1}{y}\right) = \ln x + \ln\left(\frac{1}{y}\right) = \ln x - \ln y.$$

18.2.8 Corollary 3   As $x \to 0^{+}$, $\ln x \to -\infty$.

Proof

If x is a small positive number, then x^{-1} is a large positive number. More precisely, as $x \to 0^{+}$, $x^{-1} \to \infty$ and (via 18.2.4) $\ln x^{-1} \to \infty$. Consequently $\ln x = -\ln x^{-1} \to -\infty$. (If the last step does not appear obvious, you can confirm it from the definitions.)

Our intuitive picture of the graph of ln is now reasonably complete:

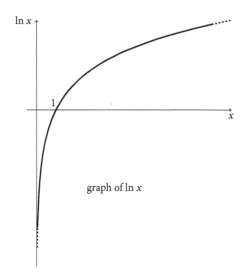

graph of ln x

and an important detail emerges: the *range* of ln is the whole real line $(-\infty, \infty)$ because, for any $x \in \mathbb{R}$, we can use the behaviour of ln near to 0 and 'near to ∞' to find a value $\ln a$ of ln that is less than x and a value $\ln b$ of ln that is greater than x. Now since ln is continuous, the IVT tells us that x itself is a value of ln.

SUMMARY: ln is a (strictly increasing and therefore) one-to-one map from $(0, \infty)$ onto $\mathbb{R} = (-\infty, \infty)$ and, of course, the vital thing about one-to-one onto maps is that they possess inverses.

18.2.9 Definition We define a function, called the *exponential function* and (for the moment) denoted by exp : $\mathbb{R} \to (0, \infty)$, by declaring exp to be the inverse of ln.

We are reluctant to reach for the notation $\ln^{-1}$ since the risk of consequent confusion is high. Instead, let's concentrate on what *inverse function* means in this case:

- For each $t > 0$ we have $\exp(\ln t) = t$.
- For each $x \in \mathbb{R}$ we have $\ln(\exp x) = x$.

At this point, look back to what we found out about inverses of continuous, differentiable, strictly increasing functions in Chapters 8 and 12 (specifically, paragraphs 8.6.8 and 12.2.16). From there, we know that exp is continuous and differentiable, and we have a formula for its derivative at each point, namely:

$$\text{at each } p \in (0, \infty), \exp'(\ln p) = \frac{1}{\ln' p}$$

which simplifies, since $\ln' p = \frac{1}{p}$, to $\exp'(\ln p) = p$. Substituting x for $\ln p$, that is, $p = \exp x$, and noting that x ranges over the whole real line as p ranges over $(0, \infty)$, that says

$$\exp'(x) = \exp x \quad \text{for every } x \in \mathbb{R}$$

and we have recovered what is possibly the most important fact about exp: that it is its own derivative. The other basic details follow easily enough:

18.2.10 Proposition

1. exp is strictly increasing.
2. $\exp x \to 0$ as $x \to -\infty$, $\exp x \to \infty$ as $x \to \infty$.
3. For each $x, y \in \mathbb{R}$, $\exp(x + y) = \exp x \times \exp y$.
4. $\exp(0) = 1$.
5. For each $y \in \mathbb{R}$, $\exp(-y) = \dfrac{1}{\exp y}$.
6. For each $x, y \in \mathbb{R}$, $\exp(x - y) = \dfrac{\exp x}{\exp y}$.

Proof

1. Its derivative (= itself) is always positive. (Alternatively, appeal to 8.6.8 or 8.6.10.)
2. Firstly, if $\varepsilon > 0$ is given, then (using part 1) $x < \ln \varepsilon$ implies $0 < \exp(x) < \exp(\ln \varepsilon) = \varepsilon$. (Recall that $\exp(x)$ is always positive.) Secondly, given $K > 0$, then $x > \ln K$ implies $\exp(x) > K$ for the same reasons.
3. Let $p = \exp x, q = \exp y$, that is, $x = \ln p, y = \ln q$. Then $x + y = \ln p + \ln q = \ln(pq)$ and therefore $\exp(x + y) = pq = \exp x \times \exp y$.
4. Because $\ln 1 = 0$.

5. $\exp(-y) \cdot \exp(y) = \exp(-y + y) = \exp(0) = 1$ (and $\exp(y) \neq 0$).

6. $\exp(x - y) = \exp(x + (-y)) = \exp(x)\exp(-y) = \exp(x)\dfrac{1}{\exp(y)} = \dfrac{\exp(x)}{\exp(y)}.$

The proposition guarantees that the overall appearance of the graph of exp is, as is widely known:

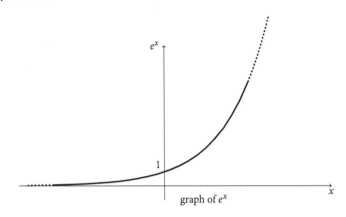

graph of e^x

As a matter of notation, it is customary to put $e = \exp(1)$ and then write not $\exp(x)$ but e^x. This is not, however, a trivial matter if we actually want to think about powers of the number e or, indeed, of non-rational powers of any positive base-number. The reader is invited to check out some details in order to understand better what is going on here:

18.2.11 EXERCISE

1. Use induction to check that exp operates through all finite sums, in the sense that $\exp(\sum_1^n x_i) = \exp(x_1)\exp(x_2)\exp(x_3)\cdots\exp(x_n)$ for all finite lists of real numbers $x_1, x_2, x_3, \cdots, x_n$.

2. Show that $\exp(n) = e^n$ for every positive integer n.

3. Show that $\exp(1/m) = e^{1/m}$ for every positive integer m.

4. Show that $\exp(r) = e^r$ for every positive rational r.

5. Show that $\exp(q) = e^q$ for every rational q.

The essence of that little investigation is that $\exp(x)$ and e^x agree at all the values of x for which common sense tells you the meaning of e^x; once x stops being rational, e^x no longer possesses a 'natural' meaning, and this is nothing to do with e, for expressions such as $2^{\sqrt{2}}$ are also beyond the grasp of basic algebra – they require proper definition before we can study them with any degree of confidence.

18.2.12 Definition   (Now that we have proper definitions of exp and ln,) for $x \in \mathbb{R}$ and $a > 0$ we define

$$a^x = \exp(x \ln a).$$

Notice first that in the special case where $a = e$, we have actually defined e^x to mean $\exp(x)$ because $\ln e = 1$ (in turn, because $\exp(1) = e$ by definition). As regards reconciling formal and informal definitions of powers of numbers other than e, you may find the following 'spiked' version of the previous exercise useful:

18.2.13 **EXERCISE**   For any $a > 0$, show that

1. $\exp(n \ln a) = a^n$ for every positive integer[3] n.

2. $\exp\left(\dfrac{1}{m} \ln a\right) = a^{1/m}$ for every positive integer[4] m.

3. $\exp(r \ln a) = a^r$ for every positive rational[5] r.

4. $\exp(q \ln a) = a^q$ for every rational[6] q.

This time, the message is that wherever a^x *can* be defined by common sense and simple algebra (that is, whenever x is merely a rational number), then that common-sense definition gives the same answer as the formal all-purpose definition set out above.

Of course, we now need to confirm that these general powers obey the familiar index laws: but this is reassuringly straightforward:

18.2.14 **Proposition**   For any $a > 0$ and real numbers x and y:

1. $a^x a^y = a^{x+y}$,

2. $\dfrac{a^x}{a^y} = a^{x-y}$,

3. $(a^x)^y = a^{xy}$.

Proof of 1.

$$
\begin{aligned}
a^x a^y &= \exp(x \ln a) \exp(y \ln a) \\
&= \exp(x \ln a + y \ln a) \\
&= \exp((x+y) \ln a) \\
&= a^{x+y}.
\end{aligned}
$$

18.2.15 **EXERCISE**   Check out the proofs of 2 and 3 above.

---

[3] The point is that we don't *need* ln and exp in order to define a^n when n is a positive integer, it just means write a down n times and multiply the lot.

[4] Again, we don't *need* ln and exp in order to define an mth root for a when m is a positive integer, it just means the (positive) number whose m^{th} power is a.

[5] Similar remarks.

[6] Similar remarks.

18.2.16 **EXERCISE**

- For a given real number a, use the one-sided version of l'Hôpital's Rule (see paragraph 16.2.6) to determine

$$\lim_{x \to 0^+} \frac{\ln(1 + ax)}{x}.$$

- Now use the sequence-based description of one-sided limits (see 10.3.6) to deduce that

$$\lim_{n \to \infty} \frac{\ln\left(1 + \frac{a}{n}\right)}{\frac{1}{n}} = a.$$

- Lastly, use continuity of the exponential function to deduce that

$$\left(1 + \frac{a}{n}\right)^n \to e^a \text{ as } n \to \infty.$$

18.2.17 **EXERCISE** Compute the Taylor series of $\ln x$ about $x = 1$, and confirm that it converges (at least) everywhere in the interval $(\frac{1}{2}, \frac{3}{2})$.

18.2.18 **EXERCISE** Compute the Taylor series of $\exp x$ about $x = 0$, and confirm that it converges everywhere on the real line.

18.3 Trigonometric functions

Surprisingly, the elementary trig functions – which most of us thought we had understood perfectly well at school – are harder to define properly than ln and exp turned out to be. Indeed, for any sort of thorough-going treatment of them, we need to start by unambiguously defining the number π. Of the two most obvious ways to do this (area of unit circle, or half-circumference of unit circle) it is the former that sits more easily with the Riemann approach to integral visualised as area under graph. The unit circle, crisply described by the equation

$$x^2 + y^2 = 1,$$

can be seen as composed of the graphs of two functions: the upper semicircle $y = f_1(x) = \sqrt{1 - x^2}$ lying above the horizontal axis, and the lower semicircle $y = f_2(x) = -\sqrt{1 - x^2}$ lying below it. By the obvious symmetry, the whole area is twice that of the region between the graph of f_1 and the horizontal axis, so

18.3.1 **Definition**

$$\pi = 2 \int_{-1}^{1} f_1$$

$$= 2 \int_{-1}^{1} \sqrt{1 - x^2} \, dx.$$

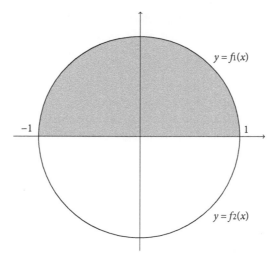

π is twice the 'shaded area'

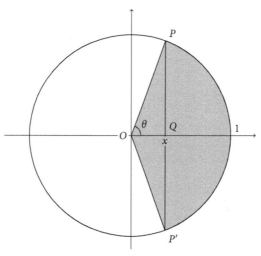

A is the 'shaded area'

18.3.2 **Comments**   To see the roughwork that lies behind the next definition, take a look at the second diagram above, where the unit circle (centre 0, radius 1) is cut by the vertical chord PP' that crosses the horizontal axis at Q (so all three points P, Q, P' have the same first coordinate x), the sector POP' (shaded) has area A and the angle POQ is designated θ. Allowing ourselves to think, for just one more paragraph, that we actually did understand basic trigonometry years ago, what is the relationship between x and A? Well, θ is the angle (within the acceptable range) whose cosine is x and, by the '$\frac{1}{2}r^2\theta$' formula for sector area, A is $\frac{1}{2}1^2(2\theta) = \theta$. Therefore

$$A = \cos^{-1} x.$$

This formula[7] would be fine for determining A from x provided we knew exactly what was meant by cosine but, at this point in the text, our reliable formal knowledge is (thanks to Riemann integration again) how to define areas. So it is really the A in that last display that we thoroughly understand, and we can now view that understanding as being an equally precise definition of $\cos^{-1}$. Once that is clear (and once we check that the idea of inverse function is legitimate here) $\cos$ can be defined as the inverse of $\cos^{-1}$, and the rest of the details about the trig functions will fall into place in a more straightforward manner.

Two more small simplifications before we proceed: firstly, the triangular area POP' could indeed be calculated via integration as twice the area lying under the straight-line graph OP, but there is no need to use such heavy machinery since *half base times perpendicular height* provides a perfectly adequate alternative; thus, the area of the triangle is $x\sqrt{1-x^2}$, and the sector area A is

$$x\sqrt{1-x^2} + 2\int_x^1 \sqrt{1-u^2}\ du.$$

(We have had to use a letter other than x for the variable of integration, since x has been assigned already to denote the first coordinate of P.)

Secondly, although the above diagram (and our thinking that it supported) tacitly assumed that θ was less than a right angle, the displayed formula for area A remains valid if θ lies between $\frac{\pi}{2}$ and π: for now, the sector area is twice the area under the graph of $x\sqrt{1-x^2}$ *minus* the triangular area POP', and that *minus* is picked up by the fact that the first x in the displayed formula is now negative (please refer to the following diagram).

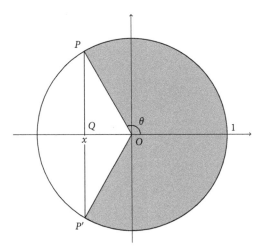

A is again the 'shaded area'

<hr>

[7] Of course, it is common practice to use the notation $\arccos x$ instead of $\cos^{-1} x$, but here we have opted for the latter in order to stress the importance of invertible functions in this approach.

18.3.3 Definition   We define the function $A : [-1, 1] \to \mathbb{R}$ via the formula

$$A(x) = x\sqrt{1 - x^2} + 2\int_x^1 \sqrt{1 - u^2}\, du, \quad -1 \leq x \leq 1.$$

18.3.4 EXERCISE

1. Verify that, on the *open* interval $(-1, 1)$, A is differentiable and its derivative is

$$A'(x) = -\frac{1}{\sqrt{1 - x^2}}.$$

(You will need to use only the product rule and the chain rule to differentiate $x\sqrt{1 - x^2}$.

Then express $\int_x^1 \sqrt{1 - u^2}\, du$ as $-\int_1^x \sqrt{1 - u^2}\, du$, and appeal to the fundamental theorem of calculus.)

2. Check that $A(-1) = \pi$ and that $A(1) = 0$, and show that

$$\lim_{x \to -1^+} A(x) = \pi \quad \text{and} \quad \lim_{x \to 1^-} A(x) = 0.$$

18.3.5 Note   The essence of the above Exercise is that A is the sort of function to which we can apply the first mean value theorem: it is continuous on $[-1, 1]$, differentiable on $(-1, 1)$ and its derivative is always less than zero here, so it is strictly decreasing and therefore one-to-one, and its range is precisely the interval $[0, \pi]$. Viewing it therefore as a map

$$A : [-1, 1] \to [0, \pi],$$

it is a bijection, and possesses an (also strictly decreasing) inverse mapping A^{-1} from $[0, \pi]$ to $[-1, 1]$. This inverse, when we have appropriately extended it over the whole real line, is our (logically watertight) definition of the cosine function.

18.3.6 Definition   The real function cos (cosine) is defined as follows (using the notation from above):

1. For $0 \leq x \leq \pi$, $\cos x = A^{-1}(x)$.

2. Then for $-\pi \leq x \leq 0$, $\cos x = \cos(-x)$.

3. Then for each integer n, $\cos(x + 2n\pi) = \cos x$.

18.3.7 Remarks

1. Although it is not strictly part of the definition, you can follow the 'evolution' of cosine through points 1, 2 and 3 above by looking at the three phases of the diagram supplied:

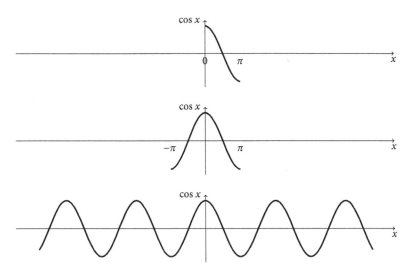

2. The inequality $-1 \leq \cos x \leq 1$ (for all real x) is built into the definition, and so is $\cos(x + 2n\pi) = \cos x$ (for all real x and integer n).

3. We also record that cos, when restricted to the interval $[0, \pi]$, has an inverse $\cos^{-1} : [-1, 1] \to [0, \pi]$ (namely, the function A) whose derivative

$$\left(\cos^{-1}\right)'(x) = -\frac{1}{\sqrt{1 - x^2}}$$

for every $x \in (-1, 1)$.

Having formally defined one of the trig functions, we can now create the others from it quite routinely:

18.3.8 Definition   The real function sin (sine) is defined as follows:

1. For $0 \leq x \leq \pi$, $\sin x = \sqrt{1 - (\cos x)^2}$.
2. Then for $-\pi \leq x \leq 0$, $\sin x = -\sin(-x)$.
3. Then for each integer n, $\sin(x + 2n\pi) = \sin x$.

18.3.9 Remarks

1. Again, it may be helpful to follow the evolution of sine through points 1, 2 and 3 above by looking at the three phases of the sketch graphs provided:

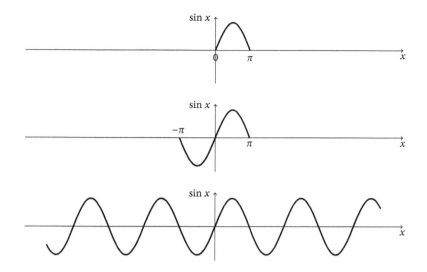

2. The inequality $-1 \le \sin x \le 1$ (for all real x) is built into the definition, as are the equalities $\sin(x + 2n\pi) = \sin x$ (for all real x and integer n) and $(\sin x)^2 + (\cos x)^2 = 1$ (for all real x).

3. It is, as you will almost certainly be aware, conventional to write $\sin x$ and $\cos x$ rather than $\sin(x)$ and $\cos(x)$ (and similarly for the other trig functions, and also for ln) provided that the argument x is a single letter, but beware of the dangers of extending this custom to cases where the argument is typographically complex. The symbol $\sin \pi x$ is capable of being interpreted either as $\sin(\pi x)$ or as $(\sin \pi)x$, so use bracketing to prevent the ambiguity. It is also conventional (so long as n is a positive integer) to denote the nth power of $\sin x$ and of $\cos x$ not as we have done above, but rather as $\sin^n x$ and $\cos^n x$. Clearly, one should carefully avoid creating confusion by doing this for $n = -1$.

18.3.10 Theorem

1. For $0 < x < \pi$, $\cos'(x) = -\sin x$.
2. For $-\pi < x < 0$, $\cos'(x) = -\sin x$.
3. For all real x except multiples of π, $\cos'(x) = -\sin x$.
4. For all real x, $\cos'(x) = -\sin x$.

Proof

1. For $0 < x < \pi$, we need only invoke once more the theorem on differentiating an inverse function to see that

$$\cos'(x) = (A^{-1})'(x) = \frac{1}{A'(\cos x)} = -\sqrt{1 - (\cos x)^2} = -\sin x.$$

2. For $-\pi < x < 0$, the result follows from part 1 because we defined cosine as an 'even' function and sine as an 'odd' function.

3. For all real x except multiples of π, the periodicity of both sine and cosine extends the validity of parts 1 and 2.

4. When x is a multiple of π, we can 'patch' the desired equality by an appeal to Theorem 12.3.20.

18.3.11 **EXERCISE** Starting with $\sin x = \sqrt{1 - \cos^2 x}$ on the interval $(0, \pi)$, show that $\sin'(x) = \cos x$ for all real x. You may expect to have to extend the result from $(0, \pi)$ to $\mathbb{R}$ in stages, as in the preceding theorem.

The addition formulas for sine and cosine (the identities for $\sin(x + y)$ and $\cos(x + y)$) are not very obvious in the present approach, but can be established indirectly by one of the nicest instances of rabbit-out-of-hat mathematics (the sort of argument that is clear *afterwards*, but completely invisible beforehand) that you are ever likely to see:

18.3.12 **Lemma 1** Suppose that a function f is twice differentiable on $\mathbb{R}$, and that

- $f + f'' = 0$ everywhere in $\mathbb{R}$,
- $f(0) = 0$, and
- $f'(0) = 0$.

Then f is the zero function (f is zero everywhere in $\mathbb{R}$).

Proof

Consider the function $g(x) = f'(x)^2 + f(x)^2$. Routine differentiation shows that $g'(x) = 0$ everywhere, so g is a constant function. Now the second and third bullet points tell us that its constant value is zero.

18.3.13 **Lemma 2** Suppose that a function f is twice differentiable on $\mathbb{R}$, and that

- $f + f'' = 0$ everywhere in $\mathbb{R}$,
- $f(0) = a$, and
- $f'(0) = b$

where a, b are constants. Then $f(x) = a \cos x + b \sin x$ everywhere in $\mathbb{R}$.

Proof

Consider the function $h(x) = f(x) - a \cos x - b \sin x$. Apply Lemma 1 to h and we find that it is the zero function.

18.3.14 **Theorem**   For all real x, y we have

1. $\sin(x + y) = \sin x \cos y + \cos x \sin y$,
2. $\cos(x + y) = \cos x \cos y - \sin x \sin y$.

Partial proof

Consider y as fixed for the moment and define f by $f(x) = \sin(x + y)$. Differentiating twice shows that $f''(x) = -\sin(x + y)$, and also notice that $f(0) = \sin y$ and $f'(0) = \cos y$. By Lemma 2, $f(x) = \sin y \cos x + \cos y \sin x$ for every real x. Since y was arbitrary, part 1 is established.

18.3.15 **EXERCISE**   Prove part 2 also.

18.3.16 **Note**   At this point, the tasks of defining and differentiating the functions tan, sec, cot and cosec, and their inverses where appropriate, are pedestrian and can safely be left unless and until there is a need for them.

18.3.17 **EXERCISE**

1. Verify that the function tan, defined (initially) on the interval $(-\frac{\pi}{2}, \frac{\pi}{2})$ by the formula

$$\tan x = \frac{\sin x}{\cos x},$$

is continuous and differentiable, with (positive) derivative $(\cos x)^{-2}$, and has range $\mathbb{R}$; also that its inverse arctan or $\tan^{-1} : \mathbb{R} \to (-\frac{\pi}{2}, \frac{\pi}{2})$ is continuous and differentiable, and that its derivative is given by

$$(\tan^{-1})'(x) = \frac{1}{1 + x^2}.$$

2. Check that the radius of convergence of the power series

$$1 - \frac{t}{3} + \frac{t^2}{5} - \frac{t^3}{7} + \frac{t^4}{9} - \cdots$$

is 1: so that, in particular, the series

$$1 - \frac{x^2}{3} + \frac{x^4}{5} - \frac{x^6}{7} + \frac{x^8}{9} - \cdots$$

and

$$x - \frac{x^3}{3} + \frac{x^5}{5} - \frac{x^7}{7} + \frac{x^9}{9} - \cdots$$

are absolutely convergent, the second one to some function $f(x)$, on (at least) the interval $(-1, 1)$.

3. Appeal to the theorem on differentiation of power series to show that

$$f'(x) = 1 - x^2 + x^4 - x^6 + x^8 - \cdots = \frac{1}{1+x^2} \quad (-1 < x < 1).$$

4. Deduce that (for $-1 < x < 1$):

$$x - \frac{x^3}{3} + \frac{x^5}{5} - \frac{x^7}{7} + \frac{x^9}{9} - \cdots = \tan^{-1}(x).$$

5. Use this to justify what we claimed in paragraph 1.6. Also use the fact that $\tan(\pi/6) = 1/\sqrt{3}$ to obtain a (numerically simple) series whose sum is $\frac{\pi}{\sqrt{12}}$.

18.3.18 **Valedictory** Thank you for travelling with us!

19 Exercises: for additional practice

These further exercises are presented broadly in line with the order in which their associated material occurs in the main text, but you should be aware that analysis is a profoundly interconnected subject, so that ideas from an earlier or a later section than the one that seems to be central to a particular question may well turn out to be valuable in crafting a good answer. Specimen solutions to these problems are available to instructors via the publishers: please visit the webpage www.oup.co.uk/companion/McCluskey&McMaster to find out how to seek access to these.

1. How far along the list of numbers

$$0.3, 0.33, 0.333, 0.3333, \ldots$$

should we go so that, from that point onwards, every number we meet is an approximation to $\frac{1}{3}$ with error smaller than 10^{-8} ?

2. How far along the list of numbers

$$1 - \frac{1}{3} - \frac{1}{9}, \ 1 - \frac{1}{5} - \frac{1}{25}, \ 1 - \frac{1}{7} - \frac{1}{49}, \ 1 - \frac{1}{9} - \frac{1}{81}, \ldots$$

can we go and be certain that, from then on, all the numbers we find are approximations to 1 whose errors are less than 10^{-6} ?

3. For the list of numbers

$$10 + \frac{1}{2}, \ 10 + \frac{2}{5}, \ 10 + \frac{3}{10}, \ 10 + \frac{4}{17}, \cdots, 10 + \frac{n}{1 + n^2}, \cdots$$

how far along should we go if we need to be sure that, from then on, each number we encounter differs from 10 by less than 0.000 003?

4. Find a stage along the list of numbers

$$\sqrt{9 + (1)^{-2}}, \ \sqrt{9 + (2)^{-2}}, \ \sqrt{9 + (3)^{-2}}, \ \sqrt{9 + (4)^{-2}} \ldots$$

Undergraduate Analysis: A Working Textbook, Aisling McCluskey and Brian McMaster 2018.
© Aisling McCluskey and Brian McMaster 2018. Published 2018 by Oxford University Press

after which we can be sure that each of these approximations to 3 will have error less than 10^{-5}.

5. Prove, by the definition of limit of a sequence, that $\dfrac{2}{n^2} - \dfrac{5}{n} + 4 \to 4$ (as $n \to \infty$).

6. Use the definition of convergence of a sequence to a limit to prove that
   (a) $\dfrac{7 - 5n^3}{2n^3} \to -\dfrac{5}{2}$,
   (b) $3 + \dfrac{4}{n} - \dfrac{5}{n^2} \to 3.$

7. Use the definition of convergence to a limit to prove that
   (a) $\dfrac{1}{5n-1} \to 0,$
   (b) $\dfrac{1}{n\sqrt{n}} \to 0,$
   (c) $\dfrac{1}{n^2 - 30\pi n} \to 0.$

8. Let $(x_n)_{n\in\mathbb{N}}$ be a given convergent sequence of real numbers whose limit is ℓ. Prove, directly from the definition of convergence, that $6x_n \to 6\ell$.

9. Show via the definition of convergence that if $x_n \geq 0$ for every $n \in \mathbb{N}$ and $\sqrt{4 + x_n} \to 2$, then $x_n \to 0$.

10. (The *arithmetic mean – geometric mean inequality*, more briefly called the *AM – GM inequality*.)
   (a) If $x \geq 0$ and $y \geq 0$, prove that $\sqrt{xy} \leq \dfrac{x+y}{2}$.
       (*Hint:* begin by noticing that $(\sqrt{x} - \sqrt{y})^2 \geq 0$.)
   (b) Use part (a) to deduce that, for every four non-negative numbers w, x, y and z, we have $\sqrt[4]{wxyz} \leq \dfrac{w+x+y+z}{4}$.
       (*Hint:* you can apply part (a) to w and x, and then to y and z. Can it then be applied to $\sqrt{wx}$ and $\sqrt{yz}$?)
   (c) Note that these are particular cases of a more general result: for any positive integer n and any list $a_1, a_2, a_3, \cdots a_n$ of non-negative numbers, we have
   $$\sqrt[n]{a_1 a_2 a_3 \cdots a_n} \leq (a_1 + a_2 + a_3 + \cdots + a_n)/n.$$

11. The *harmonic mean* of two positive numbers a and b is defined to be the reciprocal of the arithmetic mean (that is, the average) of their reciprocals. Investigate whether this is greater or smaller than their geometric mean $\sqrt{ab}$.

12. Using various parts of the algebra of limits theorem, determine the limits (as $n \to \infty$) of the sequences whose n^{th} terms are as follows:
   (a) $\dfrac{7n^3 - 4n^2 + 5}{2 + 2n - n^3}$,

(b) $\dfrac{1}{n^2 + 2}$,

(c) $\left(1 + \dfrac{1}{n} - \dfrac{2}{n^2}\right)\left(\dfrac{3n^2 + 1}{2n^2 - 1}\right)$.

13. Use the algebra of limits to determine $\lim_{n\to\infty} a_n$ and $\lim_{n\to\infty} b_n$ where

$$a_n = \dfrac{3n + 4n^2 + 5n^4}{(6 + 7n^2)^2}, \quad b_n = (23 - 7n + 2n^2)\left(\dfrac{2 - n}{5 - n^2}\right)^2.$$

14. Using various parts of the algebra of limits theorem, determine the limits (as $n \to \infty$) of the sequences whose n^{th} terms are as follows:

(a) $3 + \dfrac{5}{n}$,

(b) $\dfrac{6n + \pi^2}{5n}$,

(c) $\dfrac{2}{n^2} + \dfrac{3}{n} - 4$,

(d) $\dfrac{6n^3 + 4n^2 - 1}{17 - 7n + 2n^3}$,

(e) $\dfrac{1 + n}{1 + n + n^2}$,

(f) $\left(\dfrac{3n + 2}{4 - n}\right)^5$.

15. Put $x_n = -1 - \dfrac{3}{n} + \dfrac{2}{n^2}$ (for each positive integer n). By simplifying the difference $x_{n+1} - x_n$, show that (x_n) is an increasing sequence.

16. Show that the sequence (c_n) described by $c_n = 2 + \dfrac{1}{n} - \dfrac{1}{n^2}$ is decreasing provided that $n \geq 2$.

17. Let us denote (for each positive integer n) by x_n the number $4 + \dfrac{5}{n} - \dfrac{2}{n^2}$. Show that every x_n satisfies the inequality $-12 \leq x_n \leq +12$. By simplifying the difference $x_{n+1} - x_n$, show that (x_n) is a decreasing sequence.

18. Notice first that $2 < 3, 4 < 5, 8 < 9, 16 < 17, \ldots, 2^n < 2^n + 1$. Consequently, $1/2 > 1/3, 1/4 > 1/5, 1/8 > 1/9, 1/16 > 1/17, \ldots$, $2^{-n} > 1/(2^n + 1)$. Now put

$$x_n = \dfrac{1}{3} + \dfrac{1}{5} + \dfrac{1}{9} + \dfrac{1}{17} + \cdots + \dfrac{1}{2^n + 1}$$

for each positive integer n. Show that the sequence $(x_n)_{n\in\mathbb{N}}$ is bounded. Also check that it is increasing. Why must it converge?

19. Show that the sequence $(n - \sqrt{n})_{n\in\mathbb{N}}$ is increasing, and not bounded.

20. Explain (in terms of sequence limits) the meaning of the recurrent decimals $0.44444\ldots$ and $0.2136363636\ldots$ and express each of them as a rational number (a fraction in the usual sense of that word).

21. Explain the meaning of (and evaluate as a fraction) the recurring decimal $1.281818181\ldots$

22. Prove that a decreasing sequence that is bounded below must converge, and that its limit is the infimum of the set of all its terms.

23. Use the squeeze to find the limits of the sequences whose typical terms are:
    (a) $x_n = \sqrt[n]{4^n + 6^n}$,
    (b) $y_n = \dfrac{3n + 5\sin(n^2 + 2)}{1 - 6n}$.

24. Use the squeeze to show that the sequences whose n^{th} terms are as follows are convergent:
    (a) $\dfrac{3n}{1 - 4n} + \dfrac{(-1)^{n^2+4n}\sin(n^{12} - 2n^7)}{n}$,
    (b) $\sqrt[n]{3^n + 5^n + 8^n}$, assuming that $\sqrt[n]{a} \to 1$ for each positive constant a. (We prove this result in paragraph 6.2.3.)

25. Find the limit as $n \to \infty$ of the sequence $(\sqrt{5n + 9} - \sqrt{5n + 4})$.

26. Find the limit (as $n \to \infty$) of $\sqrt{3n + 2} - \sqrt{3n - 2}$.

27. We are given three sequences $(a_n)_{n\in\mathbb{N}}$, $(b_n)_{n\in\mathbb{N}}$ and $(c_n)_{n\in\mathbb{N}}$ and we are told only that $a_n \to \ell$, $c_n \to 0$ and $|a_n - b_n| \le |c_n|$ for every $n \in \mathbb{N}$. Prove that $b_n \to \ell$.

28. Use the squeeze to show that the sequences whose n^{th} terms are as follows are convergent:
    (a) $\pi/4 + \dfrac{(-1)^n}{\sqrt{n}}$,
    (b) $\sqrt[n]{1000 + 3^n}$. You may assume that, for any positive constant k that you choose, we have $\sqrt[n]{k} \to 1$.

29. (a) Let there be given a sequence $(x_n)_{n\in\mathbb{N}}$ for which the subsequence $(x_{2n-1})_{n\in\mathbb{N}}$ of all odd-numbered terms and the subsequence $(x_{2n})_{n\in\mathbb{N}}$ of all even-numbered terms both converge to the same limit ℓ. Prove that the entire sequence $(x_n)_{n\in\mathbb{N}}$ also converges to ℓ.

    (b) For the sequence $(a_n)_{n\in\mathbb{N}}$ described by

    $$a_n = \begin{cases} \dfrac{3 + 7n - n^2}{2n^2 + n + 12} & \text{if } n \text{ is odd,} \\[2ex] \dfrac{(0.7)^n - 1}{(0.6)^n + 2} & \text{if } n \text{ is even,} \end{cases}$$

    use part (a) to prove that it converges and to determine its limit.

30. Consider the sequence (x_n) described by $x_n = 1 - \dfrac{1}{k}$ if n is the kth prime number, $x_n = 1 + \dfrac{1}{k}$ if n is the kth non-prime positive integer. Write down the first twelve terms of the sequence (x_n). How large a value of n will guarantee that $|x_n - 1| < 0.01$?

31. Show that the sequences whose n^{th} terms are as follows are unbounded:
    (a) $(-1)^n \sqrt{n}$,
    (b) $\dfrac{n^2 + 1}{n + 5}$.

32. Show that the sequences whose n^{th} terms are as follows are unbounded:
    (a) $\sqrt[3]{n} - 1000$,
    (b) $\dfrac{1 + n^2}{5 - 8n}$.

33. Show that the following sequences are unbounded:
    (a) $\left(2n + 5 + 8\sin\left(\dfrac{n\pi}{17}\right)\right)$,
    (b) $\left(\dfrac{1 - n^2}{1 + 2n}\right)$.

34. Write down the total of the following list of numbers:

$$1 + \frac{1}{3} + \frac{1}{9} + \frac{1}{27} + \ldots + \frac{1}{3^n}$$

for an arbitrary positive integer n. Using this (or otherwise) show that the sequence (x_n) defined by the formula

$$x_n = 1 + \frac{2}{3^2} + \frac{8}{3^4} + \frac{26}{3^6} + \ldots + \frac{3^n - 1}{3^{2n}}$$

is bounded. Now use this to prove that it must be convergent.

35. Put $x_n = \dfrac{1}{1} + \dfrac{1}{3} + \dfrac{1}{5} + \dfrac{1}{7} + \ldots + \dfrac{1}{2n - 1}$ (for each positive integer n). By showing that (x_n) is not bounded, prove that it cannot be convergent.

36. Show that each of the sequences whose n^{th} terms are as follows is divergent (*hint:* consider suitable subsequences of each in turn):
    (a) $\dfrac{(-2)^n - 1}{2^n + 1}$,
    (b) $\sin\left(\dfrac{n\pi}{2}\right)$.

37. Show that the following sequences are divergent:
    (a) $\left(\cos\left(\dfrac{3n\pi}{4}\right)\right)$,
    (b) $((-1)^n + (-1)^{\lfloor n/2 \rfloor})$, where $\lfloor t \rfloor$ denotes the floor (the integer part) of t.

38. Show that each of the sequences whose n^{th} terms are as follows is divergent (*hint:* consider suitable subsequences of each in turn):

(a) $2\pi + (-1)^n \left(3 - \dfrac{2}{n+5}\right)$,

(b) $1 + \cos\left(\dfrac{(n+3)\pi}{4}\right)$.

39. Construct a sequence $(x_n)_{n\geq 1}$ such that there are ten different numbers that are limits of subsequences of (x_n).

40. The sequence $(b_n)_{n\in\mathbb{N}}$ is defined recursively by the two formulae $b_1 = 3$, $b_{n+1} = \sqrt{6b_n - 8}$. Show that
(a) $3 \leq b_n < 4$ for all $n \in \mathbb{N}$,

(b) $(b_n)_{n\in\mathbb{N}}$ is an increasing sequence,

(c) $(b_n)_{n\in\mathbb{N}}$ converges,
and determine what its limit is.

41. The sequence $(b_n)_{n\in\mathbb{N}}$ is defined recursively by the two formulae $b_1 = 6$, $b_{n+1} = \sqrt{4b_n + 32}$ (each $n \in \mathbb{N}$). Show that
(a) $6 \leq b_n < 8$ for all $n \in \mathbb{N}$,

(b) $(b_n)_{n\in\mathbb{N}}$ is an increasing sequence,

(c) $(b_n)_{n\in\mathbb{N}}$ converges;
also determine its limit.

42. The sequence (x_n) is defined recursively thus:

$$x_1 = 20, x_{n+1} = \sqrt{13x_n - 36} \ \ (n \geq 1).$$

Show that $9 < x_n \leq 20$ for all n and that (x_n) is a decreasing sequence. Then show that it converges and determine its limit.

43. Does the following sequence converge? If so, what is its limit?

$$\left(\sqrt{12}, \ \sqrt{12 + \sqrt{12}}, \ \sqrt{12 + \sqrt{12 + \sqrt{12}}}, \ \sqrt{12 + \sqrt{12 + \sqrt{12 + \sqrt{12}}}}, \ \ldots \right)$$

44. The sequence $(d_n)_{n\in\mathbb{N}}$ is defined recursively by the two formulae

$$d_1 = \frac{2}{3}, \ d_{n+1} = \frac{2 + 2d_n}{3 + d_n} \ \text{(for each } n \in \mathbb{N}\text{)}.$$

Show that
(a) $\dfrac{2}{3} \leq d_n < 1$ for all $n \in \mathbb{N}$,

(b) $(d_n)_{n\in\mathbb{N}}$ is an increasing sequence,

(c) $(d_n)_{n\in\mathbb{N}}$ converges, and determine its limit.

45. The sequence $(x_n)_{n \in \mathbb{N}}$ is defined recursively by the two formulae

$$x_1 = 2, \quad x_{n+1} = \frac{2}{1 + x_n} \quad \text{(for each } n \in \mathbb{N}\text{)}.$$

Obtain a formula for x_{n+2} in terms of x_n. Think what this tells you about the subsequence of odd-numbered terms, and about the subsequence of even-numbered terms. Use Exercises 44 and 29 to determine the limit of the sequence $(x_n)_{n \in \mathbb{N}}$.

46. Give an example of two divergent series $\sum a_n$ and $\sum b_n$ for which $\sum (a_n + b_n)$ is convergent.

47. Find (if it exists) the limit (as $n \to \infty$) of

(a) $\sqrt[n]{100 + \sin\left(\frac{n\pi}{17}\right)} v$

(b) $\dfrac{2^n}{n! + \sqrt{n!}}$

(c) $\sqrt[n]{6^{0.5n} + 3^n}$

(d) $\left(\dfrac{n^3 + n}{n^3}\right)^{2n^2 + n}$

   (*Hint:* the tricky part is to investigate $\left(1 + \dfrac{1}{n^2}\right)^n$. Once you have shown

   that $\left(1 + \dfrac{1}{n^2}\right)^{n^2}$ converges, you know that it is bounded: there is a

   constant K such that $\left(1 + \dfrac{1}{n^2}\right)^{n^2} < K$ for all n; therefore

   $\left(1 + \dfrac{1}{n^2}\right)^n < \sqrt[n]{K} \ldots$)

48. Determine the limit as $n \to \infty$ of the sequences whose n^{th} terms are as given:

(a) $\left(1 - \dfrac{4}{n^2}\right)^n$

(b) $\left(1 + \dfrac{3}{n}\right)^{2n+5}$

(c) $\dfrac{3^n}{n! + n^2}$.

(d) $\dfrac{5n^2}{3n^3} + \dfrac{5n^2 - 1}{3n^3 + 2} + \dfrac{5n^2 - 2}{3n^3 + 4} + \dfrac{5n^2 - 3}{3n^3 + 6} + \ldots + \dfrac{5n^2 - n}{3n^3 + 2n}$

49. What are the limits of the following sequences?

(a) $\left(\sqrt[3n^2+n-2]{123.47}\right)$

(b) $\left(((n!)!)^{\frac{-1}{n!}}\right)$, that is, $\left(\dfrac{1}{\sqrt[n!]{(n!)!}}\right)$

(c) $\left(\dfrac{(3\pi)^n}{n!}\right)$

(d) $\left(\dfrac{5^{2n+1}}{(2n+1)!}\right)$

(e) $\left(\left(1+\dfrac{3}{n}\right)^n\right)$

50. Determine the limits of the sequences whose typical terms are as presented below.

(a) $\left(1-\dfrac{0.5}{n}\right)^{n^2+n+10}$

(b) $\sqrt{n+7}-\sqrt{n+2}$

(c) $\sqrt[2n+3]{10n+5}$

51. Find the limit of the sequences whose n^{th} terms are as given:

(a) $\left(1+\dfrac{6}{5n}\right)^{3n+17}$,

(b) $\left(\dfrac{n}{n+2}\right)^{3n-1}$,

(c) $\dfrac{n^2}{n^3+1}+\dfrac{n^2-2}{n^3+4}+\dfrac{n^2-4}{n^3+7}+\cdots+\dfrac{n^2-2n}{n^3+3n+1}$.

52. Prove or disprove the following statements concerning a general sequence $(x_n)_{n\in\mathbb{N}}$:
   - If $|x_n|\to\sqrt{2}$ then either $x_n\to\sqrt{2}$ or $x_n\to-\sqrt{2}$;
   - $x_n^3\to 64\Rightarrow x_n\to 4$.

53. The following sequence $(x_n)_{n\in\mathbb{N}}$:

$$0,1,1,2,1,1,1,3,1,1,1,1,1,1,1,4,1,1,1,\ldots$$

is defined by specifying $x_n=1$ if n is not a power of 2 but $x_n=k$ whenever $n=2^k$. Decide (with proof) whether the sequence is convergent or divergent.

54. We are given two sequences of positive real numbers $(a_n)_{n\in\mathbb{N}}$, $(b_n)_{n\in\mathbb{N}}$ such that $(b_n)_{n\in\mathbb{N}}$ is convergent and $(a_1+a_2+a_3+\ldots+a_n)\le b_n$ for every positive integer n. Prove that $(a_n)_{n\in\mathbb{N}}$ must converge, and determine its limit.

55. The sequence $(c_n)_{n\in\mathbb{N}}$ is defined recursively by the two formulae

$$c_1=2,\; c_{n+1}=\dfrac{2+2c_n}{3+c_n}\;\text{(for each } n\in\mathbb{N}).$$

Show that
(a) $1<c_n\le 2$ for all $n\in\mathbb{N}$,

(b) $(c_n)_{n \in \mathbb{N}}$ is a decreasing sequence,

(c) $(c_n)_{n \in \mathbb{N}}$ converges,

and determine its limit.

56. Does the sequence whose n^{th} term is $\sqrt[3n+1]{4n+7}$ converge?

57. We are given a bounded sequence (y_n) such that, for every constant K, the sequence $(\sin(Ky_n))$ converges. Prove that (y_n) must also converge. (*Suggestion:* try proof by contradiction.)

58. Given that $\sum \frac{1}{n^2}$ converges, use the comparison test to prove convergence for

(a) $\sum \dfrac{5n - 1}{3n^3 + 1}$,

(b) $\sum \dfrac{5n + 1}{3n^3 - 1}$.

59. Given that the series $\sum n^{-\frac{3}{2}}$ converges, use the comparison test to show that each of the following also converges:

(a) $\sum \dfrac{3n - 2}{\sqrt{n}(2n^2 + 1)}$,

(b) $\sum \dfrac{n^2 + n + 1}{(\sqrt{n})^7 + 13}$.

60. Use the limit comparison test to decide, for each of the following series, whether it converges or diverges:

(a) $\sum n^{-2 + \frac{1}{n}}$

(b) $\sum \dfrac{100^{1/n} + (n!)^{-1/n}}{n^{1 + 2/n}}$.

61. Use the limit comparison test to decide, for each of the following series, whether it converges or diverges:

(a) $\sum \dfrac{n^3 \sqrt{n} + 5}{2n^4 - 7}$

(b) $\sum \dfrac{n^3 + 5}{2n^4 \sqrt{n} - 7}$.

62. Does the following series converge or diverge? Give reasons for your answer.

$$\sum \left(1 - \frac{1}{2n + 1}\right)^{n^2}.$$

63. For which positive values of t does

$$\sum \frac{(n + 1)! \, (2n + 1)! \, t^n}{(3n - 1)!}$$

converge, and for which does it diverge?

64. Use the nth-root test to decide whether

$$\sum \frac{\left(1 + \dfrac{1}{n}\right)^{n^2}}{\left(1 + \dfrac{4}{5n}\right)^{n^2}}$$

is convergent or divergent.

65. Use the ratio test of d'Alembert to prove the convergence of

$$\sum \frac{n!\,(2n)!\,(2\pi)^n}{(3n)!}.$$

66. Determine the real number B such that the following series converges for $0 < t < B$ but diverges for $t > B$:

$$\sum \left(\frac{4n^2 - 1}{9n^2 - 1}\right)^n t^n.$$

(Note that it is quite difficult to decide whether or not this series converges when $t = B$ exactly, and you are not asked to investigate this.)

67. Does the series

$$\sum \frac{(n!)^2 2^{2n}}{(2n)!}$$

converge or diverge?

68. For which $t > 0$ does this series converge?

$$\sum \frac{n!(2n)!(3n)!}{(6n)!} t^n$$

69. For which $t > 0$ does this series converge?

$$\sum \left(\frac{3n}{3n + 2}\right)^{n^2} t^n$$

70. For which positive values of t is the series $\sum \dfrac{((n+2)!)^2 t^n}{(2n+1)!}$ convergent, and for which is it divergent?

71. For which values of t is the series $\sum \left(\dfrac{n+5}{n}\right)^{n^2+n} |1 - t|^n$ convergent, and for which is it divergent?

72. Find a positive integer N so large that

$$\sum_{n=65}^{N} \frac{1}{n} > 100.$$

(*Suggestion:* look at the proof that the harmonic series diverges.)

73. (a) If $\sum a_n$ is a convergent series and $\sum b_n$ is a divergent series, prove that the series $\sum (a_n + b_n)$ must be divergent.

(b) Give an example of two divergent series $\sum a_n$ and $\sum b_n$ for which $\sum (a_n b_n)$ is convergent.

(c) Give an example of two convergent series $\sum c_n$ and $\sum d_n$ for which $\sum (c_n d_n)$ is divergent.

74. Show that the series

$$\sum (-1)^n \left(\frac{3n+5}{4n^2+3} \right)$$

converges.

75. Determine whether the following series converges:

$$\sum (-1)^n \left(1 + \frac{1}{n} \right)^{-n}.$$

76. Decide whether the following series converge:

(a) $\sum (-1)^n \dfrac{2n}{n^2+4}$,

(b) $\sum (-1)^n \dfrac{2n^2}{n^2+4}$.

77. (a) Does the series $\sum (-1)^n \left(\dfrac{n-5}{n} \right)^n$ converge or diverge?

(b) Does the series $\sum (-1)^n \left(\dfrac{e^2+1}{10} \right)^n$ converge or diverge?

78. We define a sequence $(x_n)_{n \in \mathbb{N}}$ by setting $x_n = \frac{4}{n+1}$ when n is odd but $x_n = \frac{2}{n}$ when n is even.

Verify that $x_n \to 0$ (as $n \to \infty$). Determine whether $\sum (-1)^{n+1} x_n$ is convergent or divergent.

79. By first factorising the bottom line and using partial fractions, find the sum of the series

$$\sum_{k=1}^{\infty} \frac{1}{4k^2 + 12k + 5}.$$

80. Let $w_n = \dfrac{1}{n^3 - n}$ for each integer $n \geq 2$. Calculate, as a rational in its lowest terms,

$$\sum_{n=2}^{59} w_n.$$

Show that the series

$$\sum_{n=2}^{\infty} w_n$$

converges, and determine the numerical value of its sum.

81. Given a sequence $(x_n)_{n \in \mathbb{N}}$ in the interval $[-10, -2]$, show (using, for example, the direct comparison test) that $\sum n^{x_n}$ is convergent.

82. There are two sequences of positive real numbers $(a_n)_{n \in \mathbb{N}}, (b_n)_{n \in \mathbb{N}}$. A subsequence $(a_{n_k})_{k \in \mathbb{N}}$ of $(a_n)_{n \in \mathbb{N}}$ satisfies the condition $a_{n_k} \geq b_k$ for every $k \geq 1$ and the series $\sum b_n$ is divergent. Prove that $\sum a_n$ is also divergent. *Suggestion:* if not, then the partial sums of $\sum a_n$ would be bounded . . .

83. Putting $t_n = -1 + \dfrac{(-1)^n}{3n+1}$, show that $\sum n^{t_n}$ is divergent. *Suggestion:* use the result of Exercise 82.

84. Identify the domains of the real functions defined by the following formulae:
    (a) $\arcsin(5 + 3x)$,
    (b) $\dfrac{x-1}{(x^2 - 49)(x^2 + 3x + 2)}$,
    (c) $\ln(6 + x - x^2)$,
    (d) $\sqrt{\ln\left(\dfrac{10}{1 + x^2}\right)}$.

85. Determine the domains of the functions described by
    (a) $\dfrac{x}{x^2 + 5x - 50}$,
    (b) $\arccos(e^x)$,
    (c) $\sqrt{\ln\left(\dfrac{x(4-x)}{3}\right)}$.

86. For the function $f(x) = \lfloor \sqrt{x} \rfloor$, where $\lfloor x \rfloor$ denotes the floor (or integer part) of x, find two sequences $(x_n), (y_n)$ such that $x_n \to 4$ and $f(x_n) \to f(4)$, but $y_n \to 4$ and $f(y_n) \not\to f(4)$.

87. For the function defined by $f(x) = \lfloor x^3 \rfloor$, find
    (a) a sequence (x_n) such that $x_n \to 0$ and $f(x_n) \to f(0)$,
    (b) a sequence (y_n) such that $y_n \to 0$ and $f(y_n) \not\to f(0)$.

88. Prove that $p(x) = x^4 - 2x^2 + 17x - 12$ defines a function p that is continuous at $x = 4$.

89. Show directly from the definition that the polynomial

$$g(x) = 2x^7 - 15x^5 + 22x - 12$$

is continuous at $x = -3$.

90. For the function f described by

$$f(x) = \begin{cases} 20 - x^2 & \text{if } x < 2, \\ 1 + 2x + 3x^2 & \text{if } x \geq 2 \end{cases}$$

find a sequence (x_n) such that $x_n \to 2$ and $f(x_n) \not\to f(2)$.

91. Prove that the function given by $f(x) = \lfloor x \rfloor^2 + \lfloor x^2 \rfloor$ is not continuous at $x = \sqrt{3}$.

92. Show that the function specified by $h(x) = \lfloor \sqrt{x} \rfloor^2$ is not continuous at 100.

93. Prove that the 'step function'

$$s(x) = \begin{cases} 0 & \text{if } x < 0, \\ 1 & \text{if } x \geq 0 \end{cases}$$

   is not continuous.

94. Verify that f is continuous at $x = 2$, where

$$f(x) = \begin{cases} 1 + x & \text{if } x \text{ is rational,} \\ 5 - x & \text{if } x \text{ is irrational.} \end{cases}$$

95. Verify that g is continuous at $x = 0$, where

$$g(x) = \begin{cases} x \sin\left(\dfrac{1}{x}\right) + x^2 \cos\left(\dfrac{1}{x}\right) & \text{if } x \neq 0, \\ 0 & \text{if } x = 0. \end{cases}$$

96. Show that the function g described by

$$g(x) = \begin{cases} x \cos\left(\dfrac{1}{\pi x}\right) & \text{if } x \neq 0, \\ 0 & \text{if } x = 0 \end{cases}$$

   is continuous at 0.

97. Show that the function h given by

$$h(x) = \begin{cases} 2 + 5x & \text{if } x \in \mathbb{Q}, \\ 10 - 3x & \text{if } x \in \mathbb{R} \setminus \mathbb{Q} \end{cases}$$

   is continuous at 1, but not continuous at -1.

98. Prove that the function
$$\frac{x^3 - x^2 + 5x - 12}{2x^2 + 3}$$
   is continuous on $\mathbb{R}$.

99. Prove that
$$f(x) = \frac{1 - 2x + 3x^2 - x^5}{2x^2 + 8x + 9}$$
   is continuous everywhere on the real line.

100. Suppose that B is a non-empty set of real numbers and that $\lambda = \inf(B)$. Show that there is a sequence (x_n) of elements of B such that $x_n \to \lambda$.

101. Show that $x^5 + 15x - 20$ has a root in $[1, 2]$.

102. Show that the polynomial $6x^4 - 8x^3 + 1$ has at least two roots in the interval $(0, 2)$.

103. Prove that $x^6 - 5x^4 + 2x + 1 = 0$ has at least four real solutions.

104. Prove that the equation $4x^5 - 8x^3 + 4x - 1 = 0$ has at least three positive solutions. It may be useful to evaluate the polynomial at $x = \frac{1}{2}$.

105. Show that the graph of the function $7 \sin x - 10 \cos x - 4x$ crosses the x-axis at least twice between 0 and π.

106. Show that the equation $x^4 + x^3 - 8x^2 + 1 = 0$ has four distinct real solutions.

107. Given that $f : [0, \pi/2] \to [0, 1]$ is continuous, prove (by considering the function $g(x) = f(x) - \sin x$) that there is a number c in $[0, \pi/2]$ such that $f(c) = \sin c$.

108. Given that $f : [0, 1] \to [0, 1]$ is continuous, prove that there exists a number $c \in [0, 1]$ such that $(f(c))^2 + 2f(c) - 4c^2 = 0$.

109. (a) Suppose that $f : [a, b] \to \mathbb{R}$ is continuous and never takes the value zero. Prove that there is $\delta > 0$ such that no value of $f(x)$ lies in the interval $[-\delta, \delta]$.

   (b) Show by example that the statement in part (a) ceases to be true if we replace $[a, b]$ by (a, b).

110. Given continuous $f : [a, b] \to \mathbb{R}$, show that there is a positive constant K such that the function $\sqrt{K + f(x)}$ is defined everywhere on $[a, b]$.

111. Given that $f : [a, b] \to \mathbb{R}$ is continuous, show that

   (a) there is a positive constant K such that the function $\ln(f(x) + K)$ is defined everywhere on $[a, b]$,

   (b) there is a positive constant A such that the function $\arcsin(Af(x))$ is defined everywhere on $[a, b]$.

112. Show by means of examples (preferably, simple ones) that

   (a) A continuous function on a bounded open interval can fail to be bounded,

   (b) A continuous function on a bounded open interval, even if it is bounded, can fail to have a maximum value and can fail to have a minimum value,

   (c) A continuous function on an unbounded closed interval can fail to be bounded,

   (d) A continuous function on an unbounded closed interval, even if it is bounded, can fail to have a maximum value and can fail to have a minimum value,

(e) A discontinuous function on a bounded closed interval can fail to be bounded,

(f) A discontinuous function on a bounded closed interval, even if it is bounded, can fail to have a maximum value and can fail to have a minimum value.

113. If $f : D \to C$ is increasing and $g : D \to C$ is decreasing, verify that the function $2f - 3g$ is increasing.

114. If $f : D \to C$ and $g : D \to C$ are both increasing, and k is a positive constant, show that $f + g$ and kf are both increasing. Also show (by presenting a suitable counterexample) that the product function fg could fail to be increasing.

115. If $h : D \to C$ and $j : C \to \mathbb{R}$ are both increasing or both decreasing, show that the composite function $j \circ h$ is increasing.

116. Determine all the limit points of $X = \{\frac{1}{n} : n \in \mathbb{N}\}$ and all the limit points of $Y = \mathbb{Q} \cap (2, 3)$.

117. Determine all the limit points of each of the following sets:
    (a) (a, b),
    (b) $[0, 2) \cup (2, 3]$,
    (c) $\mathbb{Q}$.

118. Let $g(x) = x^2(x - 1)$ and $f(x) = \sqrt{g(x)}$. What is the domain of $f(x)$? Investigate, if possible, the limit of $f(x)$ as $x \to 1$. Investigate, if possible, the limit of $f(x)$ as $x \to 0$.

119. Sketch the graph of the function $f(x) = x^2(x + 1)(x - 2)$. Now let $g(x) = \sqrt{f(x)}$:
    (a) What is the domain of g?
    (b) What is the numerical value of $g(2)$? Investigate the limit of $g(x)$ as $x \to 2$.
    (c) What is the numerical value of $g(0)$? Why can we not investigate the limit of $g(x)$ as $x \to 0$?

120. Use sequences to evaluate

$$\lim_{x \to 10} \frac{x^3 - 1000}{x^4 - 10000}.$$

121. Use sequences to evaluate

$$\lim_{x \to 3} j(x)$$

where

$$j(x) = \begin{cases} \dfrac{x^4 - 81}{x^3 - 27} & \text{while } x \neq 3, \\ -1 & \text{if } x = 3. \end{cases}$$

122. Evaluate (with proof using sequences)

   (a) $\displaystyle\lim_{x\to-4} \frac{x^2-16}{x+4}$,

   (b) $\displaystyle\lim_{x\to4} \frac{x^3-64}{x^2-16}$.

123. For the function f defined by

$$f(x) = \begin{cases} \dfrac{x^2+3x-10}{x^3+x^2-4x-4} & \text{if } x \neq 2, \\ \dfrac{1}{2} & \text{if } x = 2 \end{cases}$$

   use sequences to evaluate $\lim f(x)$ as $x \to 2$.

124. For the function f defined by

$$f(x) = \begin{cases} \dfrac{\sin(\cos(\sin x))}{2+\cos(\sin(\cos x))} & \text{if } x \leq 1, \\ x^4 - x^3 & \text{if } x > 1, \end{cases}$$

   use sequences to evaluate $\lim f(x)$ as $x \to \sqrt[3]{2}$.

125. For the function described by

$$f(x) = \begin{cases} \sin(\sin(\sin(x))) & \text{if } x \in \mathbb{Q} \cap (-\infty, 10), \\ \cos(\cos(\cos(x))) & \text{if } x \in (\mathbb{R} \setminus \mathbb{Q}) \cap (-\infty, 10), \\ 3x - x^3 & \text{otherwise} \end{cases}$$

   determine the limit of $f(x)$ as $x \to 11$.

126. The function f is defined by:

$$f(x) = \begin{cases} x & \text{if } x \text{ is rational with even denominator when in lowest terms,} \\ 2x - 2 & \text{if } x \text{ is rational with odd denominator when in lowest terms,} \\ 3x - 4 & \text{if } x \text{ is irrational.} \end{cases}$$

   Show that, in all cases, $|f(x) - 2| \leq 3|x - 2|$.
   Deduce, via the squeeze, that $f(x) \to 2$ as $x \to 2$.

127. Construct a proof of the result, that if f is a function with domain D, and p is a limit point of D, and $f(x) \to \ell$ as $x \to p$, then $|f(x)| \to |\ell|$ as $x \to p$.

128. Use sequences to prove that if $f : D \to C$ and p is a limit point of D and $f(x) \to \ell$ as $x \to p$, then $(f(x))^2 + (f(x))^3 \to \ell^2 + \ell^3$ as $x \to p$.

129. Construct a proof of the squeeze rule for limits of functions.

130. A function f is defined as follows:

(a) if x is rational then $f(x) = 2x + 7$,

(b) if x is irrational but x^2 is rational then $f(x) = 4x + 5$,

(c) in all other cases, $f(x) = 11x - 2$.

Check that, in all three cases, $|f(x) - 9| \le 11|x - 1|$ for all x. Deduce, using the squeeze, that $\lim_{x \to 1} f(x) = 9$.

131. Use the epsilon-delta description of limits to prove the following version of the squeeze: suppose that f, g, h have the same domain D, that p is a limit point of D, that

$$f(x) \le g(x) \le h(x)$$

for each $x \in D$, that $f(x) \to \ell$ as $x \to p$ and that $h(x) \to \ell$ as $x \to p$; then also $g(x) \to \ell$ as $x \to p$.

132. Use the epsilon-delta definition to prove the following: if f, g have domain D, p is a limit point of D, $f(x) \le g(x)$ for every $x \in D$ and both f and g have limits at p, then $\lim_{x \to p} f(x) \le \lim_{x \to p} g(x)$. [*Hint:* if not, then the number

$$\frac{\lim_{x \to p} f(x) - \lim_{x \to p} g(x)}{3}$$

is positive. Call it ε. Use the fact that the two limits exist to get two positive numbers δ_1 and δ_2 and put $\delta =$ the smaller of δ_1 and δ_2. Pick a point $x \ne p$ of D whose distance from p is less than δ, and look for a contradiction.]

133. Let $p(x) = 4 - 5x + 7x^2 - 2x^3$. Given a positive number ε, find a positive number δ such that $|x - 2| < \delta$ will guarantee that $|p(x) - p(2)| < \varepsilon$.

134. Let p be the polynomial described by $p(x) = 12 - 5x + x^2 + 4x^3$ and $\varepsilon > 0$ be given. Determine a positive number δ such that the condition $|x + 2| < \delta$ will guarantee that $|p(x) - p(-2)| < \varepsilon$.

135. For the polynomial function $p(x) = 3x^2 - 7x + 5$ and a given positive number ε, obtain a formula for a positive number δ such that $0 < |x - 2| < \delta$ will guarantee that $|p(x) - 3| < \varepsilon$. Why does it follow from this that p is continuous at 2?

136. The function f is defined by

$$f(x) = \begin{cases} \dfrac{3x^2 - 243}{x - 9} & \text{if } x \ne 9, \\ 27 & \text{if } x = 9. \end{cases}$$

Given $\varepsilon > 0$, get a formula for a positive number δ such that $0 < |x - 9| < \delta$ implies that $|f(x) - 54| < \varepsilon$. Is f continuous at $x = 9$?

137. Consider the function g defined by

$$g(x) = \begin{cases} \dfrac{5x^2 - 180}{x - 6} & \text{if } x \neq 6, \\ 48 & \text{if } x = 6. \end{cases}$$

For a given positive number ε, obtain a formula for a positive number δ such that $0 < |x - 6| < \delta$ will guarantee that $|g(x) - 60| < \varepsilon$. Why does it follow from this that g is not continuous at 6?

138. Show via the epsilon-delta definition of continuity that the function defined by

$$f(x) = \begin{cases} (x-1)^2 \left\lfloor \dfrac{1}{(x-1)^2} \right\rfloor & \text{if } x \neq 1, \\ 1 & \text{if } x = 1 \end{cases}$$

is continuous at 1.

139. The straight line joining $(1, 1)$ to a nearby point $(x, \sqrt{x})$ on the graph of $y = \sqrt{x}$ (assuming always $x > 0$) has gradient $\dfrac{\sqrt{x} - 1}{x - 1}$. Verify that this simplifies to $\dfrac{1}{\sqrt{x} + 1}$. Now show that

$$\left| \frac{1}{\sqrt{x} + 1} - \frac{1}{2} \right| < \frac{|1 - \sqrt{x}|}{2} < \frac{|1 - x|}{2}.$$

Then use the epsilon-delta definition of limit to determine $\lim_{x \to 1} \dfrac{\sqrt{x} - 1}{x - 1}$, that is, the gradient of the curve at $(1, 1)$.

140. Assume throughout this question that $x > 0$. Verify that

$$\left| \frac{1}{\sqrt{x} + 3} - \frac{1}{6} \right| < \frac{|3 - \sqrt{x}|}{18} < \frac{|9 - x|}{54}.$$

Now show that (if also $x \neq 9$) $\dfrac{\sqrt{x} - 3}{x - 9} = \dfrac{1}{\sqrt{x} + 3}$.

The gradient of the straight line joining the point $(9, 3)$ to a nearby point $(x, \sqrt{x})$ on the curve $y = \sqrt{x}$ is $\dfrac{\sqrt{x} - 3}{x - 9}$. Use the above roughwork and the epsilon-delta definition to evaluate the limit of this expression as $x \to 9$ (that is, the gradient of the curve itself at $x = 9$).

141. Suppose that f, g have domain D, that p is a limit point of $D \cap (p, \infty)$ and that (as $x \to p^+$) $f(x) \to \ell$ and $g(x) \to m$.
    (a) Use the epsilon-delta description of one-sided limits to prove that
    $f(x) + g(x) \to \ell + m$ (as $x \to p^+$),
    (b) Use the sequence description of one-sided limits to prove that
    $f(x) - g(x) \to \ell - m$ (as $x \to p^+$).

142. Show that left-hand limits preserve inequalities, in the following sense: if $f(x) \leq g(x)$ for all $x \in (a, b)$ and both f and g have left-hand limits at b, then

$$\lim_{x \to b^-} f(x) \leq \lim_{x \to b^-} g(x).$$

(*Suggestion:* proof by contradiction using the epsilon-delta description of limits.)

143. State and (using sequences) prove a squeeze rule for function limits as $x \to p^-$.

144. Given that

$$f(x) = \begin{cases} (1 + x - x^2)^2 & \text{if } x < 3, \\ 12 & \text{if } x = 3, \\ x^3 - x + 1 & \text{if } x > 3 \end{cases}$$

show that f has a limit at 3 but is not continuous there.

145. Verify that the function specified by

$$f(x) = \begin{cases} 1 + x + x^2 & \text{if } x < 4, \\ x^3 - 2x^2 - 3x + 1 & \text{if } x > 4, \\ 0 & \text{if } x = 4 \end{cases}$$

has a limit at $x = 4$ but is not continuous there.

146. We define $f : \mathbb{R} \to \mathbb{R}$ by the formula

$$f(x) = \begin{cases} a + bx + x^2 & \text{if } x \neq 3, \\ a + 3b & \text{if } x = 3. \end{cases}$$

Prove that f has a limit as $x \to 3$. Find the relation (in as simple a form as possible, without modulus signs!) between a and b that is equivalent to the modulus $|f|$ being continuous at 3.

147. The function f specified by

$$f(x) = \begin{cases} \dfrac{x^2 - 4x + 3}{x - 2} & \text{if } x < 1, \\ ax^2 + bx + 5 & \text{if } 1 \leq x \leq 2, \\ \dfrac{x^2 + 4x - 3}{x - 1} & \text{if } 2 < x \end{cases}$$

is continuous at 1 and at 2. Find the values of the constants a and b.

148. The formula

$$f(x) = \begin{cases} 5 - x^2 & \text{if } x < -1, \\ ax^3 + b & \text{if } -1 \le x \le 1, \\ 1 + x + x^2 & \text{if } x > 1 \end{cases}$$

is known to define a continuous function on $\mathbb{R}$. What are the numerical values of a and b?

149. The function

$$f(x) = \begin{cases} \dfrac{2 + x^3}{2 + x} & \text{if } -2 < x < -1, \\ a + bx + x^2 & \text{if } -1 \le x \le 1, \\ bx^2 - 13 & \text{if } 1 < x \end{cases}$$

is continuous at $x = -1$ and at $x = 1$. What are the numerical values of the constants a and b?

150. Given that the shared domain of f and g is not bounded above, and that the two functions tend to ℓ, m (respectively) as $x \to \infty$, use the sequence description of limits to prove that $f(x)g(x) \to \ell m$ as $x \to \infty$.

151. Given that f and g have the same domain and possess limits ℓ and m as $x \to \infty$ and that $m \ne 0$, use sequences to show that

$$\frac{f(x)}{g(x)} \to \frac{\ell}{m} \quad \text{as } x \to \infty.$$

152. Show that $\dfrac{x}{(1 + x)^2}$ tends to $-\infty$ as $x \to -1$. (You can use either the formal definition or the sequence description.)

153. Verify via the definition that $\lim_{x \to 2} \dfrac{3x + 5}{(x - 2)^2} = \infty$.

154. Let f, g have the same domain D of which p is a limit point, and suppose that (as $x \to p$) $f(x) \to \infty$ and $g(x) \to \ell \in \mathbb{R}$. Prove that $f(x) + g(x) \to \infty$ (as $x \to p$).

155. Show that $\dfrac{\sin(\pi x)}{(2x - 1)^2}$ tends to ∞ as $x \to \frac{1}{2}$. (Use the formal definition: given $K > 0$, find a positive number δ such that $|x - \frac{1}{2}| < \delta$ guarantees that $\dfrac{\sin(\pi x)}{(2x - 1)^2} > K$.)

156. Show that

$$\lim_{x \to -1} \frac{5x^2 + 3x}{(x + 1)^2} = \infty.$$

157. Use either the sequence description or the epsilon-delta description to explore the limit of $f(x) = \dfrac{2x^2 - 3}{3x^2 - 1}$ as $x \to -1$.

158. For the function f defined by $f(x) = \dfrac{1 + x + x^2}{2 + 3x + 4x^2}$, show that the limit of
$f(x)$ as $x \to \infty$ is $\frac{1}{4}$ by two different methods:
   - by the epsilon-style definition,
   - by using the sequence description of such limits.

159. Suppose that $f : (0, \infty) \to \mathbb{R}$ and that another function $g : (0, \infty) \to \mathbb{R}$ is
then defined by the formula $g(t) = f\left(\dfrac{1}{t}\right)$. Prove that the following
statements are equivalent (*suggestion:* use an epsilon-style argument):
   (a) $f(x) \to \ell$ as $x \to \infty$,
   (b) $g(t) \to \ell$ as $t \to 0^+$.

160. Differentiate (within the appropriate domain) each of the following:
   (a) $\dfrac{1 + x - x^3}{2 - x + x^2}$,
   (b) $\sin(e^x)e^{\sin x}$,
   (c) $\sqrt{x}\ln x \cos x$,
   (d) $\sin(\ln(\cos x))$.

161. Differentiate the following functions (within their domains, and assuming
that e^x, $\sin x$, $\cos x$ and $\ln x$ have their well-known derivatives). (Do not
spend a lot of time simplifying your answers.)
   (a) $f(x) = e^x \ln x \cos(2x)$,
   (b) $f(x) = \dfrac{\sin x + x^3}{\cos x - x^2}$,
   (c) $f(x) = (1 + x + e^x)^{13}$,
   (d) $f(x) = e^{\cos(\ln x)}$,
   (e) $f(x) = \dfrac{x \ln(1 + e^x)}{\sqrt{4 + x^2}}$.

162. Differentiate (with respect to x) the expressions
   (a) $\dfrac{x^2 + \sin x}{x^3 - \cos x}$,
   (b) $(x^4 - 12) \sin^9 x$,
   (c) $\ln(e^{2x} - e^{-2x})$.

163. Differentiate, with respect to x, each of the expressions
   (a) $\sin\left(\dfrac{1 + e^x}{1 - e^x}\right)$,
   (b) $(1 + x^2) \ln(1 + x^2)$,
   (c) xe^{e^x}.

164. If f is differentiable at $x = c$, show that $\dfrac{f(c + h) - f(c - h)}{2h}$ possesses a limit
as $h \to 0$.

See if you can devise a function g that is *not* differentiable at $x = 0$ and yet
$\dfrac{g(0 + h) - g(0 - h)}{2h}$ possesses a limit as $h \to 0$.

165. Suppose that f is differentiable at c. Evaluate the following:

(a) $\lim_{h \to 0} \dfrac{f(c + 2h) - f(c - 3h)}{h}$,

(b) $\lim_{h \to 0} \dfrac{(f(c + h))^2 - (f(c - h))^2}{h}$.

(Notice that the top line in (b) is the difference of two squares.)

166. If
$$f(x) = \begin{cases} 1 + x + ax^2 & \text{while } x < 4, \\ 3 + bx + x^2 & \text{while } x \geq 4 \end{cases}$$

is to define a function that is differentiable on $\mathbb{R}$, find what numerical values a and b must have.

167. Given that the function

$$f(x) = \begin{cases} ax^3 + bx & \text{if } x < 2, \\ ax^2 + 5 & \text{if } x \geq 2 \end{cases}$$

is differentiable at $x = 2$, determine the numerical values of the constants a and b.

168. The formula
$$f(x) = \begin{cases} 3x^2 + ax + b & \text{if } x < 1, \\ 2x^2 + 2bx - a & \text{if } x \geq 1 \end{cases}$$

defines a function f that is differentiable at $x = 1$. Evaluate the constants a and b.

169. Determine, if possible, the maximum and minimum values of the functions
$f(x) = \cos(2x) - \cos^2 x$ on $[0, 3\pi/4]$ and $g(x) = \dfrac{x}{x^2 + 4x + 9}$ on $\mathbb{R}$.

170. Find the maximum value and the minimum value of the expressions on the intervals indicated:

(a) $x^2 \ln x$ on $[\frac{1}{e}, 1]$,

(b) $e^x - 2e^{2x} + e^{3x}$ on $[-2, 2]$.

171. Suppose that f and g are continuous on $[a, b]$ and differentiable on (a, b), and that $f'(x) \neq g'(x)$ everywhere in (a, b). Show that the graphs of f and of g cannot intersect at two different points. (*Hint:* if they did, consider the behaviour of the function $h(x) = f(x) - g(x)$.)

172. Use Rolle's theorem on the function f given by $f(x) = x^3(1 + \sin x)$ on the interval $\left[0, \frac{3\pi}{2}\right]$, to see what it tells us about the equation

$$\frac{\cos x}{1 + \sin x} = -\frac{3}{x}.$$

173. Use Rolle's theorem to show that the equation $\tan x = \frac{2}{x}$ has a solution in the interval $(0, \pi/2)$. (*Suggestion*: consider $x^2 \cos x$.)

Show further that there is a sequence $(c_n)_{n \in \mathbb{N}}$ of numbers in the interval $(0, \pi/2)$ such that, for each positive integer n,

$$\tan c_n = \frac{n}{c_n}.$$

174. Use differentiation to show that the function $f(x) = x^2 e^x$ is decreasing on the interval $[-2, 0]$.

175. Use differentiation to show that the function

$$f(x) = (1 + x - x^2)e^{3x}$$

is decreasing on $(-\infty, -1]$, increasing on $[-1, \frac{4}{3}]$ and decreasing on $[\frac{4}{3}, \infty)$.

176. Show that

(a) $f(x) = \dfrac{4x^2 + 3x}{1 + x^2}$ is increasing on $\left[-\frac{1}{3}, 3\right]$,

(b) $g(x) = 2x \cos x - 2 \sin x - x^2$ is decreasing on $[0, \infty)$.

177. Given real constants a and b such that $a \neq 0$, show that the equation $x^5 + ax^3 + 3a^2 x + b = 0$ cannot have two distinct solutions.

178. Given a constant $a > 1$, show that the equation $x^4 + ax^3 - a = 0$ has exactly one positive solution.

179. Show that if f is decreasing on (a, b) and bounded above, then the one-sided limit $\lim_{x \to a^+} f(x)$ exists.

180. (a) Suppose we are given two sequences $(y_n)_{n \geq 1}$ and $(a_n)_{n \geq 1}$ of real numbers such that $a_n \to 0$ and $|y_m - y_n| \leq a_n$ whenever the integers n, m satisfy $0 < n < m$. Prove that $(y_n)_{n \geq 1}$ is a Cauchy sequence.

(b) (Conversely,) given that $(y_n)_{n \geq 1}$ is a Cauchy sequence, show that there is a sequence $(a_n)_{n \geq 1}$ such that $a_n \to 0$ and $|y_m - y_n| \leq a_n$ whenever $0 < n < m$.

181. Using the rather crude estimation of counting the terms and noting which one is the smallest, show that $\dfrac{1}{\sqrt{17}} + \dfrac{1}{\sqrt{18}} + \dfrac{1}{\sqrt{19}} + \cdots + \dfrac{1}{\sqrt{64}} > 6$ and that, for each positive integer n,

$$\frac{1}{\sqrt{4^{n-1} + 1}} + \frac{1}{\sqrt{4^{n-1} + 2}} + \frac{1}{\sqrt{4^{n-1} + 3}} + \cdots + \frac{1}{\sqrt{4^n}} > 3(2)^{n-2}.$$

Use this to show that the series $\sum_{n=1}^{\infty} \dfrac{1}{\sqrt{n}}$ is divergent.

182. Suppose that (x_n) and (y_n) are Cauchy sequences. Show, purely from the definition of Cauchy (and, in the case of part (c), also using the fact that Cauchy sequences are bounded) that:

(a) $(x_n + y_n)$ is Cauchy,

(b) for any constant M, (Mx_n) is Cauchy,

(c) $(x_n y_n)$ is Cauchy.

Also show by example that $\left(\dfrac{1}{x_n}\right)$ can fail to be Cauchy even in the case where $x_n > 0$ for every n.

183. Of the following two statements, just one is true in general. Give a proof for the one that is true, and find a counterexample that disproves the false one.

(a) If $(x_n)_{n\in\mathbb{N}}$ and $(y_n)_{n\in\mathbb{N}}$ are both Cauchy sequences, and there is a (strictly) positive number $\delta > 0$ such that $|y_n| \geq \delta$ for all $n \geq 1$, then the 'term-by-term quotient' sequence

$$\left(\frac{x_n}{y_n}\right)_{n\in\mathbb{N}}$$

must also be Cauchy.

(b) If $(x_n)_{n\in\mathbb{N}}$ and $(y_n)_{n\in\mathbb{N}}$ are both Cauchy sequences, and $|y_n| > 0$ for all $n \geq 1$, then the 'term-by-term quotient' sequence

$$\left(\frac{x_n}{y_n}\right)_{n\in\mathbb{N}}$$

must also be Cauchy.

184. Given a sequence (x_n), suppose we succeed in finding a positive constant K and a real number $a \in (0, 1)$ such that, for every positive integer n,

$$|x_n - x_{n+1}| < Ka^n.$$

Prove that (x_n) is Cauchy, and therefore converges.

185. Given a real constant $t \in (0, 1)$ and a sequence $(x_n)_{n\geq 1}$ that satisfies (for each positive integer n) the inequality $|x_{n+1} - x_n| < 1000t^n$, prove that $(x_n)_{n\geq 1}$ converges to some limit.

186. Given that (for each $n \geq 1$) $|x_n - x_{n+2}| < 10(0.6)^n$ and that $|x_n - x_{n+5}| < 20(0.7)^n$, show that (x_n) is Cauchy.
(*Suggestion:* $x_n - x_{n+5} + x_{n+5} - x_{n+3} + x_{n+3} - x_{n+1} = x_n - x_{n+1}$.)

187. For a sequence (x_n), the following information is known for each $n \in \mathbb{N}$:

$$|x_n - x_{n+6}| < 7(0.6)^n, |x_n - x_{n+10}| < 4(0.7)^n, |x_n - x_{n+15}| < 9(0.8)^n.$$

By estimating $|x_n - x_{n+1}|$, show that (x_n) must be convergent.

188. (a) If, for a given sequence (x_n), $|x_n - x_{n+2}| < 2^{-n}$ and $|x_n - x_{n+3}| < 3^{-n}$ for each $n \in \mathbb{N}$, does it necessarily follow that (x_n) itself is Cauchy?

(b) If, for a given sequence (x_n), the subsequence (x_{2n}) and the subsequence (x_{3n}) are both Cauchy, does it necessarily follow that (x_n) itself is Cauchy?

189. Confirm that the series

$$\sum_{k=1}^{\infty} \frac{7\sin(k^2 + 3k) - 4\cos(2k^2 - 5)}{(1.1)^k}$$

converges, by showing that the sequence of its partial sums is Cauchy.

190. Consider the sequence $(x_n)_{n\geq 1}$ defined by the formula $x_n =$

$$\frac{\sin(\cos((1 + \pi)))}{e} + \frac{\sin(\cos((1 + \pi)^2))}{e^2} + \frac{\sin(\cos((1 + \pi)^3))}{e^3} + \cdots$$
$$+ \frac{\sin(\cos((1 + \pi)^n))}{e^n}.$$

Verify that $|x_n - x_{n+1}| \leq e^{-(n+1)}$. Prove that $(x_n)_{n\geq 1}$ converges (by verifying that it is Cauchy).

191. Of the following three statements, at least one is true in general and at least one is false. Give a proof for each that is true, and find a counterexample to disprove each false one.
    (a) If $f : (a, \infty) \to \mathbb{R}$ is a continuous function on an unbounded open interval (a, ∞), and $(x_n)_{n\in\mathbb{N}}$ is any Cauchy sequence of elements of (a, ∞), then $(f(x_n))_{n\in\mathbb{N}}$ must also be Cauchy.
    (b) If $f : [a, \infty) \to \mathbb{R}$ is a continuous function on an unbounded closed interval $[a, \infty)$, and $(x_n)_{n\in\mathbb{N}}$ is any Cauchy sequence of elements of $[a, \infty)$, then $(f(x_n))_{n\in\mathbb{N}}$ must also be Cauchy.
    (c) If $f : \mathbb{R} \to \mathbb{R}$ is a continuous function on the whole real line $\mathbb{R}$, and $(x_n)_{n\in\mathbb{N}}$ is any Cauchy sequence, then $(f(x_n))_{n\in\mathbb{N}}$ must also be Cauchy.

192. A given series consists of non-negative terms. By bracketing these terms together in a particular way, we can create a new series that converges. Prove that the original series converges also (and to the same sum).

193. We consider the rearrangement of the alternating harmonic series in which the positive terms are taken in pairs followed by one negative term, thus:

$$\frac{1}{1} + \frac{1}{3} - \frac{1}{2} + \frac{1}{5} + \frac{1}{7} - \frac{1}{4} + \frac{1}{9} + \frac{1}{11} - \frac{1}{6} + \frac{1}{13} + \frac{1}{15} - \frac{1}{8} + \cdots \qquad (*)$$

Notice that if we bracket these terms together in threes, the nth bracket is

$$\left(\frac{1}{4n - 3} + \frac{1}{4n - 1} - \frac{1}{2n} \right).$$

Use this to show that (i) the bracketed series converges, and that (ii) series (*) also converges.

194. We consider the rearrangement of the alternating harmonic series in which each positive term is followed by three negative terms, thus:

$$\frac{1}{1} - \frac{1}{2} - \frac{1}{4} - \frac{1}{6} + \frac{1}{3} - \frac{1}{8} - \frac{1}{10} - \frac{1}{12} + \frac{1}{5} - \frac{1}{14} - \frac{1}{16} - \frac{1}{18} + \cdots \qquad (**)$$

Notice that if we bracket these terms together in fours, the nth bracket is

$$\left(\frac{1}{2n-1} - \frac{1}{6n-4} - \frac{1}{6n-2} - \frac{1}{6n} \right).$$

Use this to show that (i) the bracketed series converges, and that (ii) series (**) itself converges.

195. Use the ratio test of d'Alembert to find *all* values of $x \in \mathbb{R}$ for which the following series converges:

$$\sum \frac{(2-x)^k}{5^k(3k+4)}.$$

196. Find the range of values of x for which the series

$$\sum \frac{((n+1)!)^2}{(2n+1)!} 3^{2n-1} x^{2n}$$

converges.

197. If $x \in \mathbb{R}$ and (for each $k \in \mathbb{N}$)

$$a_k = \frac{k!(2k+1)!}{(3k-1)!}(1+2x)^k,$$

determine precisely the range of values of x for which the series $\sum a_k$ converges.

198. Determine the set of values of the real parameter t for which the following series is convergent:

$$\sum \frac{(n+2)^n t^n}{(3n+1)^n}.$$

199. Use the nth root test to find *all* values of $x \in \mathbb{R}$ for which the following series converges:

$$\sum \left(\frac{k}{k+1} \right)^{k^2} 2^{2k} x^k.$$

200. Find *all* values of $x \in \mathbb{R}$ for which the series $\sum y_n$ converges, where:

$$y_n = nx^n \left(1 + \frac{1}{n}\right)^{-2n^2}.$$

201. Show that if a series $\sum a_n$ is conditionally convergent (that is, convergent but not absolutely convergent) then both $\sum a_n^+$ and $\sum a_n^-$ diverge to ∞.

202. Devise a rearrangement of the alternating harmonic series that diverges to ∞.

203. Given a completely arbitrary series $\sum x_k$ that is conditionally convergent, and a completely arbitrary real number ℓ, think how you could devise a rearrangement of $\sum x_k$ that converges to ℓ.

Suggestion: the key ingredient in finding one is that both $\sum x_k^+$ and $\sum x_k^-$ diverge to infinity. You might begin by taking *just enough* of the non-negative terms of the series to make the running total greater than ℓ.

204. Let s, t be distinct numbers in the interval $(-1, 1)$. Recall that the geometric series $\sum_0^\infty s^n$ converges to $\frac{1}{1 - s}$ (and likewise for t in place of s).

Write down the Cauchy product $\sum_0^\infty c_n$ of the two series $\sum_0^\infty s^n$ and $\sum_0^\infty t^n$ and simplify the expression you obtain for its typical term c_n. From this, deduce that

$$\sum_0^\infty \frac{s^{n+1} - t^{n+1}}{s - t} \text{ converges to } \frac{1}{(1 - s)(1 - t)}.$$

205. Let $-1 < s < 1$. Write down the Cauchy product $\sum_0^\infty c_n$ of the series $\sum_0^\infty s^n$ by itself, and simplify the expression you obtain for its typical term c_n. Hence evaluate the sum of the series

$$\sum_0^\infty (n + 1)s^n.$$

206. Recall that (for all $x \in (-1, 1)$) the series $1 + x + x^2 + x^3 + x^4 + \cdots$ converges to $\frac{1}{1 - x}$ and the series $1 - x + x^2 - x^3 + x^4 - \cdots$ converges to $\frac{1}{1 + x}$. Calculate (and simplify as necessary) the Cauchy product of these two series and confirm that (as predicted by the theorem on the Cauchy product of two series) it converges to the product of the two functions.

207. Assuming the correctness of

$$e^x = 1 + x + \frac{x^2}{2!} + \frac{x^3}{3!} + \cdots = \sum_0^\infty \frac{x^n}{n!},$$

calculate (and simplify as necessary) the Cauchy product of the power series representations of e^x and of e^y, and confirm that it converges to the product of the two functions.

208. Determine the radius of convergence of each of the following:

(a) $(e^x =)\ 1 + x + \dfrac{x^2}{2!} + \dfrac{x^3}{3!} + \cdots = \sum_0^\infty \dfrac{x^n}{n!},$

(b) $(\sin x =)\ x - \dfrac{x^3}{3!} + \dfrac{x^5}{5!} - \dfrac{x^7}{7!} + \cdots = \sum_0^\infty (-1)^n \dfrac{x^{2n+1}}{(2n+1)!},$

(c) $(\cos x =)\ 1 - \dfrac{x^2}{2!} + \dfrac{x^4}{4!} - \dfrac{x^6}{6!} + \cdots = \sum_0^\infty (-1)^n \dfrac{x^{2n}}{(2n)!},$

(d) $\sum (30n + n^{30})x^n,$

(e) $\sum (n!)x^n,$

(f) $\sum \left(\dfrac{n!(n+1)!(n+2)!}{(3n+1)!} \right) x^n,$

(g) $\sum \dfrac{x^n}{\left(1 + \dfrac{3}{n}\right)^{2n^2}}.$

209. Given $a, b \in \mathbb{R}$, show that
$$f : \mathbb{R} \to \mathbb{R}, \quad x \mapsto ax + b$$
is uniformly continuous on $\mathbb{R}$.

210. Show that the function given by $f(x) = x^3$ is not uniformly continuous on any interval of the form $[a, \infty)$.

211. Decide (with proof) whether the following functions are or are not uniformly continuous on the indicated intervals.
(i) On $(0, 1), f(x) = \sin(\frac{1}{x})$.
(ii) On $(0, 1), f(x) = x\sin(\frac{1}{x})$.
(iii) On $[0, \infty), f(x) = x\sqrt{x}$.

212. Let $a \in \mathbb{R}$ be given. Find:
(i) a continuous function on $[a, \infty)$ that is not uniformly continuous,
(ii) a continuous function on (a, ∞) that is not uniformly continuous,
(iii) a continuous function on (a, b) that is not uniformly continuous.

213. An interior point b of an interval I divides it into a left portion L and a right portion R that intersect only at b. (So L takes one of the forms $(-\infty, b]$, $(a, b], [a, b]$ and R takes one of the forms $[b, \infty), [b, c), [b, c]$.)
Given a real function $f : I \to \mathbb{R}$ that is uniformly continuous on L and uniformly continuous on R, show that f is uniformly continuous on

$I = L \cup R$ also. (This is sometimes referred to as a 'gluing lemma', as we imagine gluing the two portions of the domain back together.)

214. (i) If x and y are both greater than or equal to 1, show that
$|\sqrt[3]{x} - \sqrt[3]{y}| \leq \frac{1}{3}|x - y|$.
(ii) From (i), show that the function given by $f(x) = \sqrt[3]{x}$ is uniformly continuous on $[1, \infty)$.
(iii) Now use Exercise 213 to show that $f(x) = \sqrt[3]{x}$ is uniformly continuous on $[0, \infty)$.

215. Determine whether the following real functions are uniformly continuous on the intervals indicated.
(a) On $[0, 2\pi]$ we define f by the formula $f(x) = e^{\sin(\cos(x^2))}$
(b) On $(-1, 1)$ we define f by the formula $f(x) = \dfrac{1}{1 - x^2}$.
(c) On $(0, \pi/2)$ we define f by the formula $f(x) = \dfrac{1 - \cos x}{x^2}$.
(d) On $(0, \infty)$ we define $f(x) = \ln x$.

216. Show that the function $f : [0, 1] \to \mathbb{R}$ described by $f(x) = \sqrt{1 - x^2}$ is uniformly continuous, but is not Lipschitz. (*Suggestion*: if there were a constant K such that $|f(x) - f(y)| \leq K|x - y|$ for all relevant x and y, put $x = 1$ and $y = 1 - \frac{1}{n}$ for each positive integer n and see what that tells you about K.)

217. Show that the function f described by the formula $f(x) = x - x^{-1}$ is uniformly continuous on $[1, \infty)$, but not on $(0, \infty)$.

218. (a) Suppose we know that a function (defined at least upon an interval of the form $[a, \infty)$ or (a, ∞)) possesses a limit as $x \to \infty$. Show that there is some interval of the form $[b, \infty)$ upon which this function is bounded.
(b) Show that $f(x) = 3x + \sin(\sqrt{x})$ defines a uniformly continuous function on the interval $(0, \infty)$. (*Hint*: use the result of part (a) upon the derivative of f.)

219. Show that the function f defined on $\mathbb{R}$ by the formula $f(x) = \dfrac{\sin x}{1 + x^2}$ is uniformly continuous on the real line.

220. Prove that the function specified by $f(x) = \sqrt[3]{x^2}$, $x \in \mathbb{R}$, is uniformly continuous on the real line.

221. We are given that $f : (a, \infty) \to \mathbb{R}$ is differentiable on its domain and that $f'(x) \to \infty$ as $x \to \infty$. Prove that f is not uniformly continuous on (a, ∞).

222. Show that the function $f : (0, \infty) \to \mathbb{R}$ defined by

$$f(x) = \frac{\sin x}{e^x - e^{-x}}$$

is uniformly continuous (on its domain).

223. Use the three methods indicated to show that the equation

$$\frac{\pi}{2} \cos\left(\frac{\pi x}{2}\right) = (x+1)e^{x-1}$$

has at least one solution in the interval $(0, 1)$:
(a) by applying the intermediate value theorem to
$\frac{\pi}{2} \cos\left(\frac{\pi x}{2}\right) - (x+1)e^{x-1}$,

(b) by applying the Cauchy mean value theorem to $\sin\left(\frac{\pi x}{2}\right)$ and xe^{x-1},

(c) by applying Rolle's theorem to $\sin\left(\frac{\pi x}{2}\right) - xe^{x-1}$.

224. Let n be a given positive integer. Use the Cauchy mean value theorem on the two functions f, g given by $f(x) = \sin^n x$, $g(x) = x^n$ over the interval $[0, \pi/2]$, to see what it tell us about the equation

$$\sec x = \left(\frac{\pi}{2}\right)^n \left(\frac{\sin x}{x}\right)^{n-1}.$$

225. Given that $0 < a < b$, that f is continuous on $[a, b]$ and that f is differentiable on (a, b), show using Cauchy's mean value theorem that there is a number c in the open interval (a, b) such that

$$\frac{bf(a) - af(b)}{b - a} = f(c) - cf'(c).$$

(*Hint:* consider the functions described by $\dfrac{f(x)}{x}$ and $\dfrac{1}{x}$.)

226. Assuming that $\cos t < 1$ for every $t \in (0, 2\pi)$, use Cauchy's mean value theorem to show that
(a) $\sin x < x$ for every $x \in (0, 2\pi)$,

(b) $1 - \cos x < x^2/2$ for every $x \in (0, 2\pi)$,

(c) $\sin x > x - x^3/6$ for every $x \in (0, 2\pi)$.
(*Suggestions:* for (a), try the functions described by $f(t) = \sin t$ and $g(t) = t$ on the interval $[0, x]$; for (b), try the functions described by $f(t) = 1 - \cos t$ and $g(t) = t^2/2$ on the interval $[0, x]$.)

227. The following *alleged proof* of CMVT is *incorrect*. Find out precisely why.
    'Since f satisfies the conditions of FMVT over the interval $[a, b]$, we know that
    $$\text{there exists } c \in (a, b) \text{ such that } f'(c) = \frac{f(b) - f(a)}{b - a}.$$
    'By exactly the same argument on g:
    $$\text{there exists } c \in (a, b) \text{ such that } g'(c) = \frac{g(b) - g(a)}{b - a}.$$

'Dividing one by the other (and remembering that $g(b) - g(a)$ cannot be zero, else Rolle's theorem would give $g' = 0$ somewhere, *contradiction*) we get

$$\frac{f'(c)}{g'(c)} = \frac{f(b) - f(a)}{g(b) - g(a)}$$

as desired.'

(You might wish to try running the argument of the alleged proof on a couple of simple functions such as $f(x) = x^2$ and $g(x) = x^3$ over $[0, 1]$, and observe its failure.)

228. Use l'Hôpital's rule to evaluate $\lim_{x \to 1} \dfrac{x^7 - 3x^5 + 2}{x^5 + 2x^3 - 3}$ and $\lim_{x \to 0} \dfrac{\sin x - x}{x^3}$.

229. Use l'Hôpital's rule to evaluate the following:

(a) $\lim_{x \to \pi} \dfrac{1 - e^{\sin x}}{x - \pi}$,

(b) $\lim_{x \to 0} \dfrac{1 - \cos x + x \ln(1 + x)}{\sin^2 x}$,

(c) $\lim_{x \to 0} \dfrac{e^x - 1 - x - \dfrac{1}{2}x^2}{x - \sin x}$.

230. Evaluate $\lim_{x \to -1} \left(\dfrac{e^x - (x + 2)e^{-1}}{(x + 1)^2} \right)$.

231. Evaluate $\lim_{x \to 0^+} (x \ln x)$.

(HINT: seek the limit of $f(x)/g(x)$ where $f(x) = \ln x$ and $g(x) = \frac{1}{x}$. You can assume that l'Hôpital's rule works in the 'plus or minus infinity over infinity' case just as it does in the 'zero over zero' case – see also Exercise 234.)

232. Evaluate $\lim_{x \to 0} \left(\dfrac{x - \arctan x}{x^3} \right)$.

233. Determine the following limits:

(a) $\lim_{x \to 4} \dfrac{x - \sqrt{x}}{4 - x}$,

(b) $\lim_{x \to 3} \dfrac{e^x + (2 - x)e^3}{(x - 3)^2}$,

(c) $\lim_{x \to \infty} \left\{ x \left(\dfrac{\pi}{2} - \arctan x \right) \right\}$.

234. (a) Given that $0 < \varepsilon < 1, |h - L| < \varepsilon$ and $|m - 1| < \varepsilon$, verify that

$$|hm - L| \leq (2 + |L|)\varepsilon.$$

(b) Prove the following version of l'Hôpital's rule: if f and g are both differentiable (with $g' \neq 0$) on (a, b) and, as $x \to b^-$:
   • $f(x) \to \infty$,

- $g(x) \to \infty$ and
- $\dfrac{f'(x)}{g'(x)} \to \ell$,

then also

- $\dfrac{f(x)}{g(x)} \to \ell$.

Hints for part (b):

i. (Given $\varepsilon > 0$) show that there is x_0 in (a, b) such that, for every x in (x_0, b), $\left| \dfrac{f'(x)}{g'(x)} - \ell \right| < \varepsilon$.

ii. Now show that there is x_1 in (x_0, b) such that, for every x in (x_1, b),

$$\left| \frac{1 - \dfrac{g(x_0)}{g(x)}}{1 - \dfrac{f(x_0)}{f(x)}} - 1 \right| < \varepsilon.$$

iii. For each x in (x_1, b), apply the Cauchy mean value theorem to f, g over the interval $[x_0, x]$.

235. Verify that the eighth Taylor (Maclaurin) polynomial approximating the function $\cos x$ at $a = 0$ is given by

$$p_8(x) = 1 - \frac{x^2}{2!} + \frac{x^4}{4!} - \frac{x^6}{6!} + \frac{x^8}{8!}.$$

Use Taylor's theorem, including estimation of the error or remainder, to evaluate $\cos(0.4)$ correct to six decimal places.

236. Determine a positive integer n such that there is a polynomial of degree n whose values differ from those of e^x by less than 0.001 at all points of the interval $[-1, 1]$. (You may use the Taylor/Maclaurin polynomials as determined in paragraph 16.3.3 of the text.)

237. If $|x| < \frac{1}{2}$, it is easy to check (and you may take this for granted) that $\dfrac{|x|}{1 - |x|} < 1$. Taking $f(x) = \ln(1 + x)$ and $a = 0$ in our version of Taylor's theorem, show that the kth Taylor polynomial (evaluated at x) converges to $\ln(1 + x)$ at every point of the interval $(-\frac{1}{2}, \frac{1}{2})$ by estimating the remainder at stage k.[1]

---

[1] The result is actually true on the larger interval $(-1, 1)$, but confirming this needs a slightly different method of proof.

238. Use the Taylor expansion of $\ln(1+x)$ (for small values of x) to determine whether or not the series $\sum y_n$ given by

$$y_n = \left(\frac{n+5}{n}\right)^{n^2+n} e^{-5n}$$

is convergent.

239. By considering the logarithm of the typical term and appealing to Taylor's theorem, determine whether or not the series $\sum x_n$ specified by

$$x_n = \left(\frac{3n}{3n+2}\right)^{n^2} e^{\frac{2n}{3}}$$

converges.

240. Provided that $-1 < x < 1$, what is the sum of the series

$$1 + 2x + 3x^2 + 4x^3 + \cdots + (n+1)x^n + \cdots$$

and what is the sum of the series

$$1 + 3x + 6x^2 + 20x^3 + \cdots + \frac{1}{2}(n+1)(n+2)x^n + \cdots ?$$

241. *Assuming* that it is possible to express $\arctan x$ (for $-1 < x < 1$) as the sum of a power series $a_0 + a_1 x + a_2 x^2 + a_3 x^3 + \cdots + a_n x^n + \cdots$, use the result on differentiation of power series to determine the coefficients a_n.

242. Consider the real function $f : [0,2] \to \mathbb{R}$ defined by

$$f(x) = 2 \text{ if } x \in [0,1), \quad f(1) = 0, \quad f(x) = 1 \text{ if } x \in (1,2].$$

By directly calculating the upper and lower Riemann sums for f using the partition

$$\Delta = \{0,\ 1/n,\ 2/n,\ 3/n, \cdots, (n-1)/n,\ (n+1)/n,\ (n+2)/n, \cdots, 2\}$$

where n is an arbitrary positive integer, show that f is Riemann integrable and evaluate its Riemann integral.

243. Consider the real function $f : [0,1] \to \mathbb{R}$ defined by

$$f(x) = x \text{ if } x \in [0,1), \quad f(1) = 0.$$

By directly calculating the upper and lower Riemann sums for f using the partition

$$\Delta = \{0,\ 1/n,\ 2/n,\ 3/n, \cdots, (n-1)/n,\ 1\}$$

where n is an arbitrary positive integer, show that f is Riemann integrable and evaluate its Riemann integral.

244. (a) Suppose we know that a function f (that is defined on an interval of the form $[a, b]$) takes the value 0 at every point of $[a, b]$ with one exception: there is a unique point $c \in (a, b)$ such that $f(c) \neq 0$. Show that the Riemann integral of f over $[a, b]$ exists and is zero. *Suggestion:* given $\varepsilon > 0$, choose positive $h < \frac{\varepsilon}{2|f(c)|}$ so that $a < c - h < c + h < b$ and calculate the upper and lower Riemann sums for this four-element partition.

(b) How would you modify this proof (to get the same result) if c were equal to a or to b?

245. Use the integral test to show that the series $\sum_{n=1}^{\infty} n^t$ converges if $t < -1$ and diverges if $-1 \leq t < 0$.

246. Decide whether the series $\sum_{n=3}^{\infty} \dfrac{1}{n \ln(n) \ln(\ln(n))}$ converges or diverges.

247. Use the proof of the integral test to get upper and lower estimates for
$$\sum_{n=10}^{\infty} \frac{1}{n^2}.$$

248. Consider a real function $f : [0, 1] \to \mathbb{R}$ of which we know only that it is bounded and that, for every positive integer $n \geq 2$, f is Riemann integrable over the interval $[\frac{1}{n}, 1]$. Prove that f must be integrable over $[0, 1]$ and that

$$\int_0^1 f = \lim_{n \to \infty} \int_{\frac{1}{n}}^1 f.$$

249. Evaluate
$$\int_0^1 x^2 e^x dx \quad \text{and} \quad \int_0^1 x^2 e^{x^3} dx.$$

250. The real function $f : [0, 2] \to \mathbb{R}$ is defined by the formulae $f(\frac{1}{n}) = 0$ for each positive integer n but $f(x) = 5x$ for each $x \in [0, 2]$ that is not of the form $\frac{1}{n}$ for positive integer n. Show that f is R-integrable over $[0, 2]$ and evaluate its R-integral. (*Hint:* consider 17.4.9 and 17.4.10.)

251. Give an example of an integrable function on a bounded closed interval I that is not monotonic and is discontinuous at an infinite number of points in I.

252. Show by example that the following assertion is false: if f, g are integrable over a closed bounded interval I and $g(x) > 0$ at every point x of I, then $\dfrac{f(x)}{g(x)}$ must be integrable over I also.

Think how you might slightly modify this false statement to make (and then prove) a true one about integrability of the quotient of two integrable functions.

253. Calculate the integrals of each of the following expressions over the interval indicated (assuming, where appropriate, basic properties of trig, logarithmic and exponential functions).

(a) $f(x) = x \ln x$ over $[1, e]$,

(b) $f(x) = (0.5 + xe^{x^2})(x + e^{x^2})^6$ over the interval $[0, 1]$,

(c) $f(x) = \sin^2 x \cos^2 x$ over $[0, \dfrac{\pi}{2}]$,

(d) $f(x) = e^x \sin x$ over $[0, \pi]$.

254. (a) Extend the argument of the integral test to show that, for each (fixed) positive integer n_0, if $\int_{n_0}^{n} f \to \ell$ as $n \to \infty$, then

$$\ell \le \sum_{k=n_0}^{\infty} f(k) \le \ell + f(n_0).$$

(b) Estimate the sum of the (convergent) series $\sum_1^{\infty} \frac{1}{k^5}$ with an error less than 0.001.

255. (a) For a given real number a, use the one-sided version of l'Hôpital's Rule to determine

$$\lim_{x \to 0^+} \frac{\ln(1 + ax)}{x}.$$

(b) Use the sequence-based description of one-sided limits to deduce that

$$\lim_{n \to \infty} \frac{\ln\left(1 + \dfrac{a}{n}\right)}{\dfrac{1}{n}} = a.$$

(c) Now use continuity of the exponential function to deduce that

$$\left(1 + \frac{a}{n}\right)^n \to e^a \text{ as } n \to \infty.$$

256. (a) Verify that the function tan, defined (initially) on the interval $(-\frac{\pi}{2}, \frac{\pi}{2})$ by the formula

$$\tan x = \frac{\sin x}{\cos x},$$

is continuous and differentiable, with (positive) derivative $(\cos x)^{-2}$, and has range $\mathbb{R}$; also that its inverse arctan or $\tan^{-1} : \mathbb{R} \to (-\frac{\pi}{2}, \frac{\pi}{2})$ is continuous and differentiable, and that its derivative is given by

$$(\tan^{-1})'(x) = \frac{1}{1 + x^2}.$$

(b) Check that the radius of convergence of the power series

$$1 - \frac{t}{3} + \frac{t^2}{5} - \frac{t^3}{7} + \frac{t^4}{9} - \cdots$$

is 1: so that, in particular, the series

$$1 - \frac{x^2}{3} + \frac{x^4}{5} - \frac{x^6}{7} + \frac{x^8}{9} - \cdots$$

and

$$x - \frac{x^3}{3} + \frac{x^5}{5} - \frac{x^7}{7} + \frac{x^9}{9} - \cdots$$

are absolutely convergent, the second one to some function $f(x)$, on (at least) the interval $(-1, 1)$.

(c) Appeal to the theorem on differentiation of power series to show that

$$f'(x) = 1 - x^2 + x^4 - x^6 + x^8 - \cdots = \frac{1}{1 + x^2} \quad (-1 < x < 1).$$

(d) Deduce that (for $-1 < x < 1$):

$$x - \frac{x^3}{3} + \frac{x^5}{5} - \frac{x^7}{7} + \frac{x^9}{9} - \cdots = \tan^{-1}(x).$$

(e) Use this to justify what we claimed in paragraph 2.6. Also use the fact that $\tan(\pi/6) = 1/\sqrt{3}$ to obtain a (numerically simple) series whose sum is $\frac{\pi}{\sqrt{12}}$.

Suggestions for Further Reading

Alcock, L. *How to Think About Analysis*. Oxford University Press (2014).

Appelbaum, D. *Limits, Limits Everywhere: the Tools of Mathematical Analysis*. Oxford University Press (2012).

Bryant, V. *Yet Another Introduction to Analysis*. Cambridge University Press (1990).

Burn, R.P. *Numbers and Functions: Steps into Analysis* (2nd edn). Cambridge University Press (2000).

Howie, J.M. *Real Analysis*. Springer (2001).

Spivak, M. *Calculus* (corrected 3rd edn). Cambridge University Press (2006).

Index